T0315111

A Primer on Experiments with Mixtures

A Primer on Experiments with Mixtures

JOHN A. CORNELL

University of Florida
Department of Statistics
Gainesville, FL

A JOHN WILEY & SONS, INC., PUBLICATION

For general information on our other products and services or for technical support, please contact our
Customer Care Department within the United States at (800) 762-2974, outside the United States at
(317) 572-3993 or fax (317) 572-4002.

Wiley also publishes its books in a variety of electronic formats. Some content that appears in print may
not be available in electronic formats. For more information about Wiley products, visit our web site at
www.wiley.com.

Library of Congress Cataloging-in-Publication Data

Cornell, John A., 1941-
  A primer on experiments with mixtures / John A. Cornell.
    p. cm.
  Includes bibliographical references and index.
  ISBN 978-0-470-64338-9 (cloth)
  1. Mixtures—Experiments.   2. Solution (Chemistry)—Experiments.   3. Powders—
Mixing—Experiments.   4. Mixtures—Mathematical models.   5. Solution (Chemistry)—Mathematical
models.   6. Powders—Mixing—Mathematical models.   I. Title.
  TP156.M5C67   2011
  660′.294—dc22

                                                                                2010030990

10 9 8 7 6 5 4 3 2 1

Contents

Preface

In 1981, the first edition of *Experiments with Mixtures* was published and consisted of only seven chapters and 305 pages. In 1990 and 2002, this original first edition was greatly expanded as second and third editions. Three chapters were added bringing the page counts to 632 and 649 pages, respectively.

The Prefaces to the second and third editions identified the editions as being suitable for classroom use in a one-semester advanced undergraduate or graduate-level course. Chapters 1–7 constituted the core material, and Chapters 8–10 could be selected according to need. Yet, if used as a reference for self-study, Chapters 1–2 and 4–7 provide the necessary tools for dealing with almost any type of mixture problem. Having retired from academe in 2003 but occasionally was asked to present talks and short courses on Designs and Models for Mixture Experiments, I began to question whether or not all the third edition material is really necessary, particularly as an initial offering. In my opinion, the third edition contains much more material than can be covered in a short course offered for 1, 2, or 3 days or even a one-semester graduate-level course. Even if students enrolled in a graduate-level mixtures course were previously trained in the design of experiments, or attendees at a short course on mixtures were previously educated as chemists or engineers or even in statistics, it is quite likely most of them are seeing the basic mixtures material for the very first time.

While thinking about the future of training scientists in statistics and other disciplines about mixture designs and models, I believe Dr. Wendell Smith's excellent book on *Experimental Design for Formulation* and my 3rd edition of *Experiments with Mixtures: Designs, Models, and the Analysis of Mixture Data*, should still serve mixture-knowledgeable researchers in academe, industry, and governmental positions quite nicely over the next decade. Short courses that focus on software and provide mixture designs and combined mixture components–process variables designs ought to benefit from Dr. Smith's book and my 3rd edition. What I feel is desperately needed and has taken too long a time coming is a much stronger emphasis on teaching, that is, teaching scientists and plant-level personnel how to recognize when they have encountered a mixture environment. And then the word "primer" came to mind. Taken from the French word *primaire*, my copy of Webster's dictionary defined "primer, n" as a book of elementary principles. This gave me the idea to try and lower the degree of difficulty in understanding the principles behind experimenting with mixtures by offering a book without all the unnecessary formulas that might have previously appeared in journal articles or dissertations. I contacted Wiley and asked what they thought and how they felt about my putting together *A Primer on Experiments with Mixtures*. They were very supportive of the idea and encouraged me to proceed.

Chapter 1 begins with a listing of a few simple mixture experiments such as adding sugar and/or a sweetener to a morning cup of coffee or tea to sweeten the flavor, or mixing olive oil and vinegar (and some spices) to create and flavor a salad dressing, or adding 93 octane fuel (premium unleaded) in the fuel tank of a family car that already contains some 87 octane (regular unleaded) or 89 octane fuel, in hope of getting better mileage (miles per gallon) or performance. Stated another way, the adding and/or mixing of ingredients in an effort to get a more desirable end product is something many of us do everyday, and these actions are known as *mixture experiments*. Imagine an experimental region in the shape of a simplex: with three ingredients, is an equilateral triangle, and with four components, it is a tetrahedron. This seemed easy enough to comprehend and being told and shown that mixture models are simpler in form by having fewer numbers of terms than the standard polynomials, well that was welcome news.

Chapter 2 discusses both designs and models for exploring the entire simplex factor space of the components. In fact very little material is deleted from Chapter 2 of the third edition. The section on using check points for testing a models' lack of fit has been removed and replaced with an example illustrating the use of extra points and replicated observations for testing the model's lack of fit. Scheffé's derivation of formulas for estimating the coefficients and their variances and covariances associated with having fitted the canonical form of polynomial matched up with the simplex-centroid design has been removed as well. Chapter 3, on the use of independent variables rather than the original mixture components, has been deleted. The new Chapter 3 material covers designs for when the component proportions are constrained in the form of lower and upper bounds placed on the proportions, resulting in an experimental region that may not be simplex shaped. This is where software such as DESIGN-EXPERT® and Minitab® become necessary.

Chapter 4 covers the analysis of mixture data. In an effort to cut down on the number of pages in this chapter, a criterion for selecting subsets of the terms in the Scheffé models as well as a numerical example illustrating the integrated mean-square error as the criterion in Chapter 5 of the third edition has been deleted. Measuring the slope of the response surface along the component axes along with a numerical example that illustrates the calculations for a three-component system are also omitted in order to save space.

Chapter 5 covers other mixture model forms. The inclusion of inverse terms added to terms in the Scheffé polynomials remain from the third edition along with a numerical example showing how the inverse terms pick up large values of the response near the boundaries of the experimental region. The use of ratios of components as well as Cox's polynomials for measuring component effects are included. A numerical example is provided to illustrate and compare the fits of Cox's model and Scheffé's polynomial. The chapter concludes with defining a slack variable model along with pointing out the risks in using such a model if fitting reduced model forms.

No book on mixture experiments would be complete without a chapter covering the inclusion of process variables in mixture experiments. Chapter 6 suggests designs consisting of simplex-lattices crossed with factorial arrangements. These crossed designs nicely support the fitting of combined models containing terms that include both the mixture components and the process variables. A numerical example illustrates how to measure the effects of cooking temperature and cooking time on the texture of patties made from two types of saltwater fish. If changing the amount of the mixture affects the response and the amount is allowed to vary, then a mixture-amount design and the corresponding model can be used to measure how the amount affects the response or if changing the amount affects the blending properties of components. Seeking the optimal fertilizer blend of nitrogen, phosphorous, and potassium along with the optimal fertilizer rate (low, medium, and high) for growing young (< 3 years old) citrus trees is illustrated. Since a considerable amount of interest has been shown in these expanded mixture experiments containing either process variables or varying the amount during the past decade, a section titled "Questions Raised and Recommendations Made When Fitting a Combined Model Containing Mixture Components and Other Variables," has been taken from the third edition and is included in the primer. Chapter 7 offers a brief review of least squares and the analysis of variance as well as a review of data analysis measures such as the adjusted square of the multiple correlation coefficient, the press statistic, and studentized residuals.

While putting together the material for this primer, two giants of the statistical profession passed from this world to the next. Yet I couldn't help but feel they were looking over my shoulder offering encouragement. Their names are Professor William (Bill) G. Mendenhall III and Dr. John W. Gorman. In 1963, Professor Mendenhall recruited me to become a graduate student in the Department of Statistics at the University of Florida and five years later, hired me as an Assistant Professor in the Department of Statistics. Dr. Gorman was my mentor on mixtures. I had the good fortune of working with John when he served as an Associate Editor for *Technometrics* in the early 1970s. He asked me on numerous occasions to review manuscripts, on topics in mixture experiments, that had been submitted for publication. Looking back on those days, little did I realize at the time that each manuscript I reviewed fueled my interest in learning as much as I could about mixtures. I met John for the first time at the 1975 Gordon Research Conference on Statistics in Chemistry and Chemical Engineering, and had the good fortune of being assigned his roommate, which allowed me to pick his brain that week while absorbing all I could from his experiences working with mixtures. Oh, how I miss those two wonderful gentlemen.

I am indebted to many people, some who I do not know but who offered kind words of support for my writing this primer. Steve Quigley, Associate Publisher for Wiley, has been tremendously helpful guiding me step by step during the process of selecting what should and what should not be included in the primer. Dr. Greg F. Piepel, a former student but now a colleague of mine, sent 19 pages of bibliography to me covering the years 2000 up to 2010. This enabled me to update the bibliography in Chapter 1 of the primer for which I am grateful. I am also pleased to see that research articles on mixture designs and model forms continue to be published today, some 53 years after Scheffé's pioneering article, *Experiments with Mixtures*, appeared in print. In closing, let me say that for those of you who have searched for a book containing elementary statistical principles that can assist you in performing mixture experiments, this primer is for you.

JOHN A. CORNELL
Gainesville, FL

CHAPTER 1

Introduction

Mixture experiments are performed in many areas of product development and improvement. In a mixture experiment two or more ingredients are mixed or blended together in varying proportions to form an end product. The quality characteristics of the end product are recorded for each blend to see if the quality changes or varies from one blend to the next. Some examples are as follows:

1. Recording flavor, color, and overall acceptance of a fruit punch blended from watermelon, pineapple, and orange juices.
2. Measuring the efficacy and durability of a pesticide formed by mixing several chemicals.
3. Improving brightness and durability of railroad flares, which are the product of blending proportions of magnesium, sodium nitrate, strontium nitrate, and binder.

In each of the cases above the value of the measured characteristic, or **response**, such as the **flavor** of the fruit punch in example 1), is assumed to depend only on the relative proportions of the ingredients in the mixture and **not on the amount of the mixture.** If the amount of the mixture is suspected to have an effect on the response, such as in example 2) where the efficacy of a pesticide could depend on the amount of pesticide used, then the amount is kept fixed for all blends studied. Holding the amount fixed and varying only the relative proportions of the ingredients to define different blends assures that any change in the value of the response from one blend to the next is the result of changing only the composition of the blend. This leaves us with the following definition.

Definition. In the general mixture problem the measured response depends only on the proportions of the ingredients present in the mixture and not on the amount of the mixture.

A Primer on Experiments with Mixtures, By John A. Cornell
Copyright © 2011 John Wiley & Sons, Inc. Published by John Wiley & Sons, Inc.

We shall modify this definition later in Chapter 6 when introducing mixture-amount experiments.

1.1 THE ORIGINAL MIXTURE PROBLEM

In a mixture experiment the experimenter selects a number of different blends to study and varies the proportions of two or more ingredients in each of the blends. The proportions of ingredients are measured by volume, by weight, or by mole fraction. When expressed as fractions of a mixture, the proportions are nonnegative and sum to one or unity. Thus, if we let q represent the number of ingredients (hereafter called the **components** of the mixture) in the mixture system under study, and let x_i represent the fractional proportion of the ith component in the mixture, then

$$0 \le x_i \le 1, \qquad i = 1, 2, \ldots, q \tag{1.1}$$

and

$$\sum_{i=1}^{q} x_i = x_1 + x_2 + \cdots + x_q = 1.0 \tag{1.2}$$

Take the example of formulating the fruit punch as previously mentioned. In example 1 we could let watermelon juice comprise 60% of the punch (or $x_1 = 0.60$) and let pineapple juice comprise 25% of the punch (or $x_2 = 0.25$); then orange juice must make up the remaining 15% (or $x_3 = 0.15$) of the punch. What this means is that the x_i are not independent. Why is this? Because, once the values of $x_1, x_2, \ldots, x_{q-1}$ are known, the value of x_q is automatically determined, that is, x_q must equal $1 - (x_1 + x_2 + \cdots + x_{q-1})$. Furthermore, while the x_i represent nonnegative percentages of the fruit punch, when the percentages are divided by 100%, they become fractions or proportions as in Eqs. (1.1) and (1.2).

Before we take on a numerical example, let us mention other mixture situations that we often encounter in our everyday activities. The mixture could be a morning cup of coffee or tea to which you added sugar and/or a sweetener to improve the flavor, or olive oil and vinegar with spices added to create and flavor a salad dressing, or the 87 or 89 octane fuel in the tank of the family car to which you added 93 octane fuel (premium unleaded) in hopes of getting better mileage or performance. Stated another way, the adding and/or **blending** of ingredients in an effort to get a more desirable end product is something many of us do everyday and these activities become *mixture experiments*; see Figure 1.1.

1.2 A PESTICIDE EXAMPLE INVOLVING TWO CHEMICALS

Two chemicals, Vendex and Kelthane, were mixed with three other ingredients in a 30% : 70% ratio of chemical (30%) to other ingredients (70%) pesticide experiment.

(*a*) Sweetening coffee　　　　　(*b*) Dressing a salad

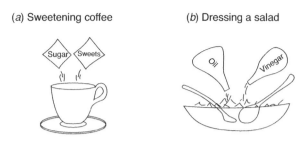

(*c*) Blending 93 octane with regular 87 octane unleaded fuel

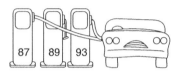

(*d*) Restoring the finish on antique furniture

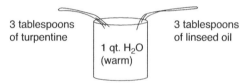

Figure 1.1 Some examples of mixture experiments.

The objective was to develop a liquid pesticide for killing mites on strawberry plants. Each chemical could be mixed alone with the other ingredients, and any combination of the two chemicals could be mixed with the other ingredients as long as the chemical made up 30% of the pesticide and the other ingredients made up the remaining 70% of the pesticide.

Tolerance to the pesticide by the mites after a period of time is what the scientist performing the experiment was trying to minimize or remove altogether. It was suspected that if the two chemicals were mixed and applied together to the plants along with the other ingredients, the efficacy or effectiveness of the pesticide would last longer than if each chemical was applied alone and two single chemical pesticides were applied sequentially, one following the other. After all, sequentially was the way chemical pesticide had been sprayed on plants in the past.

An experiment was set up consisting of different combinations of the two chemicals mixed with the other ingredients to form five different pesticide blends. The five different blends were as follows:

1. 30% Vendex plus 70% other ingredients.
2. 22.5% Vendex plus 7.5% Kelthane plus 70% other ingredients.

3. 15% Vendex plus 15% Kelthane plus 70% other ingredients.

4. 7.5% Vendex plus 22.5% Kelthane plus 70% other ingredients.

5. 30% Kelthane plus 70% other ingredients.

Each of the five pesticide blends was sprayed on a group of six strawberry plants that were uniformly infested with mites prior to applying the pesticides. Four replications of the five blends were performed in order to obtain an estimate of the experimental error variation (replication to replication variability within each particular blend). The same amount of pesticide was sprayed on each group of six plants in an effort to control as well as measure the variation among the amounts that occurred, if any. **The objective of the experiment was to try and discover a blend of the two chemicals Vendex and Kelthane that would result in the highest average percent mortality (APM) of the mites** after a specified period of time. Would one of the three middle blends 2, 3, or 4 prove to be best or would some other blend not listed above be best?

Among the five different blends above, two blends 1 and 5 consist of a single chemical. The single chemical blends are called **pure blends** or **single-component blends** (or mixtures) because the other ingredients have fixed percentages at 70% and do not change in all five blends. Furthermore, in the 70% of each blend that was contributed by the other ingredients, the relative proportions of the three individual ingredients making up the other ingredients remained fixed or unchanged from one blend to the next. This therefore is only a **two-component** mixture experiment because only the percentages of the two chemicals V and K varied from one blend to the next. When not expressed in terms of the percents of the chemicals V and K in the five pesticide blends but rather as fractional proportions of the two chemicals relative to 30% chemical, the ratios of the proportions of V : K in the five blends above are 1.0 : 0, 0.75 : 0.25, 0.50 : 0.50, 0.25 : 0.75, and 0 : 1.0, respectively. The average percent mortality (APM) among the mites of each of the five different blends when taken over $4 \times 6 = 24$ strawberry plants, along with the chemical proportions of Vendex and Kelthane in each blend are listed in Table 1.1, while a plot of the APMs across the five blends is shown in Figure 1.2.

Table 1.1   **Average Percent Mortality (APM) of Mites Across Six Plants for Each Blend of the Chemicals Vendex and Kelthane**

Chemical	Proportion of (Vendex, Kelthane)				
(V, K)	(1.0, 0)	(0.75, 0.25)	(0.50, 0.50)	(0.25, 0.75)	(0, 1.0)
Reps					
1	64.5	79.2	75.2	63.5	34.7
2	67.0	73.7	67.1	58.7	37.1
3	66.5	81.6	80.4	54.6	35.0
4	70.0	82.3	78.1	58.0	33.2
Average	67.0	79.2	75.2	58.7	35.0

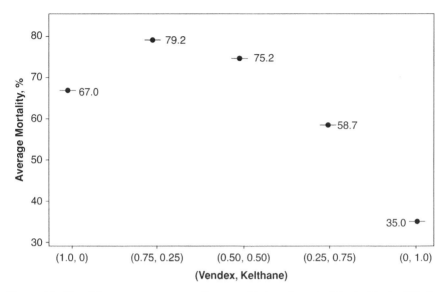

Figure 1.2 Plot of the average percent mortality across 24 plants for each of the five blends of Vendex and/or Kelthane.

From a quick look at the APMs in Figure 1.2, one may be inclined to ask:

1. Are the values of the APMs for the five blends approximately the same?
2. If you answered no, are the values of the APMs a function of the chemical composition of the pesticide?
3. If you answered 'yes' to question 2, is it possible to express the relationship between APM and chemical composition of the pesticide in the form of an equation that makes sense to us?

Suppose that we answer the questions above this way:

1. No, I do not think so because the last two blends whose APM values are 35.0% and 58.7% appear to be significantly lower than 79.2% and 75.2%.
2. Yes, it appears that upon initially replacing some of Vendex with Kelthane to produce the blend (V, K) = (0.75, 0.25) resulting in an increase in the APM value, and then adding more Kelthane, particularly above 50% results in the APM value decreasing. So with Kelthane alone as the chemical in the pesticide, the APM is 35%, which is approximately one-half of 67% and equivalent to the APM of Vendex alone in the pesticide.
3. The trend in APM values for blends starting with (V, K) = (1.0, 0) and ending with (V, K) = (0, 1.0) does not look linear (or like a straight line). Rather, the trend looks more curvilinear or quadratic, which can be modeled with a second-degree equation.

From the data values in Table 1.1 it is clear that the observed APM from using Vendex alone, with the other ingredients, is 67.0%, while when using Kelthane alone, with the other ingredients, the observed APM is 35.0%. Now, if the blending of the chemicals is strictly **additive** or linear, then we would expect the observed APM of any blend consisting of both V and K to be close to the volumetric average of the single-chemical blend averages. For example, if **additive blending** is present, the (V, K) blend (0.50, 0.50) is expected to have an APM that is (67.0% × 0.50) + (35.0% × 0.50) = 51.0%. Similarly the APM of the (0.75, 0.25) blend is expected to be (67% × 0.75) + (35% × 0.25) = 59.0% while the APM of the (0.25, 0.75) blend is expected to be (67% × 0.25) + (35% × 0.75) = 43.0%. However, from Table 1.1, since the observed APM for the (0.50, 0.50) blend is 75.2% and this value exceeds the volumetric APM of 51.0% by 24.2%, it is easy to suspect that the chemicals **Vendex and Kelthane blend in a nonadditive** manner. In fact, since the observed APM of the (0.50, 0.50) blend was higher (better) than what was expected from additive blending it might be inferred the nonlinear blending of the chemicals was *synergistic* (or beneficial) as well. If instead the observed APM of the (0.50, 0.50) blend had been lower than 51.0%, then it might be inferred that the chemicals blended *antagonistically.* Furthermore, since the observed APMs of the remaining two blends were 79.2% at (0.75, 0.25) and 58.7% at (0.25, 0.75) and these values also exceeded their respective volumetric averages of 59.0% and 43.0%, respectively, this further supports the synergistic blending of the two chemicals and thus our rejecting the hypothesis of additive blending of the two chemicals. Of course, we would have to prove or show quantitatively why we rejected the hypothesis of additive blending, and in Appendix 1A we illustrate how to test for synergistic or antagonistic blending between a pair of components when interpreting the roles played by certain terms in the fitted model.

By virtue of the constraints on the x_i's shown in Eqs. (1.1) and (1.2), the geometric description of the factor space containing the q components consists of all points on or inside the boundaries (vertices, edges, faces, etc.) of a regular $(q-1)$-dimensional simplex. For $q = 2$ components, the simplex factor space is a straight line, represented by the horizontal axis in Figure 1.2 which lists the proportions associated with the two chemicals Vendex and Kelthane, along with the other ingredients, that make up the five blends. Shown in Figure 1.3, for three components ($q = 3$), the composition space or two-dimensional simplex is an equilateral triangle, and for $q = 4$, the simplex region is a tetrahedron; see Figure 1.5. Note that since the proportions sum to unity or one as shown in Eq. (1.2), the x_i's are constrained variables, and altering the proportion of one component in a mixture will produce a change in the proportion of at least one other component in the experimental region. Additional discussion on the shape of the experimental region will be provided in Chapters 2 and 3.

The coordinate system for the values of the mixture component proportions is a *simplex coordinate* system written as $(x_1, x_2, \ldots, x_q)$ where the x_i's sum to one. With three components for example, the vertices of the triangle represent single-component mixtures $x_i = 1$, $x_j = x_k = 0$ for $i, j, k = 1, 2$ and 3, $i \neq j \neq k$ and are denoted by (1, 0, 0), (0, 1, 0), and (0, 0, 1). The interior points of the triangle represent mixtures in which none of the components are absent, that is, $x_1 > 0$, $x_2 > 0$ and $x_3 > 0$.

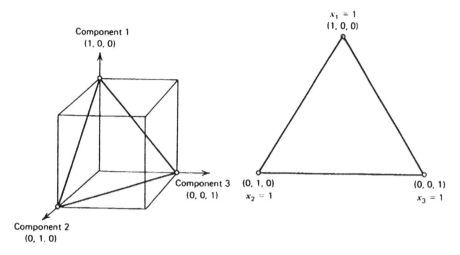

Figure 1.3 Three-component simplex region. All experimental blends must lie on or inside the triangle whose equation is $x_1 + x_2 + x_3 = 1$.

The centroid of the triangle corresponds to the mixture with equal proportions ($\frac{1}{3}$, $\frac{1}{3}$, $\frac{1}{3}$) from each of the components. Shown in Figures 1.4a,b is the three-component simplex (equilateral triangle) where the coordinates can be plotted on triangular coordinate paper that has lines parallel to the three sides of the triangle or on a hand-drawn equilateral triangle in which the blends are more easily located. Figure 1.5 shows the tetrahedron for the four components whose proportions are x_1, x_2, x_3, and x_4.

Frequently situations exist where some of the proportions x_i are not allowed to vary from 0 to 1.0. Instead, some, or all, of the component proportions are restricted by either a lower bound and/or an upper bound. In the case of component i, these constraints may be written as

$$0 \le L_i \le x_i \le U_i \le 1.0, \qquad 1 < i < q$$

where L_i is the lower bound and U_i is the upper bound. Consider the production of a commercial laundry bleach that is to be used for removing ink stains and comprised of the constituents bromine (x_1), dilute HCl (x_2) and hypochlorite powder (x_3). To be effective, the bleach must contain in solution all three of the constituents. This means that each $L_i > 0$ and each $U_i < 1.0$. To be more exact, it might be necessary to require x_2 (dilute HCl) to take values in the interval $0.05 \le x_2 \le 0.09$, which therefore forces $L_2 = 0.05$ and $U_2 = 0.09$. Then the value of $L_2 = 0.05$ forces U_1 and U_3 to be at most equal to $1 - L_2 = 0.95$. Such mixtures in which all of the components are present in nonzero proportions are called *complete* mixtures.

In the chapters that follow we will try to maintain the following points as *potential goals of a mixture experiment.* We will model the dependence of the response variable on the relative proportions of the components with some form of

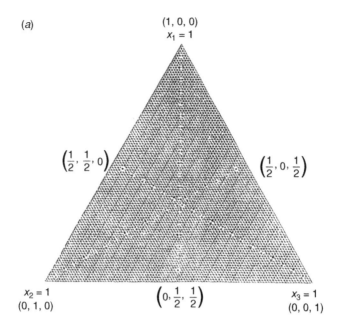

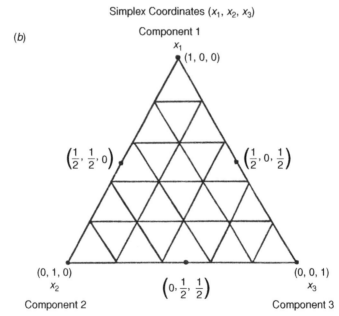

Figure 1.4 (*a*) Triangular coordinate paper. (*b*) Hand-drawn equilateral triangle where the heights of the smaller triangles inside the larger triangle are 0.20 or 20% of the height of the larger triangle.

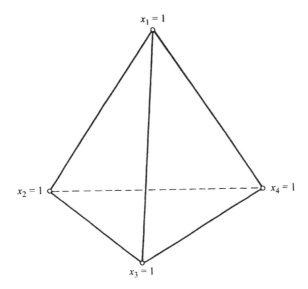

Figure 1.5 Four-component tetrahedron.

mathematical equation. Our mathematical formulations will be setup to account for the following:

1. The influence on the response of each component singly and in combination with the other components can be measured. The components with the least effect or felt to be less active might then be "screened" out, leaving only the components having the greatest effect on the response (Chapters 2 and 4).
2. Predictions of the response to any mixture or combination of the component proportions can be made.
3. Identified mixtures or blends of the components that yield desirable values of the response.

1.3 GENERAL REMARKS ABOUT RESPONSE SURFACE METHODS

In much of the experimental work involving multicomponent mixtures, the emphasis is on studying the physical characteristics, such as the shape or the highest point, of the measured response surface. Take the example of making a fruit punch by blending proportions of orange juice (x_1), pineapple juice (x_2), and grapefruit juice (x_3). The response of interest is the fruitiness flavor of the punch quantified on a 1 to 9 scale as $1 =$ not fruity, $5 =$ average, and $9 =$ extremely fruity. If the measured response (or flavor rating) to any blend of juices could be represented by the perpendicular height directly above the blend whose coordinates are located inside or on the boundaries of the triangle, then the locus of the flavor values for all one-, two-, and three-juice blends could be visualized as a surface above the triangle. One such surface, which is assumed to be continuous for all possible juice blends, is presented in Figure 1.6 and the contour

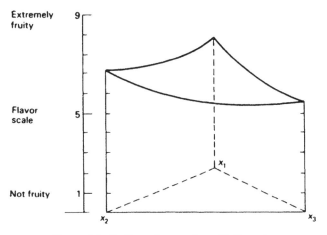

Figure 1.6 Fruitiness flavor surface of fruit punch.

plot of the estimated flavor surface is presented in Figure 1.7. Geometrically, each contour curve in Figure 1.7 is a projection onto the three-component triangle of a cross section of the flavor surface made by a plane, parallel to the triangle, cutting through the surface at a particular height. The heights of the cutting or intersecting planes that generated the contour curves in Figure 1.7 range from 5.8 up to 6.6.

The main considerations connected with the exploration of the response surface over the simplex region are (1) the choice of a proper model to approximate the surface over the region of interest, (2) the testing of the adequacy of the model in representing the response surface, and (3) a suitable design for collecting observations, fitting the

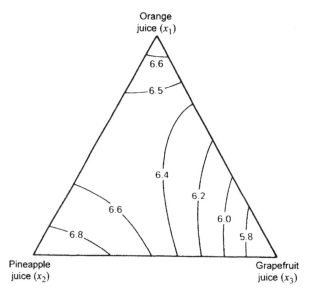

Figure 1.7 Contours of constant fruitiness flavor of the fruit punch surface.

model, and testing the adequacy of fit. To this end, we can assume that there exists some functional relationship

$$\eta = \phi(x_1, x_2, \ldots, x_q) \tag{1.3}$$

that, in theory, exactly describes the surface. We will write the quantity η to denote the true response value that is dependent on the proportions $x_1, x_2, \ldots, x_q$ of the components. A very basic assumption that we are making here is that the response surface, represented by the function ϕ, is depicted to be a continuous function in the x_i, $i = 1, 2, \ldots, q$. This assumption might be questionable for some systems such as a gaseous system whose catalytic reactions break down with the addition or deletion of components. For these systems, model forms other than the standard polynomial equations that we will work with initially are considered. Chapter 5 presents equations containing inverse terms for the purpose of modeling discontinuities of this type in the response surface.

The problem of associating the shape of the response surface with the ingredient composition centers around determining the mathematical equation that adequately represents the function $\phi(\cdot)$ in Eq. (1.3). In general, polynomial functions are used to represent $\phi(x_1, \ldots, x_q)$, the justification being that one can expand $\phi(x_1, \ldots, x_q)$ using a Taylor series, and thus a polynomial can be used also as an approximation. Normally a low-degree polynomial such as the first-degree polynomial

$$\eta = \beta_0 + \sum_{i=1}^{q} \beta_i x_i \tag{1.4}$$

or the second-degree polynomial

$$\eta = \beta_0 + \sum_{i=1}^{q} \beta_i x_i + \sum_{i \leq j}^{q} \sum^{q} \beta_{ij} x_i x_j \tag{1.5}$$

is the kind of model we believe to represent a response surface. Low-degree polynomial equations are more conveniently handled than higher degree equations because the lower degree polynomials contain a fewer number of terms and therefore require fewer observed response values in order to estimate the parameters (the β's) in the equation. On occasions when a very complicated system is being studied such as shown in Figure 1.8, we may feel the need to use a third-degree equation or some special form of a cubic or third-degree equation (especially when even a transformation of the data values does not simplify the system). Most of the time, however, we will try to be successful with at most the second-degree model.

Figure 1.8 shows contours of equal dielectric constant lines in the system $Pb(Co_{1/3}Nb_{2/3})O_3$–$PbTiO_3$–$PbZrO_3$ as estimated with a third-degree polynomial equation. As seen from the contours, the dielectric constants in the system increase with increasing proportion of $Pb(Co_{1/3}Nb_{2/3})O_3$ up to about 80%:20% of $Pb(Co_{1/3}Nb_{2/3})O_3$. Near the center of the system is a steep cliff that appears to drop off in the directions of pure $PbTiO_3$ and pure $PbZrO_3$. Contour plots as in Figures 1.7 and 1.8 are extremely helpful when studying a three-component system.

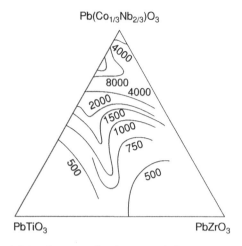

$Pb(Co_{1/3}Nb_{2/3})O_3$

$PbTiO_3$ $PbZrO_3$

Figure 1.8   Equal dielectric constant lines in system $Pb(Co_{1/3}Nb_{2/3})O_3$–$PbTiO_3$–$PbZrO_3$.

   While observing the response η during an experimental program consisting of N trials, it is natural to assume that the observed value, which we denote by y_u for the uth trial ($u = 1,2, \ldots, N$), varies about a mean of η_u with a common variance σ^2 for all $u = 1,2, \ldots, N$. The observed value contains additive experimental error ε_u:

$$y_u = \eta_u + \varepsilon_u, \qquad 1 \leq u \leq N \tag{1.6}$$

The experimental errors ε_u are assumed to be uncorrelated and identically distributed with zero mean and common variance σ^2. These properties of the errors are defined, using an expectation operator $E(\cdot)$, as

$$E(\varepsilon_u) = 0, \qquad E(\varepsilon_u^2) = \sigma^2, \qquad E(\varepsilon_u \varepsilon_{u'}) = 0,$$

$$u \neq u', \qquad u, u' = 1, 2, \ldots, N$$

Therefore the expected value for the observed value y_u is $E(y_u) = \eta_u$, for all $u = 1, 2, \ldots, N$.

   In order to approximate the functional relationship $\eta = \phi(x_1, x_2, \ldots, x_q)$ with a polynomial or with any other form of model equation, some preselected number of experimental runs are performed at various predetermined combinations of the proportions of the q components. This *set* of combinations of the proportions (or blends of the ingredients) is referred to as the *experimental design*. Once the N observations are collected, the parameters in the model are estimated by the *method of least squares*.

   As an example, suppose that we have $q = 2$ components. Coupled with the structure of y_u in Eq. (1.6), we may write

$$y_u = (\beta_1 x_1 + \beta_2 x_2)_u + \varepsilon_u \tag{1.7}$$

The absence of the parameter β_0 in Eq. (1.7) is due to the restriction $x_1 + x_2 = 1$. We discuss the derivation of the model (1.7) form in Section 2.2. With some number $N \geq 2$ of observations collected on y_u, we can obtain the estimates b_1 and b_2 of the parameters β_1 and β_2, respectively. If it is decided that the parameter estimates b_1 and b_2 are satisfactory in the sense that they are nonzero and therefore they relay information about the system we are modeling, then the unknown parameters in Eq. (1.7) are replaced by their respective estimates to give

$$\hat{y} = b_1 x_1 + b_2 x_2 \tag{1.8}$$

where $\hat{y}$ (read "y hat") denotes the predicted or estimated value of η for given values of x_1 and x_2. Of course, before any predictions are made with Eq. (1.8), we must determine that the prediction equation (1.8) does an adequate job of fitting the observed data. We discuss ways of testing the adequacy of empirical models fitted to data in Chapters 2 and 4.

The properties of the polynomials used to estimate the response function depend to a large extent on the specific program of experiments that we have called the experimental design. The experimental design also defines the range of interest of the experimenter with respect to the proportions used for each of the components. This is because the design may cover the entire simplex factor space if the experimenter's interest is with all the values of x_i ranging from 0 to 1.0 for all $i = 1, 2, \ldots, q$, or the design might cover only a subportion or smaller subspace within the simplex. This latter situation comes up in practice when additional constraints in the form of upper and/or lower bounds are placed on the component proportions, or, perhaps, when the experimenter is interested only in a group of mixtures that are located in some small region inside the simplex. Both of these cases are discussed in Chapter 3.

1.4   AN HISTORICAL PERSPECTIVE

Statistical research on mixture experiments, as represented by the number of papers that have appeared in the statistical literature, is still a relatively new activity. Almost all the theory and methodology that has emanated from the statistical community has surfaced during the last five decades. A few noticeable exceptions are the discussion of mixtures that appears in Quenouille's 1953 book, the designed experiment for administering joint dosages of hormones to mice by Claringbold in 1955, and the pioneering article on simplex-lattice designs in 1958 by H. Scheffé.

In this book we concentrate on experimental designs and techniques used in the analysis of mixture data that have evolved since the late-1950s. (In fields such as the cereal industry, the tire manufacturing industry, and the soap industry, mixture experiments date back to the turn of the century.) Table 1.2 gives a chronological sequence of papers that have appeared in the statistical literature. The list includes authors of papers that have appeared from 1953 to 2009 in journals of statistical societies and associations and in textbooks on statistics, as well as in related periodicals, such as academic technical reports, industrial bulletins, and armed services reports. Within each year the authors are listed in alphabetical order. It was not possible to include the authors of all of the works that appeared in every journal during this time period for the obvious reason of lack of space. Several of the works that are omitted in Table 1.2 but which appeared in journals like *Cereal Chemistry* and *Food Technology*, are listed in the bibliography at the end of the book.

Table 1.2   Chronological Listing of Selected Statistical Literature on Mixtures from 1953 to 2009

Year	Authors	Year	Authors	Year	Authors
1953	Quenouille	1978	Becker	1989	Agreda and Agreda	2002	Aggarwal, Sarin and Singh
1955	Claringbold*		Cornell and Gorman		Czitrom		Draper and Pukelsheim
1958	Scheffé*		Park		Donev		Kamoun et al.
1959	Quenouille		Vuchkov, Yonchev, and Damgaliev		Koons		Kowalski, Cornell, and Vining
1961	Scheffé	1979	Cornell and Khuri		Mikaeili		Myers, Montgomery. and Vining
	John and Gorman		Cornell		Murthy and Murty		Piepel, Szychowski, and Loeppky
1962	Gorman and Hinman*		Goel and Nigam		Park and Kim		Prescott et al.
	Wagner and Gorman		Hare	1990	Cornell	2003	Cornell and Gorman
1963	Kenworthy		Snee*		Lim		Dingstad, Egelandsdal, and Naes
	Scheffé*	1980	Goel		Piepel		Goldfarb, Borror, and Montgomery
	Wagner and Gorman	1981	Cornell	1991	Cornell		Heiligers and Hilgers
1964	Myers		Koons and Heasley		Cornell and Linda		White et al.
	Uranisi		Snee		Crosier	2004	Anderson-Cook et al.
1965	Bounds, Kurotori, and Cruise		Vuchkov, Damgaliev, and Yonchev	1992	Piepel		Goldfarb et al.
	Draper and Lawrence*	1982	Gorman and Cornell		Chan		Kanjilal, Majumdar, and Pal
1966	Cruise		Murthy and Murty		Czitrom		Nardia, Acchar, and Hotzad
	Box and Gardiner,		Park and Kim		Kettaneh-Wold		Prescott
	Gorman Kurotori*		Piepel	1993	Cornell		Prescott and Draper
	McLean and Anderson*		Singh, Pratap, and Das		Draper et al.		Tang et al.
1967	Diamond		Snee and Rayner		Diuneveld, Smilde, and Doornbos		White et al.
	Drew		Yonchev		Mikaeili	2005	Bjorkestol and Naes
1968	Becker	1983	Cornell		Murthy and Murty Prescott et al.		Goldfarb et al.
	Lambrakis		Cornell et al.		Vining, Cornell, and Myers		Hamada, Martz, and Steiner
	Murty and Das		Murthy and Murty	1994	Derringer		Khuri
	Thompson and Myers*		Nigam, Gupta, and Gupta		Lewis et al.		Mage and Naes
1969	Becker*		Piepel		Montgomery and Voth		Nguyen and Piepel
	Hewlett		Vuchkov, Yonchev, and Damgaliev		Piepel and Cornell		Ozol-Godfrey et al.
	Lambrakis		Ying-nan		Smith and Cornell		Piepel, Cooley, and Jones
	Watson						Rajagopal, Del Castillo, and Peterson
							Smith

Year	Author(s)
1970	Becker
	Cornell and Good
	Gorman
	Nigam
1971	Cornell
	Cox
	Keviczky
	Paku, Manson, and Nelson
	Snee
1973	Cornell*
	Nigam
	Saxena and Nigam
	Snee*
1974	Hare
	Marquardt and Snee*
	Nigam
	Snee and Marquardt
1975	Cornell
	Cornell and Ott
	Laake
	Snee
	Mendieta, Linssen, and Doornbos
1976	Snee and Marquardt
1977	Cornell
1984	Aitchison and Bacon-Shone
	Chick and Piepel
	Cornell and Gorman
	Crosier*
1985	St. John
	Ying-nan
	Yonchev
	Aitchison
	Cornell
1986	Darroch and Waller
	Gorman and Cornell
	Hare
	Piepel and Cornell
	Snee
	Ying-nan
	Cain and Price
	Cornell
1987	Crosier
	Zhu, Hu, and Chen
	Hoerl
	Piepel and Cornell
	Snee
	Sahrmann, Piepel, and Cornell
1988	Draper and St. John
	Galil and Kiefer
	Hare and Brown
	Saxena and Nigam
	Ying-nan
	Zhu, Hu, and Chen
	Chan
	Cornell
	Czitrom
	Mikaeili
	Piepel
1995	Cornell
	Heinsman and Montgomery
	Hilgers and Bauer
1996	Chen, Li, and Jackson
	Murthy and Manga
1997	Cornell and Harrison
	Liu and Neudecker
	Piepel
	Smith and Beverly
1998	Steiner and Hamada
	Chan, Guan, and Zhang
	Chan et al.
	Cornell and Ramsey
	Piepel and Redgate
1999	Chan and Sandhu
	Khuri, Harrison, and Cornell
2000	Cornell
	Anderson and Whitcomb
	Gous and Swatson
2001	Chan and Guan
	Goos and Vandebroek
2006	Chantarat et al.
	Goldfarb and Montgomery
	Goos and Donev
	Piepel
2007	Akay
	Goos and Donev
	Muteki and MacGregor
	Piepel
	Sorenson et al.
2008	Adeyeye and Oyawale
	Mandal and Pal
	Menezes et al.
	Pal and Mandal
	Piepel et al.
2009	Anderson and Whitcomb
	Borkowski and Piepel
	Piepel and Cooley
	Piepel and Landmesser
	Prescott and Draper
	Sahni, Piepel, and Naes

Note: *Refers to a paper that was referenced often. Underlined names indicate more than one paper appeared by this/these authors during the year.

15

Several of the papers cited in Table 1.2 are particularly noteworthy for their content and also because of the time at which they appeared in the statistical literature. These papers are designated with an asterisk. Each of these papers is described briefly now.

Claringbold, P. J. (1955). *Use of the simplex design in the study of the joint action of related hormones.* The first paper to introduce a design on the three-component simplex and to present the corresponding fitted model and analysis of the data.

Scheffé, H. (1958). *Experiments with mixtures.* Introduced the simplex-lattice designs and the corresponding polynomial models. This paper is probably recognized as having done more than any other toward generating interest in the areas of design and analysis of mixture experiments.

Gorman, J. W., and J. E. Hinman (1962). *Simplex-lattice designs for multicomponent systems.* A lucid presentation on the use of simplex-lattice designs and Scheffé's polynomials.

Scheffé, H. (1963). *The simplex-centroid design for experiments with mixtures.* Introduced an alternative design to the $\{q, m\}$ simplex-lattice. The first paper to consider designs and models for experiments consisting of process variables and mixture components.

Draper, N. R., and W. E. Lawrence (1965). *Mixture designs for three factors. Mixture designs for four factors.* The first paper to suggest using designs that minimized the bias in the fitted model as well as the variance through minimizing the mean square error of the estimate of the response over the simplex region.

Kurotori, I. S. (1966). *Experiments with mixtures of components having lower bounds.* The first paper to use pseudocomponents to define a pseudocomponent simplex inside the original simplex space. This paper simplified the design problem when the components' proportions are restricted by lower bounds.

McLean, R. A., and V. L. Anderson (1966). *Extreme vertices design of mixture experiments.* The first paper to recommend an algorithm to generate the coordinates of the vertices of a constrained factor space resulting from the placing of upper and lower bounds on some or all of the component proportions.

Thompson, W. O., and R. H. Myers (1968). *Response surface designs for experiments with mixtures.* The first paper to consider an ellipsoidal region of interest inside the simplex factor space and the conditions for using rotatable response surface designs for fitting polynomial models.

Becker, N. G. (1969). *Regression problems when the predictor variables are proportions.* The first paper to discuss adaptations to commonly used regression and response surface techniques, such as the method of steepest descent and reduction to canonical form, when working with proportions.

Cornell, J. A. (1973). *Experiments with mixtures: a review.* A complete review of nearly all the published statistical papers on mixture designs and models.

Snee, R. D. (1973). *Techniques for the analysis of mixture data.* A lucid discussion of several fitted model forms as well as ways of analyzing mixture data.

Marquardt, D. W., and R. D. Snee (1974). *Test statistics for mixture models.* This paper discusses the correct test statistics for testing hypotheses about the parameters in the Scheffé mixture models.

Snee, R. D. (1979). *Experimental designs for mixture systems with multicomponent constraints.* This paper provides steps for generating the coordinates of the extreme vertices of a constrained region defined by the placing of constraints on the component proportions of the form $L_1 \leq \sum_{i=1}^{q} a_i x_i \leq U_l$.

Crosier, R. B. (1984). *Mixture experiments: geometry and pseudocomponents*. Introduces upper-bound pseudocomponents and presents a formula for enumerating the number of extreme vertices, the number of edges, faces, and so on of a convex polyhedron.

REFERENCES AND RECOMMENDED READING

Cornell, J. A. (1973). Experiments with mixtures: a review. *Technometrics*, **15**, No. 3, 437–455.

Cornell, J. A. (1990a). *How to Run Mixture Experiments for Product Quality*, 2nd ed. The ASQ Basic References in Quality Control: Statistical Techniques, Vol. **5**. ASQ, Milwaukee, WI.

Gorman, J. W. and J. E. Hinman (1962). Simplex-lattice designs for multicomponent systems. *Technometrics*, **4**, No. 4, 463–487.

Hare, L. B. (1974). Mixture designs applied to food formulation. *Food Technol.*, **28**, 50–62.

Scheffé, H. (1958). Experiments with mixtures. *J. R. Statist. Soc. B*, **20**, No. 2, 344–360.

Smith, W. F. (2005). *Experimental Design for Formulation*. ASA-SIAM Series on Statistics and Applied Probability. Alexandria, VA.

Snee, R. D. (1973). Techniques for the analysis of mixture data. *Technometrics*, **15**, No. 3, 517–528.

QUESTIONS

1.1. The functional relationship between the measured response and the ingredient proportions in a mixture problem is different in several ways from the standard regression functional relationship between a dependent variable and one or more independent variables. Explain.

1.2. List several experimental situations that fall under the heading of a mixture experiment.

1.3. Pure mixtures, binary mixtures, and complete mixtures are names given to blends of one or more ingredients or components. Distinguish between the various mixture types.

1.4. Mileage figures for each of the individual fuels A and B as well as the 50%: 50% blend of the two fuels A:B in five separate cases are as follows:

Case	A	B	A:B
1	17	10	15
2	12	18	15
3	6	6	4
4	10	20	12
5	9	12	12

In which of the cases are the fuels synergistic if higher is better? In which of the cases are the fuels antagonistic and when are the fuels neither synergistic nor antagonistic? In this last situation the fuels are said to be _____.

1.5. The Scheffé second-degree model in $q = 2$ components, whose proportions are x_1 and x_2 is of the form $y = \beta_1 x_1 + \beta_2 x_2 + \beta_{12} x_1 x_2 + \varepsilon$. Recall that when the $\{2, 2\}$ simplex-lattice or design (a) below is selected as the design, the estimates of the parameters β_1, β_2, and β_{12} are calculated using the following formulas:

With design (a), the estimation formulas for b_1, b_2 and b_{12} are

$$b_1 = y_1, b_2 = y_2 \quad \text{and} \quad b_{12} = 4\left(\frac{y_3 + y_4}{2}\right) - 2(y_1 + y_2).$$

With design (b), however, the estimation formulas for the parameters are

$$b_1 = (19y_1 + 3y_2 - 3y_3 + y_4)/20, \quad b_2 = (y_1 - 3y_2 + 3y_3 + 19y_4)/20,$$

$$b_{12} = 9(y_2 + y_3 - y_1 - y_2)$$

Designs (a) and (b) below are available. To provide minimum variance estimates of the parameters, which design (a) or (b) would you choose and why, if the variance of y_i is σ^2 at $i = 1, 2, 3$, and 4?

(a) y_1 at $x_1 = 1, x_2 = 0$
  y_3, y_4 at $x_1 = x_2 = \frac{1}{2}$
  y_2 at $x_1 = 0, x_2 = 1$

(b) y_1 at $x_1 = 1, x_2 = 0$
  y_2 at $x_1 = \frac{2}{3}, x_2 = \frac{1}{3}$
  y_3 at $x_1 = \frac{1}{3}, x_2 = \frac{2}{3}$
  y_4 at $x_1 = 0, x_2 = 1$

1.6. Shown are data values at the seven blends of a three-component simplex-centroid design. High values of the response are more desirable than low values. Indicate the type of blending, linear or nonlinear, synergistic or antagonistic, that is present between the components. The model fitted is a special cubic.

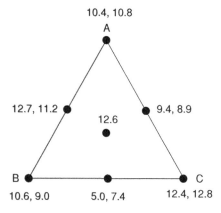

10.4, 10.8
A

12.7, 11.2   9.4, 8.9
          12.6

B   C
10.6, 9.0   5.0, 7.4   12.4, 12.8

(a) A and B:_____ and is _____.

(b) A and C:_____ and is _____.

(c) B and C:_____ and is _____.

(d) A and B and C:_____ and is _____.

Hint: $\widehat{\text{var}}(b_A) = \text{MSE}/2$, $\widehat{\text{var}}(b_{AB}) = 24(\text{MSE})/2$, $\widehat{\text{var}}(b_{ABC}) = 958.5(\text{MSE})$. Also SST = 63.79, SSR = 58.22, and SSE = 5.57.

1.7. Which particular design and model system appears simpler to work with to you? List the pro's and con's of each system.

Nonmixture or independent variables:
Pro's

Mixture or dependent variables:
Pro's

_____

_____

_____

_____

Con's

Con's

_____

_____

_____

_____

Nonmixture

Mixture

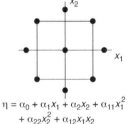

$q = 2$

$x_1 = 1$ $x_1 = 1/2$ $x_1 = 0$
$x_2 = 0$ $x_2 = 1/2$ $x_2 = 1$

$\eta = \alpha_0 + \alpha_1 x_1 + \alpha_2 x_2 + \alpha_{11} x_1^2 + \alpha_{22} x_2^2 + \alpha_{12} x_1 x_2$

$\eta = \beta_1 x_1 + \beta_2 x_2 + \beta_{12} x_1 x_2$

$q = 3$

$\eta = \alpha_0 + \alpha_1 x_1 + \alpha_2 x_2 + \alpha_2 x_3$
$+ \alpha_{11} x_1^2 + \alpha_{22} x_2^2 + \alpha_{33} x_3^2$
$+ \alpha_{12} x_1 x_2 + \alpha_{13} x_1 x_3 + \alpha_{23} x_2 x_3$

$\eta = \beta_1 x_1 + \beta_2 x_2 + \beta_3 x_3$
$+ \beta_{12} x_1 x_2 + \beta_{13} x_1 x_3 + \beta_{23} x_2 x_3$

APPENDIX 1A.   TESTING FOR NONLINEAR BLENDING OF THE TWO CHEMICALS VENDEX AND KELTHANE WHILE MEASURING THE AVERAGE PERCENT MORTALITY (APM) OF MITES

The first step in testing whether or not the chemicals blend nonlinearly is to determine if the average percent mortality (APM) values of the five blends in Figure 1.2 are different. To do so, we denote the true mean percent mortality of each of the five blends using the Greek letter mu $\mu_{(V, K)}$ and state the null hypothesis that the five means are equal as H_0: $\mu_{(1, 0)} = \mu_{(0.75, 0.25)} = \mu_{(0.5, 0.5)} = \mu_{(0.25, 0.75)} = \mu_{(0, 1)}$ versus the alternative hypothesis H_A: one or more of the $=$'s is $\neq$. The analysis begins by setting up an analysis of variance (ANOVA) table consisting of the following sources of variation among and within the five pesticide blends:

Source	d.f.	Sum of Squares	Mean Square	F	Probability
Among blends	$5 - 1 = 4$	SSAB $= 4919.07$	1229.77	1229.77/14.02 $= 87.7$	0.000
Within blends	$5(4 - 1) = 15$	SSWB $= 210.26$	14.02		
Total	$(5 \times 4) - 1 = 19$	SST $= 5129.33$			

To test H_0: $\mu_{(1, 0)} = \mu_{(0.75, 0.25)} = \ldots = \mu_{(0, 1)}$ versus H_A: one or more of the $=$ is $\neq$, the calculated value of $F = 87.7$ is compared to the table value $F(4, 15, 0.01) = 4.18$ and since $F = 87.7 > 4.18$, we reject H_0: at the 0.01 level of significance and declare one or more of the $=$ signs in H_A: is/are $\neq$, implying the means are not equal.

The next step is to show why we answered yes to the question, 'If the true APMs are not equal, does there appear to be a trend in the values of the APMs that can be defined by the chemical composition of the pesticide blends?' To answer this question, we begin by fitting the simple linear regression equation, APM = Intercept + slope K + error, or, APM $= \gamma_0 + \gamma_1 K + \varepsilon$, where the intercept ($\gamma_0$) is the value of APM when the value of Kelthane is zero and the slope (γ_1) is the change in APM for each 0.01 increase in Kelthane. The analysis begins by estimating both the intercept and the slope to produce the fitted model

$$\hat{y}(K) = 79.92 - 33.80K$$

The resulting ANOVA table is

Source	d.f.	Sum of Squares	Mean Square	F	Probability
Regression (fitted model)	1	2856.1	2856.1	2856.1/126.3 $= 22.61$	0.000[a]
Residual error	18	2273.2	126.3		
Lack of fit	3	2063.0	687.7	687.7/14.02 $= 49.06$	0.000
Pure error	15	210.3	14.02		
Total	19	5129.3			

[a]0.000 is the probability (or likelihood) of obtaining a value of F as large as 22.61 if the null hypothesis H_0 : $\mu_{(1,0)} = \cdots = \mu_{(0,1)}$ is true.

Based on the ANOVA table entries, does the fitted model $\hat{y}(K) = 79.92 - 33.80K$ adequately fit the data or does the model suffer from "lack of fit"? If the latter, then we must add one or more terms to the model. Suppose that we upgrade the model to the second degree, particularly since there appeared to be curvature in the trend of the APM values in the neighborhood of (V, K) between $(0.85, 0.15)$ and $(0.65, 0.35)$. So we increase the degree of the fitted model and rewrite it to include both Vendex and Kelthane as

$$\hat{y}(V, K) = 67.88V + 34.08K + 96.34V \times K$$

$$(1.78) \qquad (1.78) \qquad (8.08)$$

where the rightmost term stands for $V \times K$, along with the corresponding ANOVA table:

Source	d.f.	Sum of Squares	Mean Square	F	Probability
Regression	2	4886.53	2443.26	$2443.26/14.28 = 171.10$	0.000
Linear	1	2856.1	2856.10	200.01	0.000
Quadratic	1	2030.43	2030.43	142.19	0.000
Residual error	17	242.81	14.28		
Total	19	5129.3			

For the linear model above, the summary statistics R-Sq or $R^2 = 0.5568$ and R-Sq(adj) or $R_A^2 = 0.5322$. For the quadratic model above, R-Sq or $R^2 = 0.9527$ and R-Sq(adj) or $R_A^2 = 0.9471$. R^2 is the proportion of the total sum of squares that is explained by the fitted model while R_A^2 is a measure of the reduction in the estimate of the error variance based on fitting the model, or $SSE/(N - p)$, relative to the total mean square; see Section 7.4. These ANOVA tables and fitted models were produced by Minitab (1999). Plots of the two fitted models are displayed in Figure 1.9.

   Once the quadratic model $\hat{y}(V, K) = 67.88V + 34.08K + 96.34V \times K$ is selected as the better model, the null hypothesis $H_0: \beta_{12} = 0$ is tested against the alternative $H_a: \beta_{12} \neq 0$ at the $\alpha = 0.01$ level of significance. The estimate of β_{12} is 96.34 and is positive so that the test is to determine if the *synergistic blending* of Vendex with Kelthane is significant. The value of the test statistic is

$$t = \frac{96.34}{8.08} = 11.92$$

where 8.08 is the estimated standard error of the coefficient estimate (b_{vk}) whose value is 96.34; both estimates are provided by the computer software. Since the calculated value $t = 11.92$ exceeds Appendix Table A value of $t = 2.909$ at the .005 level of significance with 17 degrees of freedom, we reject H_0 and conclude the synergistic blending of Vendex and Kelthane is significant at the 0.005 level. In Figure 1.9 are shown the two fitted models where it is clear that the quadratic or nonadditive blending model fits the APMs in the data set more closely than the linear

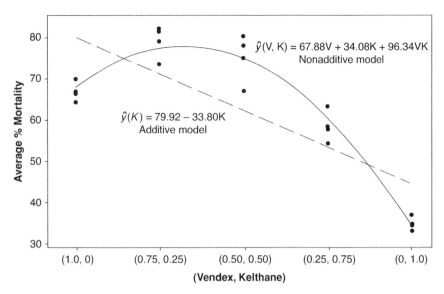

Figure 1.9 Fits of additive and nonadditive blending models across the average percent mortalities.

or additive model does. A final word on the values of Prob < 0.000 at the extreme right side of the ANOVA tables, this is saying the probability of attaining a value of F as large as each of the three values 0.000 in the table below F is less than one in a thousand.

CHAPTER 2

The Original Mixture Problem: Designs and Models for Exploring the Entire Simplex Factor Space

In this chapter we treat the most general description of the mixture problem, which is where the component proportions are to satisfy the constraints $x_i \geq 0$, $x_1 + x_2 + \cdots + x_q = 1.0$. Each component proportion x_i can take values from zero to unity, and all blends among the ingredients are possible. We concentrate on the fitting of mathematical equations to model the response surface over the entire simplex factor space, so that the empirical prediction of the response to any mixture over the entire simplex is possible.

What is meant by modeling is that a model or an equation is postulated to represent the response surface. We then choose a design at whose points we may collect observations to which the equation can be fitted (or, the coefficients in the regression equation can be estimated). Finally, the adequacy of the model is tested. This final step is to ensure that our fitted equation is a prediction tool with which we can feel comfortable.

The modeling sequence just mentioned will be altered slightly. First we discuss the simplex-lattice designs that were introduced by Scheffé in the early years (1958–1965) of the period in which research on mixture experiments was being developed. These designs are credited by many researchers to be the foundation on which the theory of experimental designs for mixtures was built, and yet these designs are still very much in use today. We then present the associated polynomial models to be fitted to data that are collected at the points of these designs.

2.1 THE SIMPLEX-LATTICE DESIGNS

To accommodate a polynomial equation to represent the response surface over the entire simplex region, a natural choice for a design would be one whose points

A Primer on Experiments with Mixtures, By John A. Cornell
Copyright © 2011 John Wiley & Sons, Inc. Published by John Wiley & Sons, Inc.

are spread evenly over the whole simplex factor space. An ordered arrangement consisting of a uniformly spaced distribution of points on a simplex is known as a *lattice.* The name lattice is used to make reference to an array of points.

A lattice may have a special correspondence to a specific polynomial equation. For example, to support a polynomial model of degree m in q components over the simplex, the lattice, referred to as a $\{q, m\}$ simplex-lattice, consists of points whose coordinates are defined by the following combinations of the component proportions. The proportions assumed by each component take the $m + 1$ *equally spaced values* from 0 to 1, that is,

$$x_i = 0, \ \frac{1}{m}, \ \frac{2}{m}, \ \ldots, \ 1 \qquad (2.1)$$

and the $\{q, m\}$ simplex-lattice consists of *all* possible combinations (mixtures) of the components where the proportions (2.1) for each component are used.

The listing of the specific component combinations comprising the $\{q, m\}$ simplex-lattice is illustrated as follows. Let us consider a $q = 3$ component system where the factor space for all blends is an equilateral triangle. Let each component assume the proportions $x_i = 0, \frac{1}{2}$, and 1 for $i = 1, 2$, and 3. Setting $m = 2$ for the proportions in Eq. (2.1), we can use a second-degree model to represent the response surface over the triangle. The $\{3, 2\}$ simplex-lattice consists of the six points on the boundary of the triangle

$$(x_1, x_2, x_3) = (1, 0, 0), (0, 1, 0), (0, 0, 1), \left(\tfrac{1}{2}, \tfrac{1}{2}, 0\right), \left(\tfrac{1}{2}, 0, \tfrac{1}{2}\right), \left(0, \tfrac{1}{2}, \tfrac{1}{2}\right)$$

The three points, which are defined as $(1, 0, 0)$ or $x_1 = 1, x_2 = x_3 = 0$; $(0, 1, 0)$ or $x_1 = x_3 = 0, x_2 = 1$; and $(0, 0, 1)$ or $x_1 = x_2 = 0, x_3 = 1$, represent single-component mixtures and these points are the three vertices of the triangle. The points $\left(\tfrac{1}{2}, \tfrac{1}{2}, 0\right)$, $\left(\tfrac{1}{2}, 0, \tfrac{1}{2}\right)$, and $\left(0, \tfrac{1}{2}, \tfrac{1}{2}\right)$ represent the binary blends or two-component mixtures $x_i = x_j = \tfrac{1}{2}, x_k = 0, k \neq i, j$, for which the nonzero component proportions are equal. The binary blends are located at the midpoints of the three edges of the triangle. The $\{3, 2\}$ simplex-lattice is shown in Figure 2.1.

Let us consider another example. We set the number of equally spaced levels (or proportions) for each component to be four, that is, $x_i = 0, \frac{1}{3}, \frac{2}{3}, 1$. If we consider all possible blends of the three components with these proportions, then the $\{3, m = 3\}$ simplex-lattice contains the following blending coordinates:

$$(x_1, x_2, x_3) = (1, 0, 0), (0, 1, 0), (0, 0, 1), \left(\tfrac{2}{3}, \tfrac{1}{3}, 0\right), \left(\tfrac{2}{3}, 0, \tfrac{1}{3}\right), \left(\tfrac{1}{3}, \tfrac{2}{3}, 0\right),$$

$$\left(\tfrac{1}{3}, 0, \tfrac{2}{3}\right), \left(\tfrac{1}{3}, \tfrac{1}{3}, \tfrac{1}{3}\right), \left(0, \tfrac{2}{3}, \tfrac{1}{3}\right), \left(0, \tfrac{1}{3}, \tfrac{2}{3}\right)$$

Note that each of the proportions of the components in every blend or mixture is either unity or a fractional number and that the sum of the fractions equals unity. When plotted as a lattice arrangement, these points represent an array of component blends that is symmetrical with respect to the orientation of the simplex (i.e., symmetrical

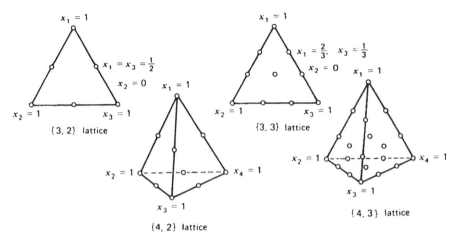

Figure 2.1 Some $\{3, m\}$ and $\{4, m\}$ simplex-lattice arrangements, $m = 2$ and $m = 3$.

with respect to the vertices and the sides of the simplex). The arrangement of the ten points of a $\{3, 3\}$ simplex-lattice is presented in Figure 2.1.

Before proceeding further, we should explain that throughout this book the coordinate system that we are using with the mixture components is called a *simplex coordinate system*. With three components, for example, the triangular coordinate system is represented by the fractional values in parentheses (x_1, x_2, x_3), where each $0 \leq x_i \leq 1$, $i = 1, 2$, and 3 and $x_1 + x_2 + x_3 = 1$. Several points of composition are chosen in the triangular system presented in Figure 2.2. When there appears to be no chance for confusion, the composition $(x_1 = a_1, x_2 = a_2, x_3 = a_3)$ is denoted by (a_1, a_2, a_3).

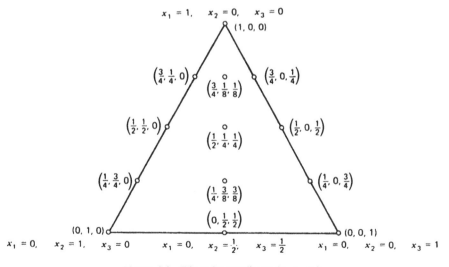

Figure 2.2 Triangular coordinates (x_1, x_2, x_3).

Table 2.1 Number of Points in the $\{q, m\}$ Simplex-Lattice for $3 \leq q \leq 10, 1 \leq m \leq 4$, where the Number of Levels for Each Component is $m + 1$

	Number of Components							
Degree of Model, m	$q = 3$	4	5	6	7	8	9	10
1	3	4	5	6	7	8	9	10
2	6	10	15	21	28	36	45	55
3	10	20	35	56	84	120	165	220
4	15	35	70	126	210	330	495	715

The number of design points in the $\{q, m\}$ simplex-lattice is $\binom{q+m-1}{m} = (q + m - 1)!/m!(q - 1)!$, where $m!$ is "m factorial" and $m! = m(m-1)(m-2) \cdots (2) (1)$. [The symbol $\binom{a}{b}$ is the combinational symbol for the number of ways a things can be taken b at a time and $\binom{a}{b} = a!/b!(a - b)!$.] In the $\{3, 2\}$ simplex-lattice, for example, the number of points is $\binom{3+2-1}{2} = 4!/2!2! = 6$, while the $\{3, 3\}$ simplex-lattice consists of $\binom{5}{3} = 10$ points. In the $\{q, m\}$ simplex-lattice then, the points correspond to pure or single-component mixtures, to binary or two-component mixtures, to ternary or three-component blends, and so on, up to mixtures consisting of at most m components. In Figure 2.1 the $\{4, 2\}$ and $\{4, 3\}$ simplex-lattices are shown. Table 2.1 lists the number of points in a $\{q, m\}$ simplex-lattice for values of q and m from $3 \leq q \leq 10, 1 \leq m \leq 4$.

2.2 THE CANONICAL POLYNOMIALS

A general form of regression function that can be fitted to data collected at the points of a $\{q, m\}$ simplex-lattice is derived using the following procedure. First, we recall that the equation of an mth-degree polynomial is written as

$$\eta = \beta_0 + \sum_{i=1}^{q} \beta_i x_i + \sum_{i \leq j}^{q} \sum^{q} \beta_{ij} x_i x_j + \sum_{i \leq j \leq k}^{q} \sum^{q} \sum^{q} \beta_{ijk} x_i x_j x_k + \cdots \tag{2.2}$$

where the terms up to the mth degree are included. The number of terms in Eq. (2.2) is $\binom{q+m}{m}$, but because the terms in Eq. (2.2) have meaning for us only subject to the restriction $x_1 + x_2 + \cdots + x_q = 1$, we know that the parameters $\beta_i, \beta_{ij}, \beta_{ijk}, \ldots$ associated with the terms are not unique. However, we may make the substitution

$$x_q = 1 - \sum_{i=1}^{q-1} x_i \tag{2.3}$$

in Eq. (2.2), thereby removing the dependency among the x_i terms, and this will not affect the degree of the polynomial. The effect of substituting Eq. (2.3) into Eq. (2.2) is that η becomes a polynomial of degree m in $q - 1$ components $x_1, x_2, \ldots,$ x_{q-1} with $\binom{q+m-1}{m}$ terms. And although the resulting formula after the substitution is simpler in form because it contains fewer components and fewer terms, the effect of component q is obscured by this substitution because the component is not included in the equation. Since we do not wish to sacrifice information on component q, we do not use Eq. (2.3). Instead we use another approach to derive an equation in place of Eq. (2.2) to represent the surface.

An alternative equation to Eq. (2.2) for a polynomial of degree m in q components, subject to the restriction on the x_i's in Eq. (1.2), is derived by multiplying some of the terms in Eq. (2.2) by the identity $(x_1 + x_2 + \cdots + x_q) = 1$ and simplifying. The resulting equation is called the "canonical" polynomial or the "canonical form of the polynomial," or simply the $\{q, m\}$ polynomial. (The name $\{q, m\}$ polynomial is given to these equations by some authors because this polynomial form is used often in conjunction with the $\{q, m\}$ simplex-lattice.) The number of terms in the $\{q, m\}$ polynomial is $\binom{q+m-1}{m}$, and this number is equal to the number of points that make up the associated $\{q, m\}$ simplex-lattice design. For example, for $m = 1$ and from Eq. (1.4),

$$\eta = \beta_0 + \sum_{i=1}^{q} \beta_i x_i$$

and upon multiplying the β_0 term by $(x_1 + x_2 + \cdots + x_q) = 1$, the resulting equation is

$$\eta = \beta_0 \left(\sum_{i=1}^{q} x_i \right) + \sum_{i=1}^{q} \beta_i x_i = \sum_{i=1}^{q} \beta_i^* x_i \tag{2.4}$$

where $\beta_i^* = \beta_0 + \beta_i$ for all $i = 1, 2, \ldots, q$. The number of terms in Eq. (2.4) is q, which is the number of points in the $\{q, 1\}$ lattice. The parameters $\beta_i^*, i = 1, 2, \ldots, q$, have simple and clear meanings that describe the shape of the response surface over the simplex region.

The general second-degree polynomial in q variables is

$$\eta = \beta_0 + \sum_{i=1}^{q} \beta_i x_i + \sum_{i=1}^{q} \beta_{ii} x_i^2 + \sum \sum_{i<j} \beta_{ij} x_i x_j \tag{2.5}$$

If we apply the identities $x_1 + x_2 + \cdots + x_q = 1$ and

$$x_i^2 = x_i \left(1 - \sum_{\substack{j=1 \\ j \neq i}}^{q} x_j \right) \tag{2.6}$$

then for $m = 2$,

$$\eta = \beta_0 \left(\sum_{i=1}^{q} x_i \right) + \sum_{i=1}^{q} \beta_i x_i + \sum_{i=1}^{q} \beta_{ii} x_i \left(1 - \sum_{j \neq i}^{q} x_j \right) + \sum_{i<j}^{q} \sum \beta_{ij} x_i x_j$$

$$= \sum_{i=1}^{q} (\beta_0 + \beta_i + \beta_{ii}) x_i - \sum_{i=1}^{q} \beta_{ii} x_i \sum_{j \neq i}^{q} x_j + \sum_{i<j}^{q} \sum \beta_{ij} x_i x_j$$

$$= \sum_{i=1}^{q} \beta_i^* x_i + \sum_{i<j}^{q} \sum \beta_{ij}^* x_i x_j \tag{2.7}$$

The number of terms in Eq. (2.7) is $q + q(q - 1)/2 = q(q + 1)/2$.

If we compare Eqs. (2.5) and (2.7), we see that the parameters in Eq. (2.7) are simple functions of the parameters in Eq. (2.5); that is, $\beta_i^* = \beta_0 + \beta_i + \beta_{ii}$ and $\beta_{ij}^* = \beta_{ij} - \beta_{ii} - \beta_{jj}$, $i, j = 1, 2, \ldots, q$, $i < j$. Furthermore Eq. (2.7) can be written in the homogeneous form as

$$\eta = \sum_{i=1}^{q} \delta_{ii} x_i^2 + \sum_{i<j}^{q} \sum \delta_{ij} x_i x_j = \sum_{i \leq j}^{q} \sum \delta_{ij} x_i x_j \tag{2.8}$$

which results from multiplying $\sum_{i=1}^{q} \beta_i^* x_i$ in Eq. (2.7) by the identity $(x_1 + x_2 + \cdots + x_q) = 1$ and then simplifying the terms.

The two models in Eqs. (2.7) and (2.8) are equivalent in the sense that one was derived from the other, Eq. (2.8) from Eq. (2.7), without changing the degree of the polynomial or reducing the number of terms. Owing to the restriction $x_1 + x_2 + \cdots + x_q = 1$ on the component proportions, an infinite number of regression functions can be derived from Eq. (2.5), and these equations are equivalent to Eqs. (2.7) and (2.8) when all of the component proportions are included. This is seen by realizing that for all functions ϕ, the linear equations $\beta_0 - \phi + \sum_{i=1}^{q} (\beta_i + \phi) x_i$ are equivalent when $\sum_{i=1}^{q} x_i = 1$.

The formula for the third-degree polynomial, or $\{q, 3\}$ polynomial, is

$$\eta = \sum_{i=1}^{q} \beta_i^* x_i + \sum_{i<j}^{q} \sum \beta_{ij}^* x_i x_j + \sum_{i<j}^{q} \sum \delta_{ij} x_i x_j (x_i - x_j) + \sum_{i<j<k}^{q} \sum \sum \beta_{ijk}^* x_i x_j x_k \tag{2.9}$$

A simpler formula for a special case of the cubic polynomial where the terms $\delta_{ij} x_i x_j$ $(x_i - x_j)$ are not considered is the special cubic polynomial

$$\eta = \sum_{i=1}^{q} \beta_i^* x_i + \sum_{i<j}^{q} \sum \beta_{ij}^* x_i x_j + \sum_{i<j<k}^{q} \sum \sum \beta_{ijk}^* x_i x_j x_k \tag{2.10}$$

Later in Section 2.15 we will encounter the special quartic model for $q = 3$, which is given by Eq. 2.50.

From this point forward, we will remove the asterisks from β_i^*, β_{ij}^*, and β_{ijk}^* and use β_i, β_{ij}, and β_{ijk} in all of the $\{q, m\}$ polynomials. The asterisks were used only to keep the parameters in the general polynomial Eq. (2.2) separate from the parameters in the derived $\{q, m\}$ polynomials. There in three components, the models of Eqs. (2.4), (2.7), (2.9), and (2.10), respectively, are to appear hereafter without asterisks, unless mentioned,

$$\eta = \beta_1 x_1 + \beta_2 x_2 + \beta_3 x_3$$

$$\eta = \beta_1 x_1 + \beta_2 x_2 + \beta_3 x_3 + \beta_{12} x_1 x_2 + \beta_{13} x_1 x_3 + \beta_{23} x_2 x_3$$

$$\eta = \beta_1 x_1 + \beta_2 x_2 + \beta_3 x_3 + \beta_{12} x_1 x_2 + \beta_{13} x_1 x_3 + \beta_{23} x_2 x_3$$
$$+ \delta_{12} x_1 x_2 (x_1 - x_2) + \delta_{13} x_1 x_3 (x_1 - x_3) + \delta_{23} x_2 x_3 (x_2 - x_3)$$
$$+ \beta_{123} x_1 x_2 x_3$$

$$\eta = \beta_1 x_1 + \beta_2 x_2 + \beta_3 x_3 + \beta_{12} x_1 x_2 + \beta_{13} x_1 x_3 + \beta_{23} x_2 x_3 + \beta_{123} x_1 x_2 x_3$$

The number of terms in the $\{q, m\}$ polynomials is a function of m, the degree of the equation, as well as the number of components q. The numbers of terms for several values of q are listed in Table 2.2.

The terms $\beta_i x_i$ and $\beta_{ij} x_i x_j$ in the $\{q, 1\}$ and $\{q, 2\}$ polynomial equations have simple interpretations. At the vertex corresponding to pure component i, for example, as in the two models of Eq. (2.4) and Eq. (2.7), we can set $x_i = 1$ to force $x_j = 0$ for all $j \neq i$; then $\eta = \beta_i$. The parameter β_i therefore represents the expected response to pure component i, and pictorially, β_i is the height of the response surface above the simplex at the vertex where $x_i = 1$ for $i = 1, 2, \ldots, q$, see Figure 2.6. (β_i heights are usually nonnegative quantities unless they represent extrapolated heights of a surface above a subregion inside the simplex as shown in Chapter 3, Figure 3.7, or unless side conditions are imposed on the β_i values. A set of conditions will be imposed on the β_i in Sections 5.7 and 5.8 when we come to Cox's polynomial model.) If Eq. (2.4) defines the response surface exactly, which is the case when the blending or mixing

Table 2.2 Number of Terms in the Canonical Polynomials

Number of Components, q	Linear	Quadratic	Special Cubic	Full Cubic
2	2	3	—	—
3	3	6	7	10
4	4	10	14	20
5	5	15	25	35
⋮	⋮	⋮	⋮	⋮
q	q	$q(q+1)/2$	$q(q^2+5)/6$	$q(q+1)(q+2)/6$

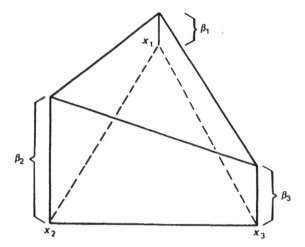

Figure 2.3 A planar surface above the three-component triangle. The surface is expressed as $\eta = \beta_1 x_1 + \beta_2 x_2 + \beta_3 x_3$, where in this case $\beta_2 > \beta_3 > \beta_1$.

among the components is strictly linear or additive, then the surface is depicted by a plane over the simplex. A planar surface for $q = 3$ is shown in Figure 2.3.

When the blending among the components is assumed to be linear (Figure 2.3), the response to the binary mixture of components i and j in the proportions x_i and x_j is given by Eq. (2.4) to be $\eta = \beta_i x_i + \beta_j x_j$, since all of the other x_k values are set to zero. If, however, the true response to the binary mixture of components i and j is more correctly represented by Eq. (2.7), which is $\eta = \beta_i x_i + \beta_j x_j + \beta_{ij} x_i x_i$, then an excess exists. The excess, which is represented by the term $\beta_{ij} x_i x_j$, is found by taking the difference between the models in Eqs. (2.7) and (2.4). If high positive values of the response are desirable, and the quantity β_{ij} is positive, the excess is called the synergism of the binary mixture, and β_{ij} is the quadratic or second-order coefficient or the binary synergism. (See Figure 2.4a.) The opposite of synergism (i.e., when β_{ij} is negative) is called antagonism of the binary mixture.

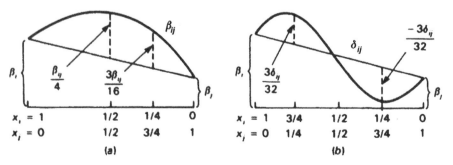

Figure 2.4 Nonlinear blending of binary mixtures, (a) Quadratic blending, $\beta_{ij} > 0$. (b) Cubic blending, $\delta_{ij} > 0$.

If the cubic formula (2.9) is the truest representation of the surface, then the excess or synergism of the binary mixture includes the additional term $\delta_{ij}x_ix_j(x_i - x_j)$, where δ_{ij} is the cubic coefficient of the binary synergism. Actually, along the $x_i - x_j$ edge, if $\delta_{ij} \neq 0$, the term $\delta_{ij}x_ix_j(x_i - x_j)$ takes on negative as well as positive values, enabling one to detect both synergistic and antagonistic blending between the components i and j. (See Figure 2.4b.) The term $\beta^*_{ijk}x_ix_jx_k$ in Eq. (2.9) represents ternary blending among the components i, j, and k in the interior of the triangle.

Another plausible way of understanding the separate terms in the first- and second-degree models in Eqs. (2.4) and (2.7) is to consider how the terms individually contribute toward the description of the shape of the mixture surface. The term β_ix_i contributes to the model only when the value of $x_i > 0$, and since β_i represents the height of the surface above the simplex at the vertex $x_i = 1$, the term β_ix_i contributes most (i.e., the value of $\beta_ix_i = \beta_i$ is greatest when $\beta_i > 0$) at $x_i = 1$. A term $\beta_{ij}x_ix_j$ in Eq. (2.7) contributes to the model everywhere in the simplex where both $x_i > 0$ and $x_j > 0$. On the edge joining the vertices corresponding to components i and j, the value of the term $\beta_{ij}x_ix_j$ is maximum when $\beta_{ij} > 0$, at $x_i = x_j = \frac{1}{2}$, where it is equal to $\beta_{ij}x_ix_j = \beta_{ij}/4$. (With the special cubic model, note that the term $\beta_{123}x_1x_2x_3$ contributes $\beta_{123}(\frac{1}{27})$ to the model, which is maximum at the centroid $(x_1, x_2, x_3, x_4, \ldots, x_q) = (\frac{1}{3}, \frac{1}{3}, \frac{1}{3}, 0, \ldots, 0)$ of the two-dimensional face of the simplex connecting the vertices $x_1 = 1, x_j = 0, j \neq 1; x_2 = 1, x_j = 0, j \neq 2;$ and $x_3 = 1, x_j = 0, j \neq 3$.)

2.3   THE POLYNOMIAL COEFFICIENTS AS FUNCTIONS OF THE RESPONSES AT THE POINTS OF THE LATTICES

As we mentioned previously, a special relationship exists between the $\{q, m\}$ simplex-lattice and the $\{q, m\}$ polynomial equation. This relationship is a one-to-one correspondence between the number of points in the lattice and the number of terms in the polynomial. As a result of this relationship, the parameters in the polynomial can be expressed as simple functions of the expected responses at the points of the $\{q, m\}$ simplex-lattice. In order to show this, we shall reintroduce the response nomenclature that was first proposed by Scheffé in his 1958 paper on mixtures.

Let the response to pure component i be denoted by η_i; the response to the binary mixture with equal proportions (0.50, 0.50) of components i and j be denoted by η_{ij}; and the response to the ternary mixture with equal proportions of components i, j, and k by η_{ijk}. In Figure 2.5 the response nomenclature is illustrated at the points of the $\{3, 2\}$ and $\{3, 3\}$ simplex-lattices, respectively.

The equations for expressing the parameters in the polynomial models in terms of η_i, η_{ij}, and η_{ijk}, are given by solving $\binom{q+m-1}{m}$ equations simultaneously. This number corresponds not only to the number of parameters in the $\{q, m\}$ polynomial equation but also to the number of lattice points and therefore to the number of expected responses η_i, η_{ij} measured at the points in the $\{q, m\}$ simplex-lattice as well. For example, if the second-degree model in Eq. (2.7) is to be used for a three-component

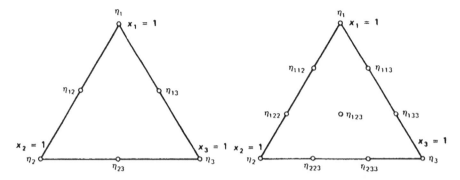

Figure 2.5 Response nomenclature at the points of the $\{3, 2\}$ and $\{3, 3\}$ simplex-lattices.

system and we have the expected responses at the points of the $\{3, 2\}$ simplex-lattice design as shown in Figure 2.5, then for the equation

$$\eta = \beta_1 x_1 + \beta_2 x_2 + \beta_3 x_3 + \beta_{12} x_1 x_2 + \beta_{13} x_1 x_3 + \beta_{23} x_2 x_3 \tag{2.11}$$

if we substitute

$$\eta_i \quad \text{at} \quad x_i = 1, \quad x_j = 0, \quad i, j = 1, 2, 3, \quad j \neq i$$
$$\eta_{ij} \quad \text{at} \quad x_i = \tfrac{1}{2}, \quad x_j = \tfrac{1}{2}, \quad x_k = 0, \quad i < j, k \neq i, j$$

into Eq. (2.11), the following $\binom{3+2-1}{2} = 6$ equations result:

$$\eta_1 = \beta_1, \quad \eta_2 = \beta_2, \quad \eta_3 = \beta_3$$
$$\eta_{12} = \beta_1(\tfrac{1}{2}) + \beta_2(\tfrac{1}{2}) + \beta_{12}(\tfrac{1}{4})$$
$$\eta_{13} = \beta_1(\tfrac{1}{2}) + \beta_3(\tfrac{1}{2}) + \beta_{13}(\tfrac{1}{4})$$
$$\eta_{23} = \beta_2(\tfrac{1}{2}) + \beta_3(\tfrac{1}{2}) + \beta_{23}(\tfrac{1}{4})$$

Solving the six equations simultaneously, and this is possible because the number of equations is equal to the number of unknown parameters, we find that the formulas for the parameters β_i and β_{ij}, $i, j = 1, 2$, and 3, $i < j$, are

$$\beta_1 = \eta_1, \quad \beta_{12} = 4\eta_{12} - 2\eta_1 - 2\eta_2$$
$$\beta_2 = \eta_2, \quad \beta_{13} = 4\eta_{13} - 2\eta_1 - 2\eta_3 \tag{2.12}$$
$$\beta_3 = \eta_3, \quad \beta_{23} = 4\eta_{23} - 2\eta_2 - 2\eta_3$$

The parameter β_i represents the response to pure component i, and β_{ij} is a contrast that compares the response at the midpoint of the edge connecting the vertices of components i and j with the responses at the vertices of components i and j. Thus, in the six-term polynomial Eq. (2.11), the sum $\beta_1 x_1 + \beta_2 x_2 + \beta_3 x_3$ represents linear or additive blending of the three components while the extra terms $\beta_{ij} x_i x_j$, $i < j$, are said to represent measures of departures from the plane of the second-degree surface resulting from the nonadditive blending of the components. Shown in Figure 2.6 are the planar and nonplanar portions of a curved surface directly above the three-component triangle. The planar surface is modeled by the three linear blending terms, while the nonplanar portion of the surface is modeled by the nonlinear blending terms of the quadratic model.

Equations (2.12) are derived using only three components for reasons of convenience. It is easy to display the expected responses at the points of the $\{3, 2\}$

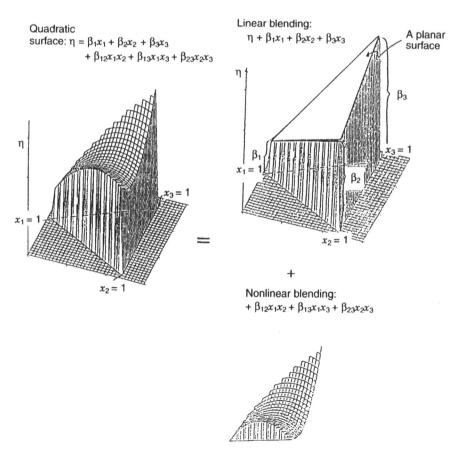

Figure 2.6 Planar and nonplanar portions of a surface exhibiting curvature along with the terms in the three-component quadratic model that describe these portions of the surface.

simplex-lattice and only six equations are necessary for setting up the formulas for the β_i and β_{ij}, $i, j = 1, 2$, and 3, $i < j$. For the general case of q components where the second-degree model of Eq. (2.7) contains $q(q+1)/2$ terms and the expected responses are positioned at the points of a $\{q, 2\}$ simplex-lattice design, the formulas for expressing the parameters β_i and β_{ij} in terms of η_i and η_{ij} are identical to Eqs. (2.12). In other words, for general q, where $i, j = 1, 2, \ldots, q$, $i < j$,

$$\beta_i = \eta_i, \quad \beta_{ij} = 4\eta_{ij} - 2(\eta_i + \eta_j) \tag{2.13}$$

For higher degree cases $m > 2$, the formulas can be derived in a manner similar to that for the second-degree model. Gorman and Hinman (1962) presented the formulas for the parameters in the cubic and quartic polynomial equations and these formulas are given in Appendix 2B.

2.4 ESTIMATING THE PARAMETERS IN THE $\{q,m\}$ POLYNOMIALS

The parameters in the $\{q, m\}$ polynomials are expressible as simple functions of the expected responses at the points of the $\{q, m\}$ simplex-lattice designs. Thus one could conjecture that to estimate the parameters in the models using observed values of the response at the lattice points, the computing formulas for b_i and b_{ij}, the estimates of β_i and β_{ij}, respectively, will be identical to Eqs. (2.13) with the observed values substituted in Eq. (2.13) in place of η_i and η_{ij}. To show that this is the case, we will consider the fitting of the three-component second-degree model in Eq. (2.11) to data values collected at the points of a $\{3, 2\}$ simplex-lattice design.

Recall from Section 1.2 that the observed value of the response in the uth trial, $1 \leq u \leq N$, denoted by y_u, is expressible in the form $y_u = \eta_u + \varepsilon_u$, where the ε_u for all $1 \leq u \leq N$ are uncorrelated and identically distributed random errors assumed to have a zero mean and a common variance σ^2. Now let us alter this notation temporarily by writing the observed response with the same nomenclature that was used for the expected response; that is, we denote the observed value of the response to the pure component i (i.e., at $x_i = 1$, $x_j = 0$, $j \neq i$) by y_i and the observed value of the response to the 50% : 50% binary mixture ($x_i = \frac{1}{2}$, $x_j = \frac{1}{2}$, $x_k = 0$ for all $i < j \neq k$) of components i and j by y_{ij}. Replacing the η_i and η_{ij} with y_i and y_{ij}, respectively, in Eq. (2.13) and letting b_i and b_{ij} denote the estimates of β_i and β_{ij}, respectively, we find that

$$b_i = y_i, \quad i = 1, 2, \ldots, q$$
$$b_{ij} = 4y_{ij} - 2(y_i + y_j), \quad i = 1, 2, \ldots, q, i < j \tag{2.14}$$

or

$$\frac{b_{ij}}{4} = y_{ij} - \frac{y_i + y_j}{2}$$

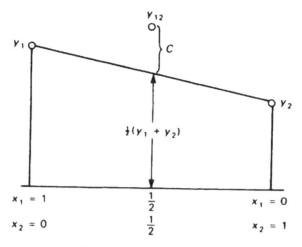

Figure 2.7   Contrast, $C = y_{12} - \frac{1}{2}(y_1 + y_2)$, and the nonlinear blending of components 1 and 2. In the quadratic model, $\beta_{12} = 4C$.

In the last contrast, $y_{ij} - (y_i + y_j)/2$, the quantity $b_{ij}/4$ represents the difference between the height of the surface observed at the blend $x_i = x_j = \frac{1}{2}$ and the average of the heights of the surface observed at the vertices corresponding to the single components i and j. This difference is shown in Figure 2.7. Furthermore, if r_i, r_j, and r_{ij} replicate observations are collected at $x_i = 1$, $x_j = 0$; at $x_i = 0$, $x_j = 1$; and at $x_i = x_j = \frac{1}{2}$, $x_k = 0$, $i < j$, $k \neq i$, j, respectively, and the averages $\bar{y}_i$, $\bar{y}_j$, and $\bar{y}_{ij}$ are calculated from the replicates, then the averages are substituted into Eq. (2.14) and the least-squares calculating formulas for the parameter estimates become

$$b_i = \bar{y}_i, \qquad i = 1, 2, \dots, q$$
$$b_{ij} = 4\bar{y}_{ij} - 2(\bar{y}_i + \bar{y}_j), \quad i, j = 1, 2, \dots, q, \quad i < j \tag{2.15}$$

Note that the scalar quantities 4 and 2 in the formula for b_{ij} do not depend on the values of r_i, r_j, and r_{ij} but rather come from the values of x_j and x_j being equal to $\frac{1}{2}$. The formulas for calculating the estimates of the parameters in the cubic model and the quartic or fourth-degree polynomial equation are presented in Appendix 2B at the end of this chapter.

Equations (2.15) for the estimates b_i and b_{ij} are the least-squares solutions to the normal equations that are shown as Eqs. (7.4) in Chapter 7. The matrix formula for the least-squares estimates is Eq. (7.5). The properties of the estimates depend on the distributional properties of the random errors ε_u. We have assumed that the errors ε_u, $1 \leq u \leq N$, are uncorrelated and identically distributed with mean zero and a variance of σ^2. Thus the means and the variances of the distributions of the estimates b_i and b_{ij}, given that the observations were collected at the points of the lattice only, are

$$E(b_i) = E(\bar{y}_i) = \beta_i$$

$$\operatorname{var}(b_i) = \operatorname{var}(\bar{y}_i) = \frac{\sigma^2}{r_i}$$

$$E(b_{ij}) = E[4\bar{y}_{ij} - 2(\bar{y}_i + \bar{y}_j)] = \beta_{ij} \qquad (2.16)$$

$$\operatorname{var}(b_{ij}) = \operatorname{var}[4\bar{y}_{ij} - 2(\bar{y}_i + \bar{y}_j)] = \frac{16\sigma^2}{r_{ij}} + \frac{4\sigma^2}{r_i} + \frac{4\sigma^2}{r_j}$$

and

$$\operatorname{cov}(b_i, b_j) = E[\bar{y}_i(\bar{y}_j)] - E(\bar{y}_i)E(\bar{y}_j) = 0, \quad i \neq j$$

$$\operatorname{cov}(b_i, b_{ij}) = E[\bar{y}_i(4\bar{y}_{ij} - 2\bar{y}_i - 2\bar{y}_j)] - E(\bar{y}_i)E(4\bar{y}_{ij} - 2\bar{y}_i - 2\bar{y}_j)$$

$$= -2E(\bar{y}_i^2) + 2(E\bar{y}_i)^2 = -\frac{2\sigma^2}{r_i} \qquad (2.17)$$

$$\operatorname{cov}(b_{ij}, b_{ik}) = \frac{4\sigma^2}{r_i}, \qquad j \neq k$$

In Eqs. (2.16) and (2.17), $E(\cdot)$ denotes expectation, $\operatorname{var}(b_i)$ represents the variance of b_i, and $\operatorname{cov}(b_i, b_{ij})$ is the covariance between b_i and b_{ij}. Knowledge of the properties of the parameter estimates and of the formulas used to calculate the variances of the estimates can be helpful when inferring whether or not the magnitudes of the parameter estimates are significantly different from zero when testing hypotheses on the parameters in the model. Furthermore, we will see in the next two sections that throughout the experimental region the variance of the predicted response is directly affected by the properties of the parameter estimates in (2.16) and (2.17). Finally, if the errors are assumed to be normally distributed, that is, if $\varepsilon_i \sim$ Normal $(0, \sigma^2)$, and if an equal number of replicates $r_i = r_{ij} = r$ is collected at each of the design points, then $b_i \sim$ Normal$(\beta_i, \sigma^2/r)$ and $b_{ij} \sim$ Normal $(\beta_{ij}, 24\sigma^2/r)$, where $\sim$ denotes "is distributed as."

Once the parameters of the second-degree model in Eq. (2.7) are estimated, and the estimates b_i and b_{ij} are substituted for the β_i and β_{ij}, respectively, the fitted equation becomes

$$\hat{y} = \sum_{i=1}^{q} b_i x_i + \sum \sum_{i<j}^{q} b_{ij} x_i x_j \qquad (2.18)$$

An estimate of the value of the response at any point $\mathbf{x} = (x_1, x_2, \ldots, x_q)'$ inside or on the boundary of the simplex is found by substituting the values of the x_i's into Eq. (2.18). We will therefore denote the estimate of η at the blend $\mathbf{x}$ by $\hat{y}(\mathbf{x})$.

2.5   PROPERTIES OF THE ESTIMATE OF THE RESPONSE $\hat{y}(\mathbf{x})$

Since the estimates b_i and b_{ij} are linear functions of random variables (the y_i and y_{ij}) and are likewise random variables, the estimate $\hat{y}(\mathbf{x})$ of the response at $\mathbf{x}$ is a random variable. When the estimates b_i and b_{ij} are unbiased, which is the case when the fitted model is of the same degree in the x_i as the true surface, then the expectation of $\hat{y}(\mathbf{x})$ is $E[\hat{y}(\mathbf{x})] = \eta$.

A formula for the variance of the estimate $\hat{y}(\mathbf{x})$ can be written in terms of the variances and covariances of the b_i and b_{ij}, which are given in Eqs. (2.16) and (2.17). An easier method for obtaining the variance of $\hat{y}(\mathbf{x})$, is to replace the parameter estimates b_i and b_{ij} by their respective linear combinations of the averages $\bar{y}_i$ and $\bar{y}_{ij}$, which are defined in Eq. (2.15). The variance of $\hat{y}(\mathbf{x})$ can then be written as a function of the variances of the $\bar{y}_i$ and $\bar{y}_{ij}$.

When treated as a function of the averages $\bar{y}_i$ and $\bar{y}_{ij}$ at the lattice points, the estimate of the response is

$$
\begin{aligned}
\hat{y}(\mathbf{x}) &= \sum_{i=1}^{q} b_i x_i + \sum \sum_{i<j} b_{ij} x_i x_j \\
&= \sum_{i=1}^{q} \bar{y}_i x_i + \sum \sum_{i<i} (4\bar{y}_{ij} - 2\bar{y}_i - 2\bar{y}_j) x_i x_j \\
&= \sum_{i=1}^{q} \bar{y}_i \left[x_i - 2x_i \left(\sum_{j \neq i}^{q} x_j \right) \right] + \sum \sum_{i<j}^{q} 4\bar{y}_{ij} x_i x_j \\
&= \sum_{i=1}^{q} a_i \bar{y}_i + \sum \sum_{i<j} a_{ij} \bar{y}_{ij}
\end{aligned}
\tag{2.19}
$$

where $a_i = x_i(2x_i - 1)$ and $a_{ij} = 4x_i x_j$, $i, j = 1, 2, \ldots, q$, $i < j$. For the coefficients a_i and a_{ij}, the values of the x_i are specified by the values in $\mathbf{x} = (x_1, \ldots, x_q)'$ and thus are fixed without error. Since the $\bar{y}_i$ and $\bar{y}_{ij}$ are averages of r_i and r_{ij} observations, respectively, then the variance of $\hat{y}(\mathbf{x})$ in Eq. (2.19) can be written as

$$
\mathrm{var}[\hat{y}(\mathbf{x})] = \sigma^2 \left\{ \sum_{i=1}^{q} \frac{a_i^2}{r_i} + \sum \sum_{i<j}^{q} \frac{a_{ij}^2}{r_{ij}} \right\}
\tag{2.20}
$$

Of course, when there is an equal number of observations, r, at each lattice point, the formula for the variance of the estimate of the response at the point $\mathbf{x}$ is simplified to

$$
\mathrm{var}[\hat{y}(\mathbf{x})] = \frac{\sigma^2}{r} \left\{ \sum_{i=1}^{q} a_i^2 + \sum \sum_{i<j}^{q} a_{ij}^2 \right\}
\tag{2.21}
$$

In Eq. (2.21) the quantity σ^2/r is dependent on the precision (through σ^2) of the experimental observations, while $\{\sum_{i=1}^{q} a_i^2 + \sum \sum_{i<j}^{q} a_{ij}^2\}$ is dependent only on the composition (through the a_i's and thus the x_i's) of the mixture at which the estimate $\hat{y}(\mathbf{x})$ is being considered. When σ^2 is unknown, it can be estimated using a sample measure, s^2, from the r_i and r_{ij} replicate observations. An estimate of var$[\hat{y}(\mathbf{x})]$ is obtained by substituting s^2 for σ^2 in Eq. (2.20) and is written as $\widehat{\text{var}}[\hat{y}(\mathbf{x})]$. Finally, if, at $\mathbf{x}$, a $(1 - \alpha) \times 100\%$ confidence interval for η is desired, then the interval is

$$\hat{y}(\mathbf{x}) - \Delta < \eta < \hat{y}(\mathbf{x}) + \Delta \qquad (2.22)$$

where $\Delta = [t_{f,\alpha/2}]\{\widehat{\text{var}}[\hat{y}(\mathbf{x})]\}^{1/2}$, f is the number of degrees of freedom associated with the sample estimate s^2 used to estimate σ^2, and $t_{f,\alpha/2}$ is the tabled t-value with f degrees of freedom at the $\alpha/2$ level of significance. The general formula for the variance of $\hat{y}(\mathbf{x})$, using matrix notation, is presented in Appendix 2A at the end of this chapter.

We now illustrate the fitting of the second-degree polynomial using data collected from a three-component yarn-manufacturing experiment. We use the example data to introduce methods for determining how well the fitted model represents the response surface.

2.6   A THREE-COMPONENT YARN EXAMPLE USING A $\{3, 2\}$ SIMPLEX-LATTICE DESIGN

Three constituents—polyethylene (x_1), polystyrene (x_2), and polypropylene (x_3)—are blended together and the resulting fiber material is spun to form yarn for draperies. Only pure blends and binary blends are studied in this example, where the response of interest is the elongation of the yarn measured in kilograms of force applied. Average values of elongation in kilograms of force are presented in Figure 2.8 at the points

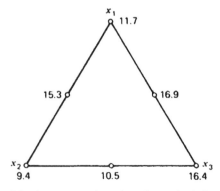

Figure 2.8   Average yarn elongation values at the design points.

Table 2.3   Observed Yarn Elongation Values

Design Point	Component Proportions			Observed Elongation Value (y_u)	Average Elongation Value $(\bar{y})$
	x_1	x_2	x_3		
1	1	0	0	11.0, 12.4	11.7
2	$\frac{1}{2}$	$\frac{1}{2}$	0	15.0, 14.8, 16.1	15.3
3	0	1	0	8.8, 10.0	9.4
4	0	$\frac{1}{2}$	$\frac{1}{2}$	10.0, 9.7, 11.8	10.5
5	0	0	1	16.8, 16.0	16.4
6	$\frac{1}{2}$	0	$\frac{1}{2}$	17.7, 16.4, 16.6	16.9

of the $\{3,2\}$ simplex-lattice design, where the averages were calculated from two replicate samples collected from each of the pure blends and three replicate samples collected on the binary or two-component blends. The mixture settings, the observed yarn elongation values, and the averages of the elongation values are presented in Table 2.3.

The fitted model in the three components is of the form

$$\hat{y}(\mathbf{x}) = b_1x_1 + b_2x_2 + b_3x_3 + b_{12}x_1x_2 + b_{13}x_1x_3 + b_{23}x_2x_3$$

where, from Eq. (2.15), the estimates are

$$b_1 = \bar{y}_1 = (11.0 + 12.4)/2 = 11.7, \quad b_2 = 9.4, \quad b_3 = 16.4$$
$$b_{12} = 4\bar{y}_{12} - 2(\bar{y}_1 + \bar{y}_2) = 4(15.3) - 2(11.7 + 9.4) = 61.2 - 2(21.1) = 19.0$$
$$b_{13} = 4\bar{y}_{13} - 2(\bar{y}_1 + \bar{y}_3) = 4(16.9) - 2(11.7 + 16.4) = 11.4, b_{23} = -9.6 \quad (2.23)$$

An estimate of the error variance σ^2 is obtained from the replicate observations at the lattice points. The estimate is

$$s^2 = \sum_{l=1}^{6} \sum_{u=1}^{2\text{ or }3} \frac{(y_{lu} - \bar{y}_l)^2}{\sum_{l=1}^{6}(r_l - 1)}$$
$$= \frac{(11.0 - 11.7)^2 + (12.4 - 11.7)^2 + (15.0 - 15.3)^2 + \cdots + (16.6 - 16.9)^2}{1 + 2 + 1 + 2 + 1 + 2}$$
$$= \frac{6.56}{9} = 0.73$$

where, in the formula for s^2, $\bar{y}_l$ is the average of the r_l observations at the lth design point: The number of degrees of freedom for the estimate s^2 is 9, and this number appears in the denominator of the quotient for s^2. Estimates of the variances of the parameter estimates in (2.23) are obtained from Eqs. (2.16) along with $s^2 = 0.73$, to give

$$\widehat{\text{var}}(b_i) = \frac{s^2}{r_i} = \frac{0.73}{2} = 0.37, \qquad i = 1, 2, \text{ and } 3$$

$$\widehat{\text{var}}(b_{ij}) = s^2 \left\{ \frac{16}{r_{ij}} + \frac{4}{r_i} + \frac{4}{r_j} \right\} = 0.73 \left\{ \frac{16}{3} + 4 \right\} = 6.81$$

The estimated standard error of each parameter estimate is the positive square root of the estimated variance of the estimate. The estimated standard errors are est. s.e. $(b_i) = \sqrt{\widehat{\text{var}}(b_i)}$ and est. s.e. $(b_{ij}) = \sqrt{\widehat{\text{var}}(b_{ij})}$, respectively. The values of the estimates of the standard errors are est. s.e.$(b_i) = 0.60$ and est. s.e.$(b_{ij}) = 2.61$. We will adopt the practice of placing the estimates of the standard errors in parentheses directly below the corresponding parameter estimate in the fitted equation.

The fitted second-degree polynomial equation is

$$\hat{y}(\mathbf{x}) = 11.7x_1 + 9.4x_2 + 16.4x_3 + 19.0x_1x_2 + 11.4x_1x_3 - 9.6x_2x_3$$
$$\phantom{\hat{y}(\mathbf{x}) = } (0.60) \quad (0.60) \quad (0.60) \quad (2.61) \quad\quad (2.61) \quad\quad (2.61) \tag{2.24}$$

If we can assume that the fitted model in Eq. (2.24) is an adequate representation of the yarn elongation surface, we can draw the following conclusions from the magnitudes of the parameter estimates:

$$b_3 > b_1 > b_2$$

Of the three single-component blends, component 3 (polypropylene) produced yarn with the highest elongation, followed by component 1 and then component 2.

$$b_{12} > 0, \quad b_{13} > 0, \quad b_{23} < 0$$

Components 1 and 2, and components 1 and 3, have binary synergistic effects; that is, the binary blends with component 1 produced higher elongation values than would be expected by simply averaging the elongation values of the pure blends. When components 2 and 3 were combined, the resulting yarn had a lower average elongation value than would be expected by averaging the elongation values of the yarn produced by the single-component blends.

Suppose that as a next step in the analysis of the yarn elongation values, we elect to test whether the binary or nonlinear blending effects are significantly different from zero, that is, we test

$$H_0 : \beta_{ij} = 0 \text{ versus } H_A : \beta_{ij} \neq 0, \qquad i, j = 12, 13, 23$$

at the $\alpha = 0.01$ level. Then the three individual t tests could be performed as follows:

$$t = \frac{b_{12}}{\sqrt{6.81}} = \frac{19.0}{2.61} = 7.28, \quad t = \frac{b_{13}}{\sqrt{6.81}} = \frac{11.4}{2.61} = 4.37,$$

$$t = \frac{b_{23}}{\sqrt{6.81}} = \frac{-9.6}{2.61} = -3.68$$

At the $\alpha = 0.01$ level and from the t-statistic Appendix Table A, $t_{0.005,9} = 3.25$, and since all three $|t|$ values above are greater than 3.25, we reject all three null hypotheses $H_0 : \beta_{ij} = 0$ in favor of $H_A : \beta_{ij} \neq 0$. More on this later at the start of Chapter 4, Section 4.2.

Our conclusions then are: if yarn with high elongation is desirable and a single-component blend is wanted, use component 3. If a binary blend is desired because either component 3 costs more than the others or because of the lack of availability of component 3, use component 1 with either of the other two components. A contour plot of the estimated elongation surface is presented in Figure 2.9.

The formula for the estimated variance of $\widehat{y}(\mathbf{x})$, at the point $\mathbf{x}$ in the three-component triangle is, from Eq. (2.20),

$$\widehat{\text{var}}[\widehat{y}(\mathbf{x})] = s^2 \left\{ \sum_{i=1}^{3} \frac{a_i^2}{r_i} + \sum_{i<j}^{3} \sum \frac{a_{ij}^2}{r_{ij}} \right\} = 0.73 \left\{ \sum_{i=1}^{3} \frac{a_i^2}{2} + \sum_{i<j}^{3} \sum \frac{a_{ij}^2}{3} \right\} \qquad (2.25)$$

where $a_i = x_i(2x_i - 1)$ and $a_{ij} = 4x_i x_j$. At the point $(x_1 = \frac{2}{3}, x_2 = \frac{1}{3}, x_3 = 0)$, for example, the coefficients are $a_1 = \left(\frac{2}{3}\right)\left(\frac{1}{3}\right) = \frac{2}{9}$, $a_2 = \left(\frac{1}{3}\right)\left(-\frac{1}{3}\right) = -\frac{1}{9}$, $a_3 = (0)$,

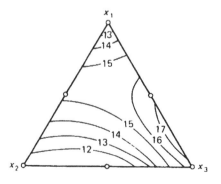

Figure 2.9 Estimated yarn elongation surface obtained with the second-degree model of Eq. (2.24).

Table 2.4　Calculated Values of $\hat{y}(x)$ and 95% Confidence Limits for the True Yarn Elongation Values of Seven Arbitrarily Selected Blends

| Blend | Component Properties | | | Lower Limit | $\hat{y}(x)$ | Upper Limit | $\{\widehat{\text{var}}[\hat{y}(x)]\}^{1/2}$ |
	x_1	x_2	x_3				
1	0.68	0.16	0.16	14.4	15.2	16.0	0.35
2	0.16	0.68	0.16	11.4	12.2	13.0	0.35
3	0.16	0.16	0.68	14.4	15.2	16.0	0.35
4	0.33	0.33	0.33	13.7	14.6	15.5	0.39
5	0.80	0.10	0.10	13.5	14.3	15.1	0.37
6	0.90	0.05	0.05	12.2	13.2	14.2	0.45
7	0.72	0.14	0.14	14.1	14.9	15.7	0.35

$a_{12} = 4 \left(\frac{2}{3}\right) \left(\frac{1}{3}\right) = \frac{8}{9}$, $a_{13} = 0$, $a_{23} = 0$, and the estimated variance of $\hat{y}\left(\frac{2}{3}, \frac{1}{3}, 0\right)$ is

$$\widehat{\text{var}}[\hat{y}(x)] = 0.73 \left\{ \frac{\left(\frac{2}{9}\right)^2 + \left(\frac{1}{9}\right)^2 + (0)^2}{2} + \frac{\left(\frac{8}{2}\right)^2 + (0)^2 + (0)^2}{3} \right\} = 0.73 \left\{ \frac{143}{486} \right\} = 0.21$$

The estimate of the response at the point $x = \left(\frac{2}{3}, \frac{1}{3}, 0\right)'$ is

$$\hat{y}(x) = 11.7 \left(\frac{2}{3}\right) + 9.4 \left(\frac{1}{3}\right) + 16.4(0) + 19.0 \left(\frac{2}{3}\right) \left(\frac{1}{3}\right) + 11.4 \left(\frac{2}{3}\right) (0)$$
$$- 9.6 \left(\frac{1}{3}\right) (0) = 15.2$$

A 95% confidence interval for η, based on Eq. (2.22) and for $\alpha = 0.05$ so that $\alpha/2 = 0.05/2 = 0.025$, is

$$15.2 - \Delta \leq \eta \leq 15.2 + \Delta$$
$$14.2 \leq \eta \leq 16.2$$

where $\Delta = [t_{9,0.025} = 2.262] \sqrt{0.21} = 1.0$. The calculated values of $\hat{y}(x)$ and $\{\widehat{\text{var}}[\hat{y}(x)]\}^{1/2}$ for arbitrarily selected blends, $x = (x_1, x_2, x_3)'$, together with the lower and upper limits for 95% confidence coefficient, are given in Table 2.4.

2.7　THE ANALYSIS OF VARIANCE TABLE

In the last section a second-degree polynomial equation was fitted to observed yarn elongation values and the objective was to try to understand how the variation or differences in the elongation values could be explained (or accounted for) in terms of the blending properties (linear and nonlinear) of the components. We now show how to partition or separate the overall variation in the elongation values into two assigned sources; the first being the variation among the averages of the elongation values attributed to the different blends, and the second being the measure of variation among the replicate samples within each blend.

The questions we will address are: Are the blends different in terms of their average yarn elongation values? If so, which blends are different and how are the differences explained by the blending properties of the three components? As an aid to us in addressing these questions, we first plot the individual data values as shown in Figure 2.10.

The variation among the six blends in the yarn elongation example is measured by computing the differences between each blend average and the overall average, squaring the differences, weighting the squared quantities by the number of replicates in each average blend, and summing the weighted quantities. Computationally we would use the equation

$$\text{Sum of squares among blends} = \sum_{l=1}^{6} r_l (\bar{y}_l - \bar{y})^2 \qquad (2.26)$$

where r_l is the number of replicate observations of the lth blend, $\bar{y}_l$ is the average of the r_l replicate observations of the lth blend, and $\bar{y}$ is the overall average of the $N = 15$ elongation values. The sum of squares in Eq. (2.26) has $6 - 1 = 5$ degrees of freedom associated with it. If there had been p different blends selected over the simplex-lattice, then $l = 1, 2, \ldots, p$ and the sum of squares among blends would have $p - 1$ degrees of freedom.

Since the $\{q, m\}$ polynomial was fitted to the data collected at the points of the $\{q, m\}$ simplex-lattice design, the number of terms in the model must equal the number of different blends defined by the design. This number is $\binom{q+m-1}{m}$, which in our example is $\binom{3+2+1}{2} = 6$. Thus the variation in the observations explained by the fitted model called the "sum of squares due to regression" or "sum of squares due to the fitted model" is the same as the sum of squares among the blends in Eq. (2.26). An alternative formula to Eq. (2.26) for calculating the sum of squares due to regression, designated as SSR, is

$$\text{SSR} = \sum_{u=1}^{N} (\hat{y}_u - \bar{y})^2 \qquad (2.27)$$

where $\hat{y}_u$ is the predicted value of y_u using the fitted model (actually $\hat{y}_u$ is the estimate of η at the uth setting of the mixture components obtained with the fitted model) and $\bar{y} = (y_1 + y_2 + \cdots + y_N)/N$ is the overall average of the observations.

The variation among the replicate observations within the blends is not accounted for (or explained) by the differences among the blends (nor by the fitted model) and is referred to as the residual variation. The formula for the residual sum of squares,

44

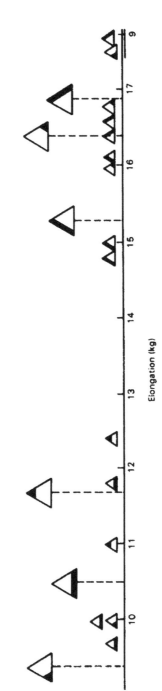

Figure 2.10 Plots of individual elongation values ($\triangle$) and the averages ($\triangle$) of each of the six blends.

Table 2.5 Analysis of Variance Table

Source of Variation	d.f.	Sum of Squares	Mean Square
Regression (fitted model)	$p - 1$	$\mathrm{SSR} = \sum_{u=1}^{N} (\hat{y}_u - \bar{y})^2$	$\mathrm{SSR}/(p - 1)$
Residual	$N - p$	$\mathrm{SSE} = \sum_{u=1}^{N} (y_u - \hat{y}_u)^2$	$\mathrm{SSE}/(N - p)$
Total	$N - 1$	$\mathrm{SST} = \sum_{u=1}^{N} (y_u - \bar{y})^2$	

denoted by SSE, is

$$\mathrm{SSE} = \sum_{u=1}^{N} (y_u - \hat{y}_u)^2 \tag{2.28}$$

and SSE has $N - p$ degrees of freedom, where p is the number of different blends or, as in this case, the number of terms in the model.

The total variation in a set of N data values is called the "sum of squares total" and is abbreviated SST. The quantity SST is computed by summing the squares of the deviations of the observed y_u from the overall average $\bar{y}$,

$$\mathrm{SST} = \sum_{u=1}^{N} (y_u - \bar{y})^2$$

The quantity SST equals the sum of SSR and SSE and SST has $N - 1$ degrees of freedom.

Using Eqs. (2.27) and (2.28), the analysis of variance table for the fitted model, containing p terms, is shown as Table 2.5. The rationale for the form of Table 2.5, which is based on testing hypotheses about the surface shape when fitting the Scheffé linear and quadratic blending models, is given in Appendix 2C. This rationale was presented by Marquardt and Snee (1974).

2.8 ANALYSIS OF VARIANCE CALCULATIONS OF THE YARN ELONGATION DATA

The analysis of variance table for the yarn elongation example of the previous section is Table 2.7. where the deviations used in calculating the sums of squares quantities are listed in Table 2.6.

Fitted model:

$$\hat{y}(\mathbf{x}) = 11.7x_1 + 9.4x_2 + 16.4x_3 + 19.0x_1x_2 + 11.4x_1x_3 - 9.6x_2x_3$$

Table 2.6 Sum of Squared Deviations SSE, SST, and SSR for Yarn Elongation Data

Observed Value (y_u)	Predicted Value ($\hat{y}_u$)	Difference or Residual ($y_u - \hat{y}_u$)	Deviation ($y_u - \bar{y}$)	Regression Deviation ($\hat{y}_u - \bar{y}$)
11.0	11.7	−0.7	−2.5	−1.8
12.4	11.7	0.7	−1.1	−1.8
15.0	15.3	−0.3	1.5	1.8
14.8	15.3	−0.5	1.3	1.8
16.1	15.3	0.8	2.6	1.8
10.0	9.4	0.6	−3.5	−4.1
8.8	9.4	−0.6	−4.7	−4.1
10.0	10.5	−0.5	−3.5	−3.0
9.7	10.5	−0.8	−3.8	−3.0
11.8	10.5	1.3	−1.7	−3.0
16.8	16.4	0.4	3.3	2.9
16.0	16.4	−0.4	2.5	2.9
17.7	16.9	0.8	4.2	3.4
16.4	16.9	−0.5	2.9	3.4
16.6	16.9	−0.3	3.1	3.4

$$\text{SSE} = \sum_{u=1}^{15} (y_u - \hat{y}_u)^2 = 6.56$$

$$\text{SST} = \sum_{u=1}^{15} (y_u - \bar{y})^2 = 134.88$$

$$\text{SSR} - \sum_{u=1}^{15} (\hat{y}_u - \bar{y})^2 = 128.32^a$$

[a]When deviations are not rounded to tenths, SST = 134.856 and SSR = 128.296.

Overall average:

$$\bar{y} = \sum_{u=1}^{15} \frac{y_u}{15} = \frac{203.1}{15} = 13.5$$

The value of the F-ratio in Table 2.7. is found by dividing the mean square for regression by the mean square for residual.

The mean squares for regression and for residual are functions of the squares of the observed y_u random variables and as such are likewise random variables. It can be shown that if the assumed model is correct, the expected values of the mean squares

Table 2.7 Analysis of Variance Table for the Yarn Elongation Example

Source of Variation	d.f.	Sum of Squares	Mean Square	F-Ratio
Regression	5	128.32	25.66	25.66/0.73 = 35.1
Residual	9	6.56	0.73	
Total	14	134.88		

for regression and for residual are

$$E(\text{mean-square residual}) = \sigma^2$$

$$E(\text{mean-square regression}) = \sigma^2 + f(\beta_1, \beta_2, \ldots, \beta_{23})$$

where the quantity $f(\beta_1, \beta_2, \ldots, \beta_{23})$ equals zero if the surface $\eta = \beta_1 x_1 + \beta_2 x_2 + \cdots + \beta_{23} x_2 x_3 = \beta_0$ where β_0 is not necessarily zero, that is, β_0 is the height of a level plane directly above the simplex (i.e., $\beta_1 = \beta_2 = \beta_3 = \beta_0, \beta_{12} = \beta_{13} = \beta_{23} = 0$) (see Figure 4.1). For the F-ratio test in Table 2.7 we assume that the errors ε_u are independent, Normal $(0, \sigma^2)$ variables. Then, if the null hypothesis $H_0: \beta_1 = \beta_2 = \beta_3 = \beta_0, \beta_{12} = \beta_{13} = \beta_{23} = 0$ is true, implying that the surface above the simplex or triangle is a level plane whose height is the same at all points, the ratio is distributed

$$\frac{(p-1) \times (\text{mean-square regression})}{\{\sigma^2 + f(\beta_1, \beta_2, \ldots, \beta_{23})\} = \sigma^2} \sim \chi^2_{(p-1)} \tag{2.29}$$

as a chi-square random variable with $p-1$ degrees of freedom, and further

$$\frac{(N-p) \times (\text{mean-square residual})}{\sigma^2} \sim \chi^2_{(N-p)} \tag{2.30}$$

and the two chi-square random variables in Eqs. (2.29) and (2.30) are independent. Since the ratio of two independent chi-squares, over their respective degrees of freedom, is an F random variable, we have

$$F = \frac{(p-1) \times (\text{mean-square regression})/(p-1)\sigma^2}{(N-p) \times (\text{mean-square residual})/(N-p)\sigma^2} \tag{2.31}$$

and the F-distribution has $(p-1) = 5$ and $(N-p) = 9$ degrees of freedom in the numerator and denominator, respectively. The value of the F-ratio in Eq. (2.31) is compared with the table value, $F_{(p-1, N-p, \alpha)}$, at the end of the book where $F_{(p-1, N-p, \alpha)}$ is the upper 100α percent point of the F-distribution with $p-1$ and $N-p$ degrees of freedom, respectively. When the F-ratio value in Eq. (2.31) is greater than or equal to the table value, the null hypothesis is rejected at the α level of significance. Since the F-value in Table 2.7 is $F = 35.1$ and exceeds the table value $F_{(5,9,\alpha=0.01)} = 6.06$, we reject $H_0: \beta_1 = \beta_2 = \beta_3 = \beta_0, \beta_{12} = \beta_{13} = \beta_{23} = 0$ at the $\alpha = 0.01$ level in favor of at least one equality being false. We will present some additional discussion on testing hypotheses in Chapter 4.

The results of the F-test performed on the mean squares in Table 2.7 prompt us to reject the hypothesis that the elongation surface is a planar surface of constant height above the triangle. It is not clear by the rejection of the level plane whether or not the surface is a plane. It might be a plane that is tilted such as shown in Figure 2.3, or the surface may possess curvature as shown in Figure 2.6 and not be a plane at all. If the surface is a tilted plane whose heights above the three vertices are unequal, then $H_0: \beta_1 = \beta_2 = \beta_3 = \beta_0, \beta_{12} = \beta_{13} = \beta_{23} = 0$ is rejected in favor of

$H_A: \beta_{12} = \beta_{13} = \beta_{23} = 0$ and $\beta_1 \neq \beta_2 = \beta_3$ or $\beta_1 = \beta_2 \neq \beta_3$ or $\beta_1 \neq \beta_2 \neq \beta_3$. If the surface possesses curvature, then $H_0: \beta_1 = \beta_2 = \beta_3 = \beta$, $\beta_{12} = \beta_{13} = \beta_{23} = 0$ is rejected in favor of $H_A: \beta_{12} = \beta_{13} = \beta_{23} \neq 0$ and $\beta_1 = \beta_2 = \beta_3$ or $\beta_1 \neq \beta_2 \neq \beta_3$ or $\beta_1 = \beta_2 \neq \beta_3$ or $\beta_1 = \beta_2 \neq \beta_3$. Also another statistic that we will find useful quite often when checking the closeness of the fitted model to the observed data values is the square of the adjusted multiple correlation coefficient R_A, where $R_A^2 = 1 - [(N-1) \text{SSE}/(N-p)\text{SST}]$; see Section 7.4. With the yarn elongation data, the value of R_A^2 with the fitted second-degree model is $R_A^2 = 1 - [(14)6.56/(9)134.86] = 0.924$.

2.9 THE PLOTTING OF INDIVIDUAL RESIDUALS

A measure of the closeness of the mixture surface predicted by the fitted model to observed values of the response at the design points can be obtained by computing the differences $e_u = y_u - \hat{y}_u$, $u = 1, 2, \ldots, N$, where y_u is the observed value of the uth response and $\hat{y}_u$ is the predicted value of the response for the uth trial. These differences are called *residuals*. An approximate average of the squares of the residuals was used to obtain an estimate of the observation or error variance when SSE in Eq. (2.28) was divided by $N - p$, but often it is equally important to examine the residuals individually as well as jointly for detecting inadequacies in the model. This is because the sizes of the residuals can be made large by factors that are not accounted for in the proposed model as well as by not utilizing the strengths of the important components to the correct degree with the fitted model.

Residuals may be plotted various ways, using different types of plotting paper. First, a general inspection of the signs and of the sizes is carried out with just a dot diagram on standard grid paper. (See Box, Hunter, and Hunter, 1978, ch. 6.) If the assumption is true that the errors are independently and identically distributed Normal $(0, \sigma^2)$, the residual plot will have roughly the appearance of a sample from a normal distribution centered at zero. Of course, unless the number of residuals is sufficiently large (e.g. $N \geq 30$ for a rough rule of thumb), it is difficult to envision any shape tendencies in the appearance of the distribution. Daniel and Wood (1971) suggest several techniques for studying residual patterns but state that none of the techniques can be expected to work well with fewer than 20 observations.

We present a plot of the residuals from Table 2.6 in Figure 2.12. As we will see, the use of the residual plot can be an effective way of comparing the fits of competing models. In fact we will compare the fit of the second-degree model of Eq. (2.24) to the elongation data with the fit of a first-degree model to the same data by noting the reduction in the sizes or spread of the residuals with the two models.

Plots of residuals $e_u = y_u - \hat{y}_u$ versus the individuals predicted values $\hat{y}_u$, or values of e_u versus the size of the individual values of the x_i or linear combinations of the x_i, are other techniques that are useful for detecting an inadequate model or for checking assumptions made about the errors. In particular, to check the normality assumption of the errors, some computer software packages offer plots of the studentized residuals on normal probability paper. The individual studentized residuals are also plotted against the individual predicted values to check for constancy of the error variance across the range of values of $\hat{y}_u$. Still another statistic, the C_p statistic, can be used to compare the fits of competing models. This statistic measures the sum of the squared biases, plus the square residuals (the e_u^2 terms) at all data points.

2.10   TESTING THE DEGREE OF THE FITTED MODEL: A QUADRATIC MODEL OR PLANAR MODEL?

The fitting of the second-degree yarn elongation model of Eq. (2.24) provided information on each of the components individually as well as on pairs of components. At the conclusion of the example, component 1 was said to blend in a synergistic manner with each of components 2 and 3, while components 2 and 3 were said to blend in an antagnositic way. These claims were made because of the signs and relative magnitudes of the estimates b_{12}, b_{13}, and b_{23}.

The question we ask ourselves now is: If we refit the data using only a first-degree model in x_1, x_2, and x_3, would the first-degree model fit the 15 elongation values as well as the second-degree equation (2.24) did? Stated another way: Is there sufficient evidence to reject $H_0: \beta_{12} = \beta_{13} = \beta_{23} = 0$ because of the falsehood of one or more of the equality signs?

The least-squares estimates of the parameters in the first-degree model $\eta = \beta_1 x_1 + \beta_2 x_2 + \beta_3 x_3$ are not as easily calculated as illustrated in the formulas of Eq. (2.15) when the data are collected at the points of the $\{3, 2\}$ simplex-lattice. This is because there is no longer a one-to-one correspondence between the number of design points (= 6) and the number of terms (= 3) in the model. However, if r_1 observations are collected at the vertices of the triangle and r_2 observations are collected at the midpoints of the three edges, where in our case $r_1 = 2$ and $r_2 = 3$, then the formulas for the estimates b_1, b_2, and b_3 of the parameters β_1, β_2, and β_3 in the fitted first-degree model are

$$b_1 = A\left[r_1\bar{y}_1 + \frac{r_2(\bar{y}_{12} + \bar{y}_{13})}{2}\right] + B\left[r_1(\bar{y}_2 + \bar{y}_3) + \frac{r_2(\bar{y}_{12} + \bar{y}_{13} + 2\bar{y}_{23})}{2}\right]$$

$$b_1 = A\left[r_1\bar{y}_2 + \frac{r_2(\bar{y}_{12} + \bar{y}_{23})}{2}\right] + B\left[r_1(\bar{y}_1 + \bar{y}_3) + \frac{r_2(\bar{y}_{12} + \bar{y}_{23} + 2\bar{y}_{13})}{2}\right]$$

$$b_3 = A\left[r_1\bar{y}_3 + \frac{r_2(\bar{y}_{13} + \bar{y}_{23})}{2}\right] + B\left[r_1(\bar{y}_1 + \bar{y}_2) + \frac{r_2(\bar{y}_{13} + \bar{y}_{23} + 2\bar{y}_{12})}{2}\right]$$

where $A = (4r_1 + 3r_2)/[(r_1 + r_2)(4r_1 + r_2)]$ and $B = -r_2/[(r_1 + r_2)(4r_1 + r_2)]$. With the 15 yarn elongation values in Table 2.3, $A = 17/55$ and $B = -3/55$ and $b_1 = (17/55)[71.7] + (-3/55)[131.4] = 14.994$, $b_2 = 9.831$, and $b_3 = 15.795$, so that the first-degree fitted model is

$$\hat{y}(\mathbf{x}) = 15.0x_1 + 9.8x_2 + 15.8x_3$$
$$\phantom{\hat{y}(\mathbf{x}) = } (1.4) \quad\;\; (1.4) \quad\;\; (1.4) \tag{2.32}$$

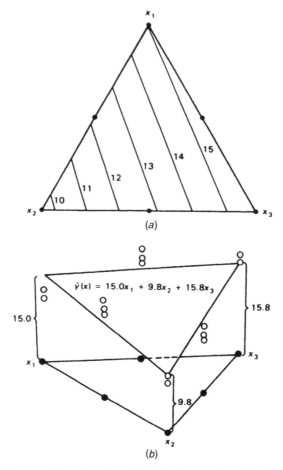

Figure 2.11 (a) Contours of the estimated yarn elongation planar surface of Eq. (2.32). (b) A plot of the heights of the yarn elongation values relative to the planar surface of Eq. (2.32).

where the numbers in parentheses below the parameter estimates are the estimated standard errors or estimated standard deviations of the estimates. The surface contours generated with the fitted model (2.32) are shown in Figure 2.11.

The sums of squares quantities associated with the fitted first-degree model of Eq. (2.32) are

$$SSR = \sum_{u=1}^{N=15} (\hat{y}_u - \bar{y})^2 = 57.63$$

$$SSE = \sum_{u=1}^{N=15} (y_u - \hat{y}_u)^2 = 77.23$$

$$SST = \sum_{u=1}^{N=15} (y_u - \bar{y})^2 = 134.86$$

A measure of the proportion of the total variation in the data values, SST, explained by the fitted model is the *coefficient of determination*

$$R^2 = \frac{\text{SSR}}{\text{SST}}$$

The value of R^2 for the fitted first-degree model of Eq. (2.32) is $R^2 = 57.63/134.86 = 0.4273$. Other model fitting summary statistics are PRESS, R_A^2, and R_{PRESS}^2 (see Sections 7.4 and 7.5). With Eq. (2.32), PRESS $= 120.81$, $R_A^2 = 0.3319$, and $R_{\text{PRESS}}^2 = 0.1041$. For the second-degree model of Eq. (2.24), these quantities are SSR $= 128.30$, SSE $= 6.56$, SST $= 134.86$, $R^2 = 0.9514$, PRESS $= 18.30$, $R_A^2 = 0.9243$, and $R_{\text{PRESS}}^2 = 0.8643$. A test that compares the first- and second-degree models in terms of how well they account for or explain the variation in the response values uses the residual sum of squares associated with each of the models. The test statistic is an F-ratio of the form (see Section 7.6)

$$F = \frac{(\text{SSE}_{\text{reduced}} - \text{SSE}_{\text{complete}})/r}{\text{SSE}_{\text{complete}}/(N - p)} \tag{2.33}$$

where

$\text{SSE}_{\text{reduced}} =$ the residual sum of squares associated with the reduced model (the first-degree equation of Eq. (2.32)),

$\text{SSE}_{\text{complete}} =$ the residual sum of squares associated with the complete model (the second-degree equation of Eq. (2.24) and Table 2.7)

$r =$ the difference in the number of parameters in the complete and reduced models ($r = 6 - 3 = 3$), and p is the number of parameters in the complete model ($p = 6$).

In the F-test of Eq. (2.33), the numerator is the reduction in the sum of squared residuals (or unexplained variation) that is achieved when going from the first-degree to the second-degree model, divided by the number of additional terms in the second-degree model, while the denominator is the estimate of the observation or error variance found with the complete second-degree model. With the yarn elongation data, the value of the F-ratio in Eq. (2.33) is

$$F = \frac{(77.23 - 6.56)/3}{6.56/9} = 32.2 \tag{2.34}$$

Since $F = 32.2$ exceeds $F_{(3, 9, 0.01)} = 6.99$, we conclude that the drop or decrease in the residual or unexplained sum of squares when the fitted model is quadratic is large enough, relative to $s^2 = 0.73$, to justify using the second-degree model to describe the yarn elongation surface. The decrease in the magnitudes of the individual residuals from fitting the second-degree model is illustrated in Figure 2.12, where the dot diagram is a plot of the residuals from both models. Also the differences between the shapes of the two estimated surfaces is seen by comparing the contour plots of Figures 2.9 and 2.11a.

Finally, Table 2.8a lists the typical software output from a fitted model in the form of the actual observed values, predicted values, residuals, studentized residuals,

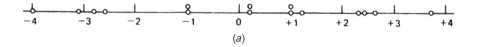

(a)

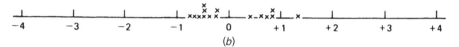

(b)

Figure 2.12 Plots of the sizes of the residuals with the yarn elongation data arising from the fit of the first-degree model (o) and the fit of the second-degree model ($\times$). Sizes of the residuals $= y_u - \hat{y}_u$.

Cook's distance, and outlier t values. This output was provided by DESIGN-EXPERT of Stat-Ease (1998). Listing the outputs from the fitted linear blending model of Eq. (2.32) and the quadratic blending model of Eq. (2.24) in a single table enables us to compare the fits of the two models visually by contrasting the values of the residuals, studentized residuals, and so on. As mentioned later in Section 4.2, rule-of-thumb values for the studentized residuals, Cook's distance, and outlier t for accepting the model fit are less than 3.0, less than 1.0, and less than 3.5, respectively. While both models are acceptable by these rules, our choice is the quadratic model of Eq. (2.24).

Having decided from the F-test in Eq. (2.34) or from the plot of the residuals in Figure 2.12 or from the increase in the values of R^2, R_A^2, and R_{PRESS}^2 with Eq. (2.24) that the second-degree fitted model does a better job of describing the shape of the elongation surface (or in explaining the variation in the elongation values) than does the fitted first-degree model, we might next decide to perform tests of hypothesis on the individual parameters β_{12}, β_{13}, and β_{23}. Such tests determine the significance of the individual parameters, as shown in Appendix 1A and in Section 2.6.

We have seen that the $\{q, m\}$ simplex-lattice design, for $m = 2$ and $m = 3$, gives an equally spaced distribution of points over the simplex and that it provides just enough points to enable the associated $\{q, m\}$ polynomial in the x_i to be fitted. A possible objection to the simplex-lattice design is that, while it is generally intended for the prediction of the response to mixtures of q components, of $q - 1$ components, and of $q - 2$ components at least, the model is fitted to observations collected on mixtures of at most m components. This means that the parameters in the second-degree model ($m = 2$) are estimated from data collected on mixtures consisting of at most pairs of components. Furthermore the $\{q, 2\}$ simplex-lattice is strictly a points-on-the-boundary-only design. A partial solution to the use of too few component mixtures is provided by the simplex-centroid design, which is introduced in Section 2.13.

Displayed in Table 2.8b is the printout from DESIGN-EXPERT showing the measured yarn elongation values along with both the fitted first- and second-degree models and their associated ANOVA tables. Following the fitted second-degree model are the elongation data values, the predicted elongation values using the second-degree model, the residuals and their leverage values that are discussed later in Section 4.7, along with other model adequacy statistics discussed in Sections 7.4 and 7.5.

**Table 2.8a   Typical Output with Software Packages Such as
DESIGN-EXPERT of Stat-Ease (1998)**

		Linear Blending Model Eq. (2.32)			
Actual Value	Predicted Value	Residual	Studentized Residual	Cook's Distance	Outlier t
11.0	14.99	−3.99	−1.894	0.535	−2.166
12.4	14.99	−2.59	−1.230	0.226	−1.260
15.0	12.41	2.59	1.092	0.058	1.101
14.8	12.41	2.39	1.007	0.049	1.008
16.1	12.41	3.69	1.556	0.118	1.667
8.8	9.83	−1.03	−0.489	0.036	−0.473
10.0	9.83	0.17	0.080	0.001	0.077
10.0	12.81	−2.81	−1.187	0.068	−1.210
9.7	12.81	−3.11	−1.313	0.084	−1.359
11.8	12.81	−1.01	−0.427	0.009	−0.412
16.8	15.79	1.01	0.477	0.034	0.461
16.0	15.79	0.21	0.097	0.001	0.093
17.7	15.39	2.31	0.973	0.046	0.970
16.4	15.39	1.01	0.424	0.009	0.409
16.6	15.39	1.21	0.509	0.013	0.492

		Quadratic Model Eq. (2.24)			
Actual Value	Predicted Value	Residual	Studentized Residual	Cook's Distance	Outlier t
11.0	11.70	−0.70	−1.160	0.224	−1.185
12.4	11.70	0.70	1.160	0.224	1.185
15.0	15.30	−0.30	−0.430	0.015	−0.410
14.8	15.30	−0.50	−0.717	0.043	−0.696
16.1	15.30	0.80	1.148	0.110	1.171
8.8	9.40	−0.60	−0.994	0.165	−0.993
10.0	9.40	0.60	0.994	0.165	0.993
10.0	10.50	−0.50	−0.717	0.043	−0.696
9.7	10.50	−0.80	−1.148	0.110	−1.171
11.8	10.50	1.30	1.865	0.290	2.245
16.8	16.40	0.40	0.663	0.073	0.641
16.0	16.40	−0.40	−0.663	0.073	−0.641
17.7	16.90	0.80	1.148	0.110	1.171
16.4	16.90	−0.50	−0.717	0.043	−0.696
16.6	16.90	−0.30	−0.430	0.015	−0.410

Table 2.8b   Least Squares Regression and ANOVA Table

3-Component Yarn Elongation Example

Constituents: Polyethylene (x_1), polystyrene (x_2), polypropylene (x_3).

Measured response: Elongation of the yarn.

Data:

x_1	x_2	x_3	Y_u	$\bar{Y}$
1	0	0	11.0, 12.4	11.7
1/2	1/2	0	15.0, 14.8, 16.1	15.3
0	1	0	8.8, 10.0	9.4
1/2	0	1/2	17.7, 16.4, 16.6	16.9
0	0	1	16.8, 16.0	16.4
0	1/2	1/2	10.0, 9.7, 11.8	10.5

Design-Expert ANOVA Table for Linear Blending Model

Elongation $= \beta_1 x_1 + \beta_2 x_2 + \beta_3 x_3$

ANOVA for Mix Linear Model

SOURCE	SUM OF SQUARES	DF	MEAN SQUARE	F VALUE	PROB > F
MODEL	57.6	2	28.91	4.48	0.035
RESIDUAL	77.2	12	6.44		
Lack Of Fit	70.7	3	23.56	32.32	<0.001
Pure Error	6.6	9	0.73		
COR TOTAL	134.9	14			

ROOT MSE	2.54	R-SQUARED	0.4273
DEP MEAN	13.54	ADJ R-SQUARED	0.3319
C.V.	18.74%	PRED R-SQUARED	0.1041

Predicted Residual Sum of Squares (PRESS) = 120.8

COMPONENT	COEFFICIENT ESTIMATE	DF	STD ERROR	t FOR H0 COEF=0 PROB > |t|		VIF
A-p. ethylene	14.99	1	1.41	Not Applicable		1.08
B-p. styrene	9.83	1	1.41	Not Applicable		1.08
C-p. propylen	15.79	1	1.41	Not Applicable		1.08

ANOVA Table and Output for Quadratic Blending Model

Elongation $= \beta_1 x_1 + \beta_2 x_2 + \beta_2 x_2 + \beta_{12} x_1 x_2 + \beta_{13} x_1 x_3 + \beta_{23} x_2 x_3$

ANOVA for Mix Quadratic Model

SOURCE	SUM OF SQUARES	DF	MEAN SQUARE	F VALUE	PROB > F
MODEL	128.3	5	25.66	35.20	<0.001
RESIDUAL	6.6	9	0.73		
Lack Of Fit	0	0			
Pure Error	6.6	9	0.73		
COR TOTAL	134.9	14			

ROOT MSE	0.85	R-SQUARED	0.9514
DEP MEAN	13.54	ADJ R-SQUARED	0.9243
C.V.	6.31%	PRED R-SQUARED	0.8643

Predicted Residual Sum of Squares (PRESS) = 18.3

COMPONENT	COEFFICIENT ESTIMATE	DF	STD ERROR	t FOR H0 COEF=0 PROB > |t|		VIF
A-p. ethylene	11.70	1	0.60	Not Applicable		1.75
B-p. styrene	9.40	1	0.60	Not Applicable		1.75
C-p. propylen	16.40	1	0.60	Not Applicable		1.75
AB	19.00	1	2.61	7.28 <	0.001	1.75
AC	11.40	1	2.61	4.37	0.002	1.75
BC	-9.60	1	2.61	-3.68	0.005	1.75

Obs Ord	ACTUAL VALUE	PREDICTED VALUE	RESIDUAL	LEVER	STUDENT RESID	COOK'S DIST	OUTLIER t	Run Ord
1	11.00	11.70	-0.70	0.500	-1.16	0.224	-1.19	1
2	12.40	11.70	0.70	0.500	1.16	0.224	1.19	6
3	15.00	15.30	-0.30	0.333	-0.43	0.015	-0.41	4
4	14.80	15.30	-0.50	0.333	-0.72	0.043	-0.70	5
5	16.10	15.30	0.80	0.333	1.15	0.110	1.17	3
6	8.80	9.40	-0.60	0.500	-0.99	0.165	-0.99	2
7	10.00	9.40	0.60	0.500	0.99	0.165	0.99	15
8	17.70	16.90	0.80	0.333	1.15	0.110	1.17	14
9	16.40	16.90	-0.50	0.333	-0.72	0.043	-0.70	13
10	16.60	16.90	-0.30	0.333	-0.43	0.015	-0.41	12
11	16.80	16.40	0.40	0.500	0.66	0.073	0.64	11
12	16.00	16.40	-0.40	0.500	-0.66	0.073	-0.64	10
13	10.00	10.50	-0.50	0.333	-0.72	0.043	-0.70	9
14	9.70	10.50	-0.80	0.333	-1.15	0.110	-1.17	8
15	11.80	10.50	1.30	0.333	1.86	0.290	2.24	7

2.11 TESTING MODEL LACK OF FIT USING EXTRA POINTS AND REPLICATED OBSERVATIONS

In the previous section we compared and tested the fit of a first-degree model with that of a second-degree model using the data collected at the points of a $\{3, 2\}$ simplex-lattice design. This test of the equivalence of the two fitted model forms was actually a test to determine whether the data from two-component blends reflected linearity in the blending properties of the components (which is the case if $\beta_{12} = \beta_{13} = \beta_{23} = 0$ in model (2.7) for $q = 3$). Since the F-test in Eq. (2.34) was significant, we concluded that one or more of the quadratic blending terms in the second-degree model was necessary to pick up the nonlinear blending between one or more pairs of the three components.

A slightly different approach that could have been used in developing an appropriate model form would have been to initially fit the first-degree model and then ask: Does the model adequately fit the observed response values? In this approach we assume initially that the components blend linearly and, upon fitting the first-degree model, question whether there is evidence that is contrary to this assumption. If there is evidence that suggests the fitted first-degree model is not adequate, then either additional experiments must be performed that might improve the fit or the form of the model must be changed. We selected the latter course of action by fitting the second-degree model, since the $\{3, 2\}$ simplex-lattice supports the fitting of a second-degree model.

There are several approaches that can be taken to test lack of fit of a fitted model. One approach requires that replicate observations be taken at one or more design points, where the number of distinct design points exceeds the number of terms in the fitted model. When this happens, the residual sum of squares from the analysis of the fitted model can be partitioned into two sums of squares: the sum of squares due to lack of fit of the model and the sum of squares due to pure error, where the latter sum of squares is calculated by using the replicates. These sums of squares, when divided by their respective degrees of freedom, are then compared in the form of an F-ratio as shown in Eq. (2.33). Also see Draper and Smith (1981), Draper and Herzberg (1971), or Khuri and Cornell (1998). When replicate observations are not available, other authors (Green, 1971; Daniel and Wood, 1971; Shillington, 1979) have proposed grouping values of the response that are observed at similar (i.e., close to but not the same) settings of the regressor variables. Such grouped values are called "pseudoreplicates" or "near-neighbor observations."

2.11.1 A Second Numerical Example Illustrating the Use of Extra Points and Replicated Observations for Test 1 Model Lack Fit

A three-component experiment was set up to determine whether an artificial sweetner could be used in a popular athletic-sports drink. The sweeteners were glycine, saccharin, and an enhancer. The amount of sweetener was held fixed in all blends at 4% of the total volume (250 mL).

Initially six combinations of the three sweeteners were selected for testing and are listed as blends 1 to 6 in Table 2.9. The data values represent an "intensity of

Table 2.9   Data from the Artificial Sweetener Experiment

Blend	Glycine x_1	Saccharin x_2	Enhancer x_3	Average Intensity of Sweetness Aftertaste Score $(y)^a$ from 20 Respondents	Average Score ($\bar{y}$)
1	1	0	0	10.1, 10.7	10.40
2	0	1	0	5.8, 6.5	6.15
3	0	0	1	4.2, 3.6	3.90
4	1/2	1/2	0	14.5, 15.4, 15.0	14.97
5	1/2	0	1/2	12.9, 12.0, 11.6	12.17
6	0	1/2	1/2	11.6, 13.0, 12.2	12.27
Check points					
7	1/3	1/3	1/3	8.2	
8	2/3	1/6	1/6	17.0	
9	1/6	2/3	1/6	6.0	
10	1/6	1/6	2/3	7.2	

[a]Each sweetness aftertaste score is an average of 20 respondents.

sweetness aftertaste" score for each blend, where each data point is an average of 20 replies from a large consumer population. Each respondent in the population was asked to rate the intensity of sweetness aftertaste using a scale of 1 (positively no aftertaste) to 30 (very extreme aftertaste). Low scores (<15) are considered desirable. Overall, 300 people participated in the scoring. Forty respondents, comprising two replications of 20 respondents each, scored each of the three single sweeteners while 60 respondents, three replicates of 20 respondents, scored each of the two-sweetener blends. The data values in Table 2.9 are rounded to one decimal place to simplify the computations.

The second-degree model fitted to the 15 data values at the first six blends (1–6) of Table 2.9 is

$$\hat{y}(\mathbf{x}) = 10.40x_1 + 6.15x_2 + 3.90x_3 + 26.77x_1x_2 + 20.07x_1x_3 + 28.97x_2x_3$$
$$\quad\;\;(0.40)\quad\;\;(0.40)\quad\;(0.40)\quad\;(1.73)\quad\quad\;(1.73)\quad\quad\;(1.73)$$

$$(2.35)$$

where the quantities in parentheses are the estimated standard errors of the coefficient estimates. The estimate of the error variance σ^2 obtained from the replicates at the six design points is $\hat{\sigma}^2_{\text{ext}} = 0.3206$ with nine degrees of freedom. Predictions of the mean response at the six blends with model (2.35) would yield the mean responses, or average scores, listed at the six blends in Table 2.9 because the fitted model contains the same number of terms as there are points in the design.

To check on the adequacy of the fitted model (2.35) *at blends inside the three-sweetener triangle*, four additional blends were chosen and are listed as check points 7 to 10 in Table 2.9. The check points are the centroid of the triangle (point 7) as well as points midway between the centroid and each of the three vertices (points 8, 9, and

10). As a group, the four check-point locations represent positions that are farthest away distancewise from the original six design points and that when used to test the adequacy of the model would appear to maximize the power of the test.

At each of the four check-point blends, 20 people were asked to score the intensity of aftertaste and the average scores recorded by the additional 80 persons are listed in Table 2.9. Using the fitted model (2.35), predicted values of the response at the points 7 to 10 are 15.24, 14.62, 13.23, and 11.55, respectively. The test for model lack of fit inside the triangle then is a test on the magnitudes of the differences in the 4×1 vector $\mathbf{d}$:

$$d_l = y_l^* - \hat{y}_l^*, \quad l = 7, 8, 9, 10$$

$$\mathbf{d} = \begin{bmatrix} 8.2 - 15.24 \\ 17.0 - 14.62 \\ 6.0 - 13.23 \\ 7.2 - 11.55 \end{bmatrix} = \begin{bmatrix} -7.04 \\ 2.38 \\ -7.23 \\ -4.35 \end{bmatrix}$$

An F-test of the differences is shown in Cornell (2002, ch. 2), but we will provide an equivalent F-test in Eq. (2.37) since the replicated observations occur at the initial six design points only and not at the check points.

Fitted to the complete set of 19 data values in Table 2.9, the second-degree model is

$$\hat{y}(\mathbf{x}) = 11.52x_1 + 5.80x_2 + 3.99x_3 + 20.39x_1x_2 + 13.99x_1x_3 + 21.91x_2x_3$$
$$\quad (1.92) \qquad (1.92) \qquad (1.92) \qquad (8.07) \qquad \quad (8.07) \qquad \quad (8.07)$$

$$(2.36)$$

where the residual sum of squares is 99.6567 with $18 - 5 = 13$ degrees of freedom. If we partition the residual sum of squares into pure-error sum of squares (SS) and lack-of-fit sum of squares, we find

Pure-error (due to replicates) sum of squares $= 2.8854$ with 9 d.f.

Lack-of-fit sum of squares $=$ residual sum of squares $-$ pure-error sum of squares

$$= 99.6567 - 2.8854$$

$$= 96.7713 \qquad \text{with } 13 - 9 = 4 \text{ d.f.}$$

A test of zero lack of fit of the model (2.36) gives

$$F = \frac{\text{lack-of-fit SS}/4}{\text{pure-error SS}/9}$$

$$= \frac{96.7713/4}{2.8854/9} = 75.46 \qquad (2.37)$$

which, when compared to $F_{(4,9,0.01)} = 6.42$, causes us to reject zero lack of fit.

Since the quadratic model (2.36) possesses significant lack of fit, one suggestion for improving the form of the model would be to upgrade it by adding the special cubic term, $\beta_{123}x_1x_2x_3$, to it. The special cubic model, fitted to the 19 values in Table 2.9 is

$$\hat{y}(\mathbf{x}) = 11.25x_1 + 5.54x_2 + 3.73x_3 + 26.93x_1x_2 + 20.52x_1x_3$$
$$\phantom{\hat{y}(\mathbf{x}) = } (1.39) \quad\ (1.39) \quad\ (1.39) \quad\ (6.09) \quad\quad\ (6.09)$$
$$+\ 28.44x_2x_3 - 180.68x_1x_2x_3 \tag{2.38}$$
$$ (6.09) \quad\quad\ (50.9)$$

Even though each of the cross-product or binary coefficient estimates, as well as the ternary blending coefficient estimate, appears to be significantly different from zero (tested by dividing each estimate by its standard error and comparing the quotient to a tabled value of $t_{12,0.005} = 3.055$), the model (2.38) with a residual sum of squares equal to 44.9238 still has significant lack of fit and therefore additional terms must be added to it. We will discuss the addition of terms like $\beta_{1123}x_1^2x_2x_3$, $\beta_{1223}x_1x_2^2x_3$, and $\beta_{1233}x_1x_2x_3^2$ to the quadratic model to form the special quartic model for three, components later in Section 2.15.

2.12   THE SIMPLEX-CENTROID DESIGN AND ASSOCIATED POLYNOMIAL MODEL

In a q-component simplex-centroid design, the number of distinct points is $2^q - 1$. These points correspond to q permutations of $(1, 0, 0, \ldots, 0)$ or q single-component blends, the $\binom{q}{2}$ permutations of $\left(\frac{1}{2}, \frac{1}{2}, 0, \ldots, 0\right)$ or all binary mixtures, the $\binom{q}{3}$ permutations of $\left(\frac{1}{3}, \frac{1}{3}, \frac{1}{3}, 0, \ldots, 0\right), \ldots$, and so on, with finally the overall centroid point $(1/q, 1/q \ldots, 1/q)$ or q-nary mixture. In other words, the design consists of every (nonempty) subset of the q components, but only with mixtures in which the components that are present appear in equal proportions. Such mixtures are located at the centroid of the $(q - 1)$-dimensional simplex and at the centroids of all the lower di-mensional simplices contained within the $(q - 1)$-dimensional simplex. Presented in Figure 2.13 are the three-component and four-component simplex-centroid designs.

At the points of the simplex-centroid design, data on the response are collected and a polynomial is fitted that has the same number of terms (or parameters to be estimated) as there are points in the associated design. The polynomial equation is

$$\eta = \sum_{i=1}^{q} \beta_i x_i + \sum\sum_{i<j}^{q} \beta_{ij}x_ix_j + \sum\sum\sum_{i<j<k}^{q} \beta_{ijk}x_ix_jx_k + \cdots + \beta_{12\ldots q}x_1x_2\cdots x_q$$

$$(2.39)$$

As in the case of the previous models, which were expressed in the canonical forms in Eqs. (2.4) and (2.7), the parameter β_i in Eq. (2.39) represents the expected response to the pure component i and is called the linear blending value of component i, and

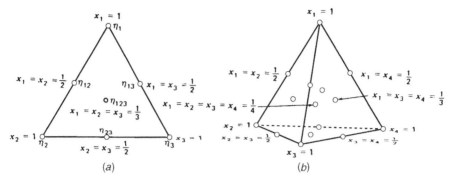

Figure 2.13   Simplex-centroid designs for (a) three components and (b) four components.

β_{ij} is the coefficient of the nonadditive blending of components i and j. The other β_{ijk} parameters are defined similarly.

The formulas for the estimates of the first two sets of parameters in Eq. (2.39) are

$$b_i = \bar{y}_i, \qquad i = 1, 2, \ldots, q$$

$$b_{ij} = 4\bar{y}_i - 2(\bar{y}_i + \bar{y}_j) \tag{2.40}$$

which are identical to Eq. (2.14) or Eq. (2.15), used previously to calculate b_{ij} with the simplex-lattice design. The equality of the formulas in Eqs. (2.40) and (2.14) means that with the more elaborate simplex-centroid design, an observation that is collected at the centroid of the face of the simplex-centroid design is not used to estimate the binary coefficient β_{ij} nor is it used to estimate β_i or β_j.

Some examples of the sizes of the variances of the parameter estimates and covariances between pairs of estimates in the $q = 3$ component case are

$$
\begin{aligned}
b_i : \quad & i = 1, 2, 3 : \operatorname{cov}(b_i, b_j) = \sigma^2 && i = j \\
& && = 0 && i \neq j \\
b_{ij} : \quad & i < j : \operatorname{cov}(b_i, b_{i'j}) && = -2\sigma^2 && i = i' \\
& && = 0 && i \neq i' \\
& \operatorname{cov}(b_{ij}, b_{i'k}) && = 4\sigma^2 && i = i', j < k \\
& && = 24\sigma^2 && i = i', j = k \\
& && = 0 && i \neq i', j \neq k \\
b_{123} : \quad & \operatorname{cov}(b_i, b_{123}) && = 3\sigma^2 && i = 1, 2, \text{ or } 3 \\
& \operatorname{cov}(b_{ij}, b_{123}) && = -60\sigma^2 && i = 1, 2, j = 2, 3 \\
& \operatorname{cov}(b_{123}, b_{123}) && = 1188\sigma^2
\end{aligned}
\tag{2.41}
$$

2.13 AN APPLICATION OF A FOUR-COMPONENT SIMPLEX-CENTROID DESIGN: BLENDING CHEMICAL PESTICIDES FOR CONTROL OF MITES

Four chemical pesticides—Vendex (x_1), Omite (x_2), Kelthane (x_3), and Dibron (x_4)—were sprayed on strawberry plants in an attempt to control the mite population. Each chemical was applied individually and in combination with each of the others to comprise the four single-component blends, six binary blends, four ternary blends, and the four chemicals together. Each of the 15 chemical treatments was sprayed on three plants in each of four blocks of 45 plants. Seven days after spraying, the total number of mites on 10 leaves sampled from each plant was recorded. This number was divided by the total number of mites recorded from 10 leaves of the same plant just prior to spraying and then multiplied by 100% to approximate the mite mean percentage survival per plant. An average was taken across the three plants that received the same treatment and the average value was used as a datum.

The average relative percentages (averaged across the four replications) and the component proportions are presented in Table 2.10. The fitted model is

$$\hat{y}(\mathbf{x}) = 1.8x_1 + 25.4x_2 + 28.6x_3 + 38.5x_4 - 34.8x_1x_2 - 48.4x_1x_3 + 34.2x_1x_4$$

$$(11.8) \qquad\qquad\qquad (58.0)$$

$$-94.4x_2x_3 + 21.8x_2x_4 - 91.4x_3x_4 + 624.6x_1x_2x_3 - 584.7x_1x_2x_4$$

$$(408.4)$$

$$-238.5x_1x_3x_4 - 40.8x_2x_3x_4 - 1464.8x_1x_2x_3x_4 \qquad\qquad (2.42)$$

$$(4076.9)$$

and the estimate of the error variance was $s^2 = 561.5$ with 42 d.f. The quantities in parentheses below the parameter estimates in Eq. (2.42) are the estimated standard errors of the parameter estimates. The estimated standard errors are the positive square roots of the variance estimates, where the latter are

$$\widehat{\mathrm{var}}(b_i) = \frac{s^2}{4} = 140.4$$

$$\widehat{\mathrm{var}}(b_{ij}) = \frac{24s^2}{4} = 3369.0$$

$$\widehat{\mathrm{var}}(b_{ijk}) = \frac{1188s^2}{4} = 16.7 \times 10^4$$

$$\widehat{\mathrm{var}}(b_{1234}) = \frac{118400s^2}{4} = 16.6 \times 10^6$$

Table 2.10 Average Percentage of Mites per Plant Relative to Initial Numbers 7 Days after Spraying the Chemical Treatment

| Chemical Proportions | | | | Blend | Average |
x_1	x_2	x_3	x_4	Designation	Percentage $(\bar{y})^a$
1	0	0	0	V	1.8
0	1	0	0	O	25.4
0	0	1	0	K	28.6
0	0	0	1	D	38.5
0.5	0.5	0	0	VO	4.9
0.5	0	0.5	0	VK	3.1
0.5	0	0	0.5	VD	28.7
0	0.5	0.5	0	OK	3.4
0	0.5	0	0.5	OD	37.4
0	0	0.5	0.5	KD	10.7
0.33	0.33	0.33	0	VOK	22.0
0.33	0.33	0	0.33	VOD	2.6
0.33	0	0.33	0.33	VKD	2.4
0	0.33	0.33	0.33	OKD	11.1
0.25	0.25	0.25	0.25	VOKD	0.8

^{a}The 15-term model was fitted also to the arcsin $(\sqrt{\bar{y}})$ values with similar results. The percentages are used here simply for purposes of illustrating the fitting of model (2.42)

In this experiment, where the response is the average percentage of mites on the plants 7 days after treatment relative to the number counted on the plants just prior to spraying, treatments with low ($\bar{y}_u \leq 10$) average response values are preferred to (or are considered more effective than) treatments with high ($\bar{y}_u \geq 20$) average response values. According to the average percentage values in Table 2.10 as well as several of the parameter estimates in the fitted model of Eq. (2.42), the following inferences might be made about the chemical blends used in the experiment. With the 7-day posttreatment data, there is some evidence that the most effective blends are:

Particular Blend	Reason for Inference
V is the most effective single chemical	The estimate $b_1 = 1.8$ appears to be significantly lower than the other single-component estimates
VO, VK, and OK are the most effective	$\bar{y}_{VO} = 4.9$, $\bar{y}_{VK} = 3.1$, and $\bar{y}_{OK} = 3.4$ appear to be significantly smaller in magnitude than the other $\bar{y}_{ij}$
VOD and VKD are the most effective	$\bar{y}_{VOD} = 2.6$ and $\bar{y}_{VKD} = 2.4$ appear to be significantly smaller in magnitude than $\bar{y}_{VOK} = 22.0$ and $\bar{y}_{OKD} = 11.1$
VOKD is an effective blend	$\bar{y}_{VOKD} = 0.8$ is the lowest average response and appears to be significantly smaller than all the other responses except $\bar{y}_V = 1.8$

To validate these inferences, we would need to perform tests of hypotheses on the model's parameters using the parameter estimates in Eq. (2.42). Hypothesis tests are discussed in Sections 2.6 and 4.1.

Plotting estimated response contours for systems with four or more components is not an easy exercise. This is because to represent the surface described by the fitted model in Eq. (2.42) graphically in two dimensions, one of the component proportions, say, x_i, must be fixed so that the values of the other, x_j, can be varied over the range $0 \le x_j \le 1 - x_i$. For example, let us assume that surface contours across the values of the component proportions x_1, x_2, and x_3 are desired at three levels 0, 0.2, and 0.6 of x_4. The estimated response equation (2.42) is, with each of these cases,

$$x_4 = 0 \qquad \hat{y}(\mathbf{x}) = 1.8x_1 + 25.4x_2 + 28.6x_3 - 34.8x_1x_2 - 48.4x_1x_3$$
$$-94.4x_2x_3 + 624.6x_1x_2x_3 \qquad (2.43)$$

$$x_4 = 0.2 \qquad \hat{y}(\mathbf{x}) = 1.8x_1 + 25.4x_2 + 28.6x_3 + 38.5(0.2) - 34.8x_1x_2$$
$$(0 \le x_j \le 0.8) \qquad\quad -48.4x_1x_3 + 34.2x_1(0.2) - 94.4x_2x_3$$
$$j = 1, 2, 3 \qquad\qquad +21.8x_2(0.2) - 91.4x_3(0.2) + 624.6x_1x_2x_3$$
$$-584.7x_1x_2(0.2) - 238.5x_1x_3(0.2)$$
$$-40.8x_2x_3(0.2) - 1464.8x_1x_2x_3(0.2)$$
$$= 7.7 + (1.8 + 6.8)x_1 + (25.4 + 4.4)x_2$$
$$+(28.6 - 18.3)x_3 - (34.8 + 116.9)x_1x_2$$
$$-(48.4 + 47.7)x_1x_3 - (94.4 + 8.2)x_2x_3$$
$$+(624.6 - 293.0)x_1x_2x_3$$
$$= 7.7 + 8.6x_1 + 29.8x_2 + 10.3x_3 - 151.7x_1x_2$$
$$-96.1x_1x_3 - 102.6x_2x_3 + 331.6x_1x_2x_3 \qquad (2.44)$$

$$x_4 = 0.6 \qquad \hat{y}(\mathbf{x}) = 23.1 + 22.3x_1 + 38.5x_2 - 26.2x_3 - 385.6x_1x_2$$
$$(0 \le x_j \le 0.4) \qquad\quad -191.5x_1x_3 - 118.9x_2x_3 - 254.3x_1x_2x_3$$
$$j = 1, 2, 3 \qquad\qquad\qquad\qquad\qquad\qquad\qquad\qquad\qquad (2.45)$$

The plots of the surface contours for Eqs. (2.43) through (2.45) are presented in Figure 2.14.

2.14   AXIAL DESIGNS

The $\{q, m\}$ simplex-lattice and q-component simplex-centroid designs are boundary designs in that, with the exception of the overall centroid, the points of these designs are positioned on the boundaries (vertices, edges, faces, etc.) of the simplex factor space. Axial designs, however, are designs consisting mainly of complete mixtures or q-component blends where most of the points are positioned inside the simplex. Axial designs have been recommended for use when component effects are to be measured

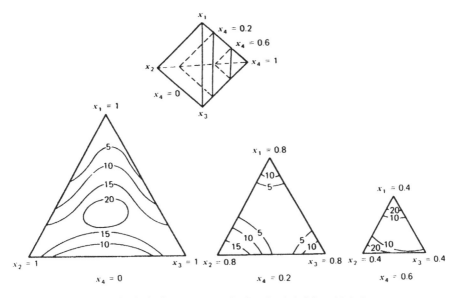

Figure 2.14 Surface contours at the three levels 0, 0.2, and 0.6 of x_4.

and in screening experiments (see Section 4.5) particularly when first-degree models are to be fitted. To define an axial design, we state the following.

Definition. The axis of component i is the imaginary line extending from the base point $x_i = 0$, $x_j = 1/(q-1)$ for all $j \neq i$, to the vertex where $x_i = 1$, $x_j = 0$ all $j \neq i$.

The base point is the centroid of the $(q-2)$-dimensional boundary (sometimes called a $(q-2)$-flat), which is opposite the vertex $x_i = 1$, $x_j = 0$, all $j \neq i$. The length of the axis is the shortest distance from the opposite $(q-2)$-dimensional boundary to the vertex. This distance is defined in the simplex coordinate system as one unit. Figure 2.15 presents the axes for components 1, 2, and 3 in the three-component triangle.

An *axial* design's points are positioned only on the component axes. (With three components, both the $\{3, 2\}$ simple-lattice and simplex-centroid designs are outer extreme point axial designs.) The simplest form of axial design is one whose points are positioned equidistant from the centroid $(1/q, 1/q, \ldots, 1/q)$ toward each of the vertices. The distance from the centroid, measured in the units of x_i, is denoted by Δ and the maximum value for Δ is $(q-1)/q$. Such a design has been suggested in Cornell (1975). A three-component axial design is shown in Figure 2.16.

Let us write the matrix form of the first-degree model $\eta = \sum_{i=1}^{q} \beta_i x_i$ as $\eta = \mathbf{x}'\boldsymbol{\beta}$, where $\mathbf{x} = (x_1, x_2, \ldots, x_q)'$ and $\boldsymbol{\beta}$ is a $q \times 1$ vector of the β_i, $i = 1, 2, \ldots, q$. With an axial design of the form shown in Figure 2.16, if r observations are collected at each of the points, the form of the variance-covariance matrix of the vector $\mathbf{b} =$

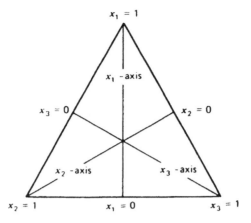

Figure 2.15 The x_i axes, $i = 1, 2,$ and 3.

$(b_1, b_2, \ldots, b_q)'$ of estimates of the parameters in the model is

$$\text{var}(\mathbf{b}) = \{d\mathbf{I} + e\mathbf{J}\}\sigma^2 \tag{2.46}$$

where

$$d = \frac{(q-1)^2}{r\Delta^2 q^2}, \quad e = \frac{\Delta^2 q^2 - (q-1)^2}{r\Delta^2 q^3} \tag{2.47}$$

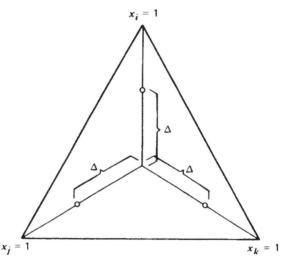

Figure 2.16 Three-component axial design where the distance from the center of the simplex to the points on the axes is Δ.

I is an identity matrix of order q and **J** is a $q \times q$ matrix of 1's. Given the form of the matrix var(**b**) in Eq. (2.46), the variance of each parameter estimate b_i is

$$\text{var}(b_i) = (d + e)\sigma^2 = \left\{\frac{(q-1)^3 + \Delta^2 q^2}{r\Delta^2 q^3}\right\}\sigma^2 \qquad (2.48)$$

and the covariance between pairs of estimates b_i and b_j is $\text{cov}(b_i, b_j) = e\sigma^2$. Now, the larger the value of Δ, the smaller are the values of var(b_i) and cov(b_i, b_j). In fact, if the points are positioned at the vertices, then $\Delta = (q-1)/q$, cov(b_i, b_j) = 0, and

$$\text{var}(b_i) = \left\{\frac{(q-1)^3 + (q-1)^2}{r(q-1)^2 q}\right\}\sigma^2 = \frac{\sigma^2}{r}$$

If the points are positioned midway between the centroid of the simplex and each of the vertices so that $\Delta = (q-1)/2q$, then cov(b_i, b_j) = $-3\sigma^2/rq$, and

$$\text{var}(b_i) = \frac{(4q-3)}{rq}\sigma^2$$

In Table 2.11, values of var(b_i)r/σ^2 and cov(b_i, b_j)r/σ^2 are listed for increasing values of Δ for $q = 3, 4, 5, 6$, and 7. Incremental values of $\Delta = a/q$ range from $a = 1$ to $a = q - 2$. With the larger values of q or larger numbers of components, the faster the values of var(b_i) and cov(b_i, b_j) decrease for increasing values of Δ. This means that when fitting the first-degree model to an axial design of the type considered, the greater the number of components, the more spread the design should be in order to increase the precision of the parameter estimates and reduce the correlations between the estimates. Here precision refers to the reciprocal of the variance and correlation between pairs of estimates is directly related to the covariance between the pairs. Additional discussion on the use of axial designs is presented in Chapters 4 and 5 when the topic of screening the components is discussed.

Table 2.11 Variances and Covariances of Parameter Estimates Associated with the Simple Axial Design, $\Delta = a/q$

Number of Components $q =$	3	4	4	5	5	5	6	6	6	6	7	7	7	7	7
Size of $\Delta = a/q$, where $a =$	1	1	2	1	2	3	1	2	3	4	1	2	3	4	5
var(b_i)r/σ^2	3	7	$\frac{31}{16}$	13	$\frac{17}{5}$	$\frac{73}{45}$	21	$\frac{43}{8}$	$\frac{67}{27}$	$\frac{47}{32}$	31	$\frac{55}{7}$	$\frac{25}{7}$	$\frac{58}{28}$	$\frac{241}{175}$
cov (b_i, b_j)r/σ^2	-1	-2	$-\frac{5}{16}$	-3	$-\frac{3}{5}$	$-\frac{7}{45}$	-4	$-\frac{7}{8}$	$-\frac{8}{27}$	$-\frac{3}{32}$	-5	$-\frac{8}{7}$	$-\frac{3}{7}$	$-\frac{5}{28}$	$-\frac{11}{175}$

2.15   COMMENTS ON A COMPARISON MADE BETWEEN AN AUGMENTED SIMPLEX-CENTROID DESIGN AND A FULL CUBIC LATTICE FOR THREE COMPONENTS WHERE EACH DESIGN CONTAINS TEN POINTS

For the sake of simplicity, in setting up the experimental blends as well as ease in calculating the model coefficient estimates, it can be argued that the $\{q, m\}$ simplex-lattices and simplex-centroid designs, along with their corresponding model forms, are among the most popular and appealing designs and models used for exploring the entire simplex factor space. One need only review the literature from allied disciplines such as the food industry, the rubber industry, petroleum and textile industries, and many other fields where mixture experiments are performed daily, to find these designs being used on a regular basis. Nevertheless, in thinking that many mixture surfaces, when viewed over the entire three-component triangle, are actually more complicated in shape than those that can be modeled with a first-or second-degree Scheffé polynomial, Cornell (1986) suggested the use of and compared the two 10-point designs shown in Figure 2.17. The rationale for suggesting the designs in Figure 2.17 was (1) often the reported quadratic-shaped surface would have been detected and reported as being cubic in shape had the design supported the fitting of a cubic model, and (2) unless the design consists of a greater number of points than there are terms in the fitted model, one cannot test for the adequacy (or lack of fit) of the fitted model.

In order to compare the two designs in Figure 2.17, specific design criteria must be defined. The following is a list of four properties of a good design that seem to be particularly relevant when fitting mixture surfaces and thus can be used as design criteria for comparing the two designs. The design should:

1. Generate a satisfactory distribution of information throughout the experimental region (the triangle).

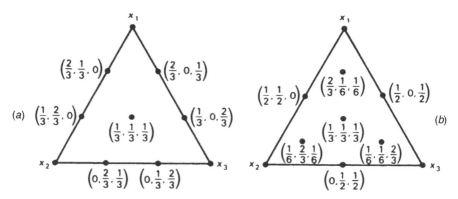

Figure 2.17   Three-component 10-point designs, (*a*) A $\{3,3\}$ simplex-lattice. (*b*) A simplex-centroid design augmented with three interior points.

2. Ensure that the fitted model predicts a value, $\hat{y}(\mathbf{x})$, at all points in the experimental region that is as close as possible to the true value of the response.

3. Give good detectability of model lack of fit.

4. Provide an internal estimate of the error variance.

In this section we refer to the designs in Figures 2.17a,b as designs A and B, respectively. Design A is the $\{3, 3\}$ simplex-lattice and supports the fitting of the cubic model

$$\eta = \beta_1 x_1 + \beta_2 x_2 + \beta_3 x_3 + \beta_{12} x_1 x_2 + \beta_{13} x_1 x_3 + \beta_{23} x_2 x_3 + \beta_{123} x_1 x_2 x_3$$

$$+ \delta_{12} x_1 x_2 (x_1 - x_2) + \delta_{13} x_1 x_3 (x_1 - x_3) + \delta_{23} x_2 x_3 (x_2 - x_3) \qquad (2.49)$$

while design B does not. The last three terms in Eq. (2.49) are useful for studying changes in the shape of the response surface of a degree higher than two (i.e., cubic) for binary blends of pairs of components. An illustration of cubic blending between the components i and j is shown in Figure 2.4b. For example, if we denote the responses to the four blends along the edge connecting the $x_1 = 1$ and $x_2 = 1$ vertices by $\eta_1, \eta_{112}, \eta_{122}$, and η_2 and consider the cubic model in the two components whose proportions are denoted by x_1 and x_2,

$$\eta = \beta_1 x_1 + \beta_2 x_2 + \beta_{12} x_1 x_2 + \delta_{12} x_1 x_2 (x_1 - x_2)$$

then the coefficients in the above equation, expressed in terms of the responses, are

$$\beta_1 = \eta_1, \quad \beta_2 = \eta_2, \quad \beta_{12} = (9/4)[(\eta_{112} + \eta_{122}) - (\eta_1 + \eta_2)],$$

$$\delta_{12} = (9/4)[3(\eta_{112} - \eta_{122}) - (\eta_1 - \eta_2)]$$

The quantity $2\beta_{12}/9$ is a measure of the difference between the average of the responses to the binary blends, and the average of the responses to the single components. The quantity $4\delta_{12}/27$ is a comparison of the difference between the binary responses $(\eta_{112} - \eta_{122})$ with the differences between the responses to the pure mixtures $(\eta_1 - \eta_2)/3$, where the latter difference, $(\eta_1 - \eta_2)/3$, is taken over an interval three times as wide as the interval used for measuring $\eta_{112} - \eta_{122}$.

Design B provides a more uniform distribution of information about the surface inside the triangle than does design A. With design B, the ratio of mixture types (pure : binary : ternary) is 3 : 3 : 4 compared to 3 : 6 : 1 with design A. With a more uniform distribution of mixture types and coupled with the fact that each component is studied at six levels, $x_i = 0, \frac{1}{6}, \frac{1}{3}, \frac{1}{2}, \frac{2}{3}$, and 1, design B supports the fitting of the *special quartic model*

$$\eta = \beta_1 x_1 + \beta_2 x_2 + \beta_3 x_3 + \beta_{12} x_1 x_2 + \beta_{13} x_1 x_3 + \beta_{23} x_2 x_3$$

$$+ \beta_{1123} x_1^2 x_2 x_3 + \beta_{1223} x_1 x_2^2 x_3 + \beta_{1233} x_1 x_2 x_3^2 \qquad (2.50)$$

Equation (2.50) is especially useful for detecting curvature of the surface in the interior of the triangle that cannot be picked up by terms in Eq. (2.49).

In terms of the criterion 1 then, design A contributes more information than design B in terms of how the response behaves for binary blends measured along the boundaries of the triangle. Design B provides more information than A about the response to complete mixtures, that is, to blends of all three components inside the triangle.

With respect to criterion 2, comparisons can be made between designs A and B in terms of the properties $\hat{y}(\mathbf{x})$, namely, the var[$\hat{y}(\mathbf{x})$] and the bias of $\hat{y}(\mathbf{x})$. Quantification of these properties is possible by calculating the *average variance of* $\hat{y}(\mathbf{x})$ and the *average squared bias of* $\hat{y}(\mathbf{x})$, where "average" is taken to mean averaged over the experimental region, or triangle in this case. For the average variance of $\hat{y}(\mathbf{x})$, a value can readily be obtained once the form of the fitted model has been specified. A value for the average squared bias of $\hat{y}(\mathbf{x})$, however, requires not only specifying the form of the fitted model but also specifying the form of the assumed true response equation, which generally is of higher degree than the fitted model.

Rather than work through the derivations of the properties of $\hat{y}(\mathbf{x})$ that are presented in Cornell (1986), we simply report below the results by stating the recommendations given by Cornell (1986) concerning the desirability of designs A and B in Figure 2.17.

- Design A, consisting of nine points equispaced along the boundaries of the triangle, is particularly useful for studying the three binary systems, x_1 with x_2, x_1 with x_3, and x_2 with x_3. The corresponding cubic model (2.49) centers one's attention on the detection and estimation of the binary blending between pairs of components through the coefficients of the terms $\beta_{ij}x_ix_j$ and $\delta_{ij}x_ix_j(x_i-x_j)$.

- Design B is the recommended design to use if one is interested initially in fitting a low-degree (first- or second-degree) mixture polynomial but is uncertain about the shape of the surface above the triangle. When sequentially developing a model for describing the shape of the surface, one begins by fitting the simplest model form first, testing the model for adequacy of fit, augmenting the model if necessary by adding terms to it, fitting the augmented model and testing for adequacy of fit, and so on. Then, of course, design B is the more powerful arrangement to use. The power meant here is to reject the zero lack of fit of the fitted model whose true surface is more complicated in shape than can be described by the terms in the fitted model.

Furthermore design B is an axial design of the type discussed in the previous section, where the proportions for component i along its axis are $x_i = 0, \frac{1}{3}, \frac{2}{3}$, and 1. Additional discussion on the use of axial designs is presented in Chapters 4 and 5 with relation to topics such as the screening of components and the measuring of component effects using both Scheffé's model and Cox's mixture models.

2.16 REPARAMETERIZING SCHEFFÉ'S MIXTURE MODELS TO CONTAIN A CONSTANT (β_0) TERM: A NUMERICAL EXAMPLE

Many standard regression programs do not provide the correct analysis of a fitted Scheffé-type mixture model for one of two reasons:

1. The fitted model is required to contain a constant term (β_0) and the Scheffé models do not.
2. Or, if a no-constant term option is available in the program, then the regression or fitted model sum of squares and the total sum of squares in the analysis of variance table are not corrected for the overall mean.

As a result of point 2, summary statistics such as the F-test, R^2, and R_A^2, which almost always accompany the analysis of variance (ANOVA) table and are computed using the uncorrected (for the mean) regression sum of squares and total sum of squares, are incorrect. In general, the values of F, R^2, and R_A^2 are inflated, giving the wrong impression that the fitted model is better than it really is.

A simple technique that can be used to alleviate both the need for the model to contain a constant term and the calculation of the corrected regression and total sum of squares in the ANOVA table is to write the Scheffé models with a constant term. This is done most simply without having to impose a restriction on the values of the parameters in the linear blending portion of the model, by deleting one of the $\beta_i x_i$ terms, say, $\beta_q x_q$. In other words, instead of writing the Scheffé first- or second-degree models as

$$\eta = \beta_1 x_1 + \beta_2 x_2 + \cdots + \beta_q x_q \tag{2.51}$$

or

$$\eta = \beta_1 x_1 + \beta_2 x_2 + \cdots + \beta_q x_q + \sum \sum_{i<j}^{q} \beta_{ij} x_i x_j \tag{2.52}$$

one writes

$$\eta = \beta_0 + \beta_1^* x_1 + \beta_2^* x_2 + \cdots + \beta_{q-1}^* x_{q-1} \tag{2.53}$$

or

$$\eta = \beta_0 + \sum_{i=1}^{q-1} \beta_i^* x_i + \sum \sum_{i<j}^{q} \beta_{ij} x_i x_j \tag{2.54}$$

and only the parameters in the linear blending portion of the models change in their meaning. The parameter β_0 assumes the role of β_q in (2.51) or (2.52) and the

remaining $q - 1$ parameters, β_i^*, represent the differences $\beta_i - \beta_q$, $i = 1, 2, \ldots$
$q - 1$.

Let us illustrate the model form (2.54) using the yarn elongation data of Table 2.3. The fitted second-degree Scheffé model with the estimated standard errors in parentheses was shown as Eq. (2.24) to be

$$\hat{y}(\mathbf{x}) = 11.7x_1 + 9.4x_2 + 16.4x_3 + 19.0x_1x_2 + 11.4x_1x_3 - 9.6x_2x_3 \quad (2.55)$$
$$(0.60) \qquad (0.60) \qquad (0.60) \qquad (2.61) \qquad (2.61) \qquad (2.61)$$

Now, if we fit the model (2.54) where $q = 3$, the result is

$$\hat{y}(\mathbf{x}) = b_0 + b_1^*x_1 + b_2^*x_2 + b_{12}x_1x_2 + b_{13}x_1x_3 + b_{23}x_2x_3$$
$$= 16.4 - 4.7x_1 - 7.0x_2 + 19.0x_1x_2 + 11.4x_1x_3 - 9.6x_2x_3 \quad (2.56)$$
$$(0.60) \quad (0.85) \quad (0.85) \quad (2.61) \qquad (2.61) \qquad (2.61)$$

Comparing the model forms (2.55) and (2.56), we see that

(2.56) equals	(2.55)
b_0	$b_3 = 16.4$
b_1^*	$b_1 - b_3 = 11.7 - 16.4 = -4.7$
b_2^*	$b_2 - b_3 = 9.4 - 16.4 = -7.0$
b_{12}	$b_{12} = 19.0$
b_{13}	$b_{13} = 11.4$
b_{23}	$b_{23} = -9.6$

As for the estimated standard errors of the coefficients in (2.56), s.e.$(b_0) =$ s.e.(b_3), and s.e.$(b_{ij}) =$ s.e.(b_{ij}). However, for those estimates in (2.56) that are differences,

$$\text{s.e.}(b_i^*) = \{\text{var}(b_i) + \text{var}(b_3) - 2\text{cov}(b_i, b_3)\}^{1/2}$$
$$= \{0.73(0.5) + 0.73(0.5) - 2(0.73)(0)\}^{1/2} = 0.85$$

where $i = 1$ and 2.

The ANOVA table for the fitted model (2.56) containing the constant term is the same as Table 2.7

Source	d.f.	Sum of Squares	Mean Squares	F
Regression	5	128.32	25.66	35.20
Residual	9	6.56	0.73	
Total	14	134.88		

Summarizing then, any of the Scheffé mixture models (2.4), (2.7), (2.9), and (2.10) can be reparameterized to contain a constant term (β_0) by simply deleting any one of the $\beta_i x_i$ terms in the linear blending portion of the model and then adding the constant term. If cross-product terms are present in the model, they remain in the reparameterized model. Furthermore, if after fitting the reparameterized model one desires to express the fitted model in the Scheffé form, the estimated constant coefficient is substituted for the deleted coefficient estimate and the remaining linear blending coefficients are obtained by adding $b_i^* + b_0$ to produce b_i. For example, if $\beta_q x_q$ is deleted from the second-degree model to produce

$$\hat{y}(\mathbf{x}) = b_0 + b_1^* x_1 + b_2^* x_2 + \cdots + b_{q-1}^* x_{q-1} + \sum_{i<j}^{q} b_{ij} x_i x_j$$

then to get the Scheffé second-degree canonical polynomial, we write

$$\hat{y}(\mathbf{x}) = (b_1^* + b_0)x_1 + (b_2^* + b_0)x_2 + \cdots + (b_{q-1}^* + b_0)x_{q-1} + b_0 x_q + \sum_{i<j}^{q} b_{ij} x_i x_j$$

$$= b_1 x_1 + b_2 x_2 + \cdots + b_q x_q + \sum_{i<j}^{q} b_{ij} x_i x_j$$

2.16.1 SAS Input Statements and ANOVA Printout for a Fruit Punch Experiment

Watermelon (x_1), pineapple (x_2), and orange (x_3) juice concentrates were used as primary ingredients of a fruit punch. Ten blends of the three-juice concentrates were evaluated for overall general acceptance by a sensory panel. The ingredient proportions and the average acceptance values scored on a scale of 1 (extremely poorer than reference) to 9 (extremely better than reference) for three replications of each blend are listed in Table 2.12. Each general acceptance value in Table 2.12

Table 2.12   Fruit Punch General Acceptance Ratings

Blend	Watermelon %W	$(x_1)^a$	Pineapple %P	(x_2)	Orange %O	(x_3)	General Acceptance (y_u)	Average $\bar{y}$
1	100	1.0	0	0	0	0	4.3, 4.7, 4.8	4.60
2	65	0.5	35	0.5	0	0	6.3, 5.8, 6.1	6.07
3	30	0	70	1.0	0	0	6.5, 6.2, 6.3	6.33
4	30	0	35	0.5	35	0.5	6.2, 6.2, 6.1	6.17
5	30	0	0	0	70	1.0	6.9, 7.0, 7.4	7.10
6	65	0.5	0	0	35	0.5	6.1, 6.5, 5.9	6.17
7	54	0.34	23	0.33	23	0.33	6.0, 5.8, 6.4	6.07
8	80	0.72	10	0.15	10	0.14	5.4, 5.8, 6.6	5.93
9	40	0.14	40	0.57	20	0.29	5.7, 5.0, 5.6	5.43
10	40	0.14	20	0.29	40	0.57	5.2, 6.4, 6.4	6.00

$^a x_1 = \%W - 30/70, \; x_2 = \%P/70, \; x_3 = \%O/70.$

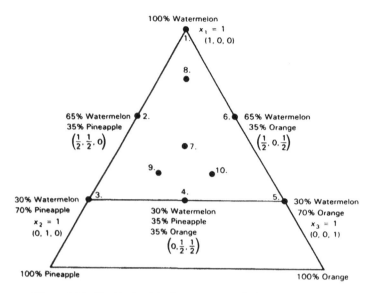

Figure 2.18   Ten blends of the three-component fruit punch experiment.

represents an average of eight panelists' scores. Blends 1 to 6 comprise the {3, 2} simplex-lattice design while blends 7 to 10 are interior blends. The design is shown in Figure 2.18.

The second-degree model fitted to the entire set of 30 acceptance values in Table 2.12 is

$$\hat{y}(\mathbf{x}) = 4.77x_1 + 6.27x_2 + 7.11x_3 + 2.15x_1x_2 + 1.10x_1x_3 - 3.54x_2x_3$$
$$\quad (0.24) \quad (0.25) \quad (0.25) \quad (1.13) \quad (1.13) \quad (1.02) \tag{2.57}$$

The numbers in parentheses directly below the coefficient estimates are the estimated standard errors (positive square root of the variance) of the coefficient estimates. These standard error values were taken from the computer printout of the analysis shown in Table 2.14.

The SAS version 6.03, input statements for generating the Scheffé second-degree model without and with a constant (INTERCEPT) term are listed in Table 2.13. The computer printout of the analysis of the 30 general acceptance values that generated Eq. 2.57 is listed in Table 2.14. Several entries on the printout are worth mentioning at this point. Specifically, since the second-degree model that was fitted and is shown as Eq. (2.57) does not contain a constant term β_0, the NOINT (or NO INTercept) option was used in the model statement, which is the first model statement listed in Table 2.13. As a result, with this model form, the degrees of freedom (d.f.) and the sum of squares for the "Model" and "Uncorrected total" entries in Table 2.14 are not corrected for the overall mean of the 30 data values. Consequently the d.f. for both the model and uncorrected total in Table 2.14 are inflated by 1 and the sum of

Table 2.13   SAS Program Statements Used to Produce Tables 2.14 and 2.15

```
options Is = 76 ps = 62 nodate nonumber;
data;
    input dpoint x1 x2 x3 y1 y2 y3;
    array ys{3} yl − y3;
    do i = 1 to 3;
        y = ys{i};
        output;
    end;
    keep x1 x2 x3 y;
    cards;
1 1 0 0 4.3 4.7 4.8
2 .5 .5 0 6.3 5.8 6.1
3 0 1 0 6.5 6.2 6.3
4 0 .5 .5 6.2 6.2 6.1
5 0 0 1 6.9 7 7.4
6 .5 0 .5 6.1 6.5 5.9
7 .34 .33 .33 6 5.8 6.4
8 .72 .14 .14 5.4 5.8 6.6
9 .14 .57 .29 5.7 5 5.6
10 .14 .29 .57 5.2 6.4 6.4;
    run;
proc glm;
    model y = x1 x2 x3 x1*x2 x1*x3 x2*x3 / NOINT;
proc glm;
    model y = x1 x2 x1*x2 x1*x3 x2*x3;
    estimate 'beta1' intercept 1 x1 1;
    estimate 'beta2' intercept 1 x2 1;
    estimate 'beta3' intercept 1;
    run;
```

squares for the model and uncorrected total are also inflated. The correct d.f. and sum of squares for the model and corrected total entries are listed in Table 2.15.

Since the sum of squares entries for the model and uncorrected total are inflated in Table 2.14 so also are the values of the F-test (F-value $= 955.14$) and of R^2 (R-square $= 0.995830$). The correct values of the F-test and of R^2 are listed in Table 2.15.

The printout for Table 2.15 was generated by deleting one of the linear blending terms from the second-degree model and allowing SAS to include a constant term (or intercept) in the fitted model. For our example $\beta_3 x_3$ was deleted. The deletion of one of the $\beta_i x_i$ terms in the model is necessary because SAS automatically inserts a constant term (β_0) in the model unless one specifies NOINT in the model statement, and it is not possible to estimate the intercept β_0 and all of the linear coefficients β_i, $i = 1, 2, \ldots, q$ owing to the restriction $x_1 + x_2 + \cdots + x_q = 1$. Consequently, with

**Table 2.14   Fruit Punch Data: Computer Output from Analysis of the
30 General Acceptance Values**

General Linear Models Procedure

Dependent variable: Y

Source	d.f.	Sum of Squares	Mean Square	F-Value	$Pr > F$
Model	6	1084.49825	180.74971	955.14	0.0001
Error	24	4.54175	0.18924		
Uncorrected total	30	1089.04000			

	R-Square	C.V.	Root MSE		Y Mean
	0.995830	7.266430	0.43502		5.9866667

Source	d.f.	Type I SS	Mean Square	F-Value	$Pr > F$
X1	1	480.255527	480.255527	2537.82	0.0001
X2	1	364.276843	364.276843	1924.95	0.0001
X3	1	236.866735	236.866735	1251.68	0.0001
X1*X2	1	0.650804	0.650804	3.44	0.0760
X1*X3	1	0.169700	0.169700	0.90	0.3531
X2*X3	1	2.278635	2.278635	12.04	0.0020

Source	d.f.	Type III SS	Mean Square	F-Value	$Pr > F$
X1	1	76.171495	76.171495	402.51	0.0001
X2	1	121.609914	121.609914	642.62	0.0001
X3	1	156.473002	156.473002	826.85	0.0001
X1*X2	1	0.679641	0.679641	3.59	0.0702
X1*X3	1	0.177607	0.177607	0.94	0.3423
X2*X3	1	2.278635	2.278635	12.04	0.0020

Parameter	Estimate	T for H_0: Parameter $= 0$	$Pr > \|T\|$	Standard Error of Estimate
X1	4.773601512	20.06	0.0001	0.23793379
X2	6.266368008	25.35	0.0001	0.24719381
X3	7.108060450	28.76	0.0001	0.24719381
X1*X2	2.148058023	1.90	0.0702	1.13347590
X1*X3	1.098086662	0.97	0.3423	1.13347590
X2*X3	−3.536609727	−3.47	0.0020	1.01919177

Note: The model used is Eq. (2.57) with the NOINT option included in the model statement.

Table 2.15 Fruit Punch Data: Computer Output from Analysis of the 30 General Acceptance Values

General Linear Models Procedure

Dependent variable: Y

Source	d.f.	Sum of Squares	Mean Square	F-Value	$Pr > F$
Model	5	9.29291245	1.85858249	9.82	0.0001
Error	24	4.54175421	0.18923876		
Corrected total	29	13.83466667			

	R-Square	C.V.	Root MSE		Y Mean
	0.671712	7.266430	0.43502		5.9866667

Source	d.f.	Type I SS	Mean Square	F-Value	$Pr > F$
X1	1	5.11961353	5.11961353	27.05	0.0001
X2	1	1.07415889	1.07415889	5.68	0.0255
X1*X2	1	0.65080425	0.65080425	3.44	0.0760
X1*X3	1	0.16970035	0.16970035	0.90	0.3531
X2*X3	1	2.27863544	2.27863544	12.04	0.0020

Source	d.f.	Type III SS	Mean Square	F-Value	$Pr > F$
X1	1	8.88687702	8.88687702	46.96	0.0001
X2	1	1.10482246	1.10482246	5.84	0.0237
X1*X2	1	0.67964140	0.67964140	3.59	0.0702
X1*X3	1	0.17760739	0.17760739	0.94	0.3423
X2*X3	1	2.27863544	2.27863544	12.04	0.0020

Parameter	Estimate	T for H_0: Parameter $= 0$	$Pr > \lvert T \rvert$	Standard Error of Estimate
beta1	4.77360151	20.06	0.0001	0.23793379
beta2	6.26636801	25.35	0.0001	0.24719381
beta3	7.10806045	28.76	0.0001	0.24719381

Parameter	Estimate	T for H_0: Parameter $= 0$	$Pr > \lvert T \rvert$	Standard Error of Estimate
INTERCEPT	7.108060450	28.76	0.0001	0.24719381
X1	−2.334458938	−6.85	0.0001	0.34065742
X2	−0.841692441	−2.42	0.0237	0.34834804
X1*X2	2.148058023	1.90	0.0702	1.13347590
X1*X3	1.098086662	0.97	0.3423	1.13347590
X2*X3	−3.536609727	−3.47	0.0020	1.01919177

Note: The model used is Eq. (2.59). Coefficients for X1, X2, and X3 in a Scheffé-type model are obtained by "Estimate" statements.

the addition of the constant term (or INTERCEPT) in the model and the deletion of one of the linear terms, the DF and sum of squares for the model and corrected total entries are adjusted for the overall mean.

The estimates of the coefficients in the second-degree model of Eq. (2.57) are listed at the bottom of Table 2.14. Listed also are the estimated standard errors of the coefficient estimates and the values of the corresponding t-test statistic:

$$t = \frac{\text{estimate}}{\text{s.e. of estimate}} \tag{2.58}$$

The level of significance of the t test in (2.58) is given under $Pr > |T|$.

The model that was fitted to provide the printout in Table 2.15 is of the form

$$
\begin{aligned}
\hat{y}(\mathbf{x}) &= b_0 + b_1^* x_1 + b_2^* x_2 + b_{12} x_1 x_2 + b_{13} x_1 x_3 + b_{23} x_2 x_3 \\
&= 7.11 - 2.33 x_1 - 0.84 x_2 + 2.15 x_1 x_2 + 1.10 x_1 x_3 - 3.54 x_2 x_3 \\
&\quad (0.25) \;\; (0.34) \quad\;\; (0.35) \quad\;\; (1.13) \quad\quad\;\; (1.13) \quad\quad\;\; (1.02)
\end{aligned} \tag{2.59}
$$

The quantities in parentheses below the coefficient estimates are the estimated standard errors of the coefficient estimates. These estimated standard errors are listed in Table 2.15 in the bottom rightmost column.

As we discussed in the previous section, the difference between the fitted models of Eqs. (2.57) and (2.59) exists only with the first three terms in the two models. In Eq. (2.59) above, the intercept b_0 is equal to b_3 in Eq. (2.57). The b_i^*, $i = 1, 2$ in Eq. (2.59) represent differences between the b_i and b_3 in Eq. (2.57). More specifically,

Eq. (2.59)	equals	Eq. (2.57)
$b_1^* = -2.33$		$b_1 - b_3 = 4.77 - 7.11$
$b_2^* = -0.84$		$b_2 - b_3 = 6.27 - 7.11$

Thus, if one fits the model of Eq. (2.59) to obtain the correct analysis of variance table as shown in Table 2.15 but wishes to have a Scheffé-type model of the form shown in Eq. (2.57), it is easy to rewrite Eq. (2.59) as

$$
\begin{aligned}
\hat{y}(\mathbf{x}) &= (b_1^* + b_0) x_1 + (b_2^* + b_0) x_2 + b_0 x_3 + b_{12} x_1 x_2 + b_{13} x_1 x_3 + b_{23} x_2 x_3 \\
&= (-2.33 + 7.11) x_1 + (-0.84 + 7.11) x_2 + 7.11 x_3 + 2.15 x_1 x_2 \\
&\quad + 1.10 x_1 x_3 - 3.54 x_2 x_3 \\
&= 4.77 x_1 + 6.27 x_2 + 7.11 x_3 + 2.15 x_1 x_2 + 1.10 x_1 x_3 - 3.54 x_2 x_3
\end{aligned}
$$

which is exactly the model of Eq. (2.57). The estimates of the coefficients for the first three terms (linear blending terms) are listed in Table 2.15 in the "Parameter" column and are labeled beta1, beta2, and beta3, respectively.

2.17　QUESTIONS TO CONSIDER AT THE PLANNING STAGES OF A MIXTURE EXPERIMENT

At the planning stages prior to performing a mixture experiment, there are generally several questions that need to be answered concerning not only the types of blends that will be studied but also the method that is to be used in analyzing the data when collected. Listed below are questions that can aid an experimenter who is planning to run a mixture experiment. Also listed are the specific chapters in this book where the information surrounding the question can be found.

Definition. A feasible blend of ingredients (or components) is one from which an acceptable end product will result.

　　Questions:　　　　　　　　　　　　　　　　　　　　　　　　　　**Chapter**

1. For my particular situation, how are feasible mixtures defined? Are they blends consisting of:
   (a) The individual ingredients alone as well as any combination of the q ingredients?　　　　　　　　　　　　　　　　　2
   (b) At least $2 \leq r \leq q$ of the components with nonzero proportions? Some components may be absent ($x_i = 0$) but not more than $q - r$ can be absent in any blend.　　　　　　　**2, 3**
   (c) All of the components? In other words, none of the components can be absent in any blend (forcing the lower bounds $0 < L_i \leq x_i$ for all $i = 1, 2, \ldots, q$).　　　　　　　3
2. What are the objectives of the experiment? Do I know which components are the most active and which are least active? If not, should I run a screening experiment first? If I know which components are the major players, do I know what their blending properties are (i.e., how they blend together)? Do I know how each affects the responses of interest both individually as well as together?　　　　　　　　　　　　　　　　　　　　　**2, 4**
3. Is it reasonable to assume that the shape of the mixture surface is smooth so that some type of polynomial model can be fit? If so, can I fit a Scheffé-type mixture model in the component proportions or must I fit a standard form of polynomial model in a set of independent variables? If not a polynomial, what other forms of mixture models can I consider?　　　　　　　**2, 5**
4. How is the experimental region of feasible blends defined? (See question 1.) Is it the whole simplex region and, if so, can a lattice design be used? If the region is not the whole simplex but rather is only a subregion of the simplex, how do I choose which blends to use? Should I consider the use of pseudocomponents?　　　　　　　　　　　　　　　　　**2, 3**

5. Are there other factors such as process variables or the amount of the mixture that might have some influence on the blending properties of the components and thus ought to be considered? If so, how do we include these factors in the design and in the model to be fitted?   **6**

6. In planning the total number of experiments (blends) to run, should additional blends other than the basic design points be considered? If so, where should these points be taken? Can we replicate some or all of the blends? Do I need to consider blocks or groups of runs?   **2, 6**

2.18 SUMMARY

In this chapter were presented designs for exploring the entire simplex factor space. We began by introducing the $\{q, m\}$ simplex-lattice designs and then presented the class of simplex-centroid designs. Accompanying both classes of designs were their associated canonical form of polynomial model.

Formulas for estimating the parameters in the canonical polynomial models were presented where the response was observed at the simplex design points. Data from a three-component yarn experiment, where a $\{3, 2\}$ simplex-lattice was set up, were used to illustrate the calculations required in obtaining the second-degree fitted model coefficients as well as to show how to calculate the sums of squares quantities in the analysis of variance table. A discussion on choosing between a fitted first-degree model and a fitted second-degree model was presented and a plot of the residuals from both fitted models was constructed to aid in the decision making.

Axial designs, consisting of points positioned on the component axes, were also presented. For a three-component system, one specific axial design was defined as the simplex-centroid design augmented with three interior points. Recommendations were given based on a comparison made between the strengths of a three-component augmented simplex-centroid design and the $\{3, 3\}$ simplex-lattice for fitting mixture models over the three-component triangle.

In the next-to-last section of the chapter, it was shown that one can re-parameterize the Scheffé-type mixture models so that the model contains a constant term (β_0) by deleting one of the linear blending terms. The inclusion of the constant term is necessary for many least-squares regression programs in order that the regression and total sums of squares in the ANOVA table are calculated correctly. Finally, it was shown that once the model containing the constant term is fitted, one can easily reexpress the fitted surface with a canonical polynomial model of the Scheffé type.

REFERENCES AND RECOMMENDED READING

Agreda, C. L., and V. H. Agreda (1989). Designing mixture experiments. *CHEMTECH*, **19**, 573–575.

Box, G. E. P., W. G. Hunter, and J. S. Hunter (1978). *Statistics for Experimenters. An Intro to Design, Data Analysis, and Model Building*. Wiley, New York.

Cornell, J. A. (1975). Some comments on designs for Cox's mixture polynomial. *Technometrics*, **17**, 25–35.

Cornell, J. A. (1986). A comparison between two ten-point designs for studying three-component mixture systems. *J. Qual. Technol.*, **18**, 1–15.

Cornell, J. A. (1990). *How to Run Mixture Experiments for Product Quality*, 2nd ed. The ASQ Basic References in Quality Control: Statistical Techniques, Vol. 5. American Society for Quality, Milwaukee, WI, 1–71.

Cornell, J. A. (1990). Mixture experiments. In Subir Ghosh, ed., *Statistical Design and Analysis of Industrial Experiments*. Marcel Dekker, New York, pp. 175–209.

Cornell, J. A. (2002). *Experiments with Mixtures: Designs, Models, and the Analysis of Mixture Data*, 3rd ed. Wiley, New York.

Cornell, J. A., J. T. Shelton, R. Lynch, and G. F. Piepel (1983). Plotting three-dimensional response surfaces for three-component mixtures or two-factor systems. *Bulletin 836, Agricultural Experiment Station*, Institute of Food and Agricultural Sciences, University of Florida, Gainesville, 1–31.

Cornell, J. A., and S. B. Linda (1991). Models and designs for experiments with mixtures. Part I: Exploring the whole simplex region. *Bulletin 879, Agricultural Experiment Station*, Institute of Food and Agricultural Sciences, University of Florida, Gainesville, 1–49.

Draper, N. R., and A. M. Herzberg (1971). On lack of fit. *Technometrics*, **13**, 231–241.

Draper, N. R., and H. Smith (1981). *Applied Regression Analysis*, 2nd ed. Wiley, New York.

Gorman, J. W., and J. E. Hinman (1962). Simplex-lattice designs for multicomponent systems. *Technometrics*, **4**, 463–487.

Khuri, A. I., and J. A. Cornell (1998). Lack of fit revisited. *J. Combinatorics Inf. Syst. Sci.*, **23**, 193–208.

Marquardt, D. W., and R. D. Snee (1974). Test statistics for mixture models. *Technometrics*, **16**, 533–537.

SAS Institute (1982). *SAS/GRAPH User's Guide*. Version 6.03. SAS Institute, Inc., Cary, NC.

Scheffé, H. (1958). Experiments with mixtures. *J. R. Stat. Soc.* **B20** (2), 344–360.

Scheffé, H. (1963). Simplex-centroid design for experiments with mixtures. *J. R. Stat. Soc.* **B25** (2), 235–263.

Shelton, J. T., A. I. Khuri, and J. A. Cornell (1983). Selecting check points for testing lack of fit in response surface models. *Technometrics*, **25**, 357–365.

Smith, W. F. (2005). *Experimental Design for Formulation*. ASA-SIAM Series on Statistics and Applied Probability. ASA, Alexandria, VA.

Snee, R. D. (1971). Design and analysis of mixture experiments. *J. Qual. Technol.*, **3**, 159–169.

Snee, R. D. (1973). Techniques for the analysis of mixture data. *Technometrics*, **15**, 517–528.

Snee, R. D. (1979). Experimenting with mixtures. *CHEMTECH*, **9**, 702–710.

QUESTIONS

2.1. List the component proportions comprising the blends in a two-component $\{2,4\}$ simplex-lattice. List the 20 blends that comprise a $\{4,3\}$ simplex-lattice.

2.2. The following data represent average numbers of mites $(\times 10^{-2})$ counted on plants layed out in triplicate where two types (A and B) of chemical pesticides were sprayed on the plants.

Pesticide Blend		Average Number of Mites $(\times 10^{-2})$			
$A(x_1)$	$B(x_2)$	Y_1	Y_2	Y_3	$\bar{Y}$
100% (1)	0% (0)	3.8	3.0	3.7	3.5
50% $\frac{1}{2}$	50% $\frac{1}{2}$	1.8	2.2	2.0	2.0
0% (0)	100% (1)	4.0	4.7	4.8	4.5

(a) Plot the response average mite number versus each pesticide blend.

(b) Calculate the coefficient estimates for the fitted model

$$\hat{Y}(x) = b_1 x_1 + b_2 x_2 + b_{12} x_1 x_2$$

(c) The within blend estimate of σ^2 is $s^2 = 0.14$. Is there evidence of nonlinear-blending from using the two pesticides? *Hint:* Test H_0: $\beta_{12} = 0$ verses $H_A : \beta_{12} \neq 0$ at the 0.05 level of significance. $t_{6,.05} = 1.943$, $t_{6,.025} = 2.447$.

2.3. Show for $q = 3$ that when the special cubic model in Eq. (2.10) is fitted to the expected responses at the points of a design that is the $\{3, 2\}$ simplex-lattice augmented with a point at the centroid $x_1 = x_2 = x_3 = \frac{1}{3}$ of the triangle, then $\beta_{123} = 27\eta_{123} - 12(\eta_{12} + \eta_{13} + \eta_{23}) + 3(\eta_1 + \eta_2 + \eta_3)$, where η_{123} is the expected response at the centroid. How do the formulas for the parameters β_i and β_{ij} differ from Eq. (2.13), if at all, by the addition of the centroid response η_{123}?

2.4. Refer to the three-component yarn elongation example in Section 2.6. Predict the elongation value of yarn produced from the blend whose proportions are $x = (0.50, 0.20, 0.30)'$ with Eq. (2.24) and set up a 95% confidence interval for the true elongation value η at x.

2.5. The following contour plots were generated from six models fitted to data collected at the points of a simplex-centroid design in three components.

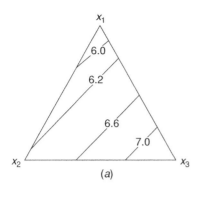

(a)

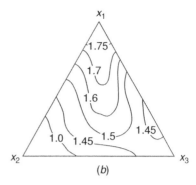

(b)

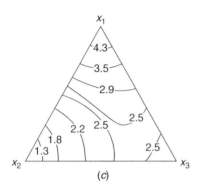

(c)

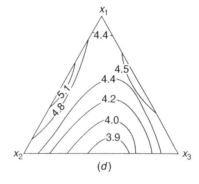

(d)

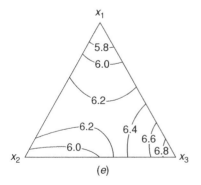

(e)

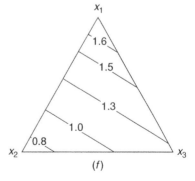

(f)

The models produced the six contour plots. Match each model with its corresponding contour plot and define the surface as planar, second-degree, or special cubic.

Model	Plot
1. $\hat{y}(\mathbf{x}) = 5.4x_1 + 5.9x_2 + 7.2x_3 + 2.7x_1x_2 + 0.1x_1x_3 - 1.9x_2x_3$	_____
2. $\hat{y}(\mathbf{x}) = 5.6x_1 + 6.2x_2 + 7.5x_3$	_____
3. $\hat{y}(\mathbf{x}) = 1.84x_1 + 0.67x_2 + 1.51x_3 + 0.14x_1x_2 - 1.01x_1x_3$ $+ 0.27x_2x_3 + 8.68x_1x_2x_3$	_____
4. $\hat{y}(\mathbf{x}) = 4.85x_1 + 1.28x_2 + 2.74x_3 - 3.80x_1x_2$ $- 3.66x_1x_3 + 0.58x_2x_3$	_____
5. $\hat{y}(\mathbf{x}) = 1.69x_1 + 0.68x_2 + 1.29x_3$	_____
6. $\hat{y}(\mathbf{x}) = 4.29x_1 + 4.57x_2 + 4.43x_3 + 2.84x_1x_2 + 0.84x_1x_3$ $- 2.56x_2x_3 - 11.19x_1x_2x_3$	_____

2.6. The following parameter estimates were calculated from multiple observations collected at the points of a $\{3, 2\}$ simplex-lattice.

$$b_1 = 5.5, \qquad b_2 = 7.0, \qquad b_3 = 8.0$$
$$\tfrac{1}{4}b_{12} = 8.75, \quad \tfrac{1}{4}b_{13} = 2.25, \quad \tfrac{1}{4}b_{23} = 4.5$$

(a) What were the values of the average responses at the lattice points that produced the parameter estimates above?

(b) Suppose that three observations were collected at each binary blend ($x_i = x_j = 0.50$, $x_k = 0$) and four observations were collected at each vertex. Which of the quadratic coefficients are significantly ($\alpha = 0.05$) greater than zero if an estimate of the variance of each observation is $s^2 = 3.0$?

2.7. It is possible to perform 12 experiments in order to study the three-component system where x_1, x_2, and x_3 represent the component proportions. It is suspected (but not guaranteed) that only a quadratic model is necessary:

$$y = \beta_1 x_1 + \beta_2 x_2 + \beta_3 x_3 + \beta_{12} x_1 x_2 + \beta_{13} x_1 x_3 + \beta_{23} x_2 x_3 + \varepsilon$$

Three designs are proposed where the circled points indicate replicate observations are to be collected:

(a) Six points each replicated twice.

(b) Seven points with the midedge points replicated twice and centroid replicated three times.

(c) Nine points with the centroid replicated three times.

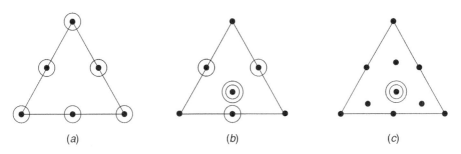

(a)   (b)   (c)

Which design would you choose and why? *Hint:* Suppose that you fit the quadratic model to data collected at the points of each design. Fill in the degrees of freedom column in the ANOVA table for each design below. What information does each design provide that the other designs do not?

Source	d.f. (a)	d.f. (b)	d.f. (c)
Regression (fitted model)	_____	_____	_____
Residual (LOF + pure error)	_____	_____	_____
Total	11	11	11
List	Design (a)	Design (b)	Design (c)
Good properties	_____	_____	_____
Bad properties	_____	_____	_____

2.8. In the special cubic model

$$\hat{y}(\mathbf{x}) = \sum_{i=1}^{3} b_i x_i + \sum_{i<j}^{3} b_{ij} x_i x_j + b_{123} x_1 x_2 x_3$$

the size of the quadratic coefficient b_{ij} must be approximately how many times the size of the planar coefficient b_i in order for the term $b_{ij} x_i x_j$ to contribute as much as the term $b_i x_i$, to the model at the lattice points $x_i = x_j = \frac{1}{2}$? The magnitude of the ratio of the cubic coefficient to the planar coefficient b_{123}/b_i must be approximately how large for the terms to contribute equally to the model at the centroid point? How large must the ratio b_{123}/b_{ij} be for the special cubic term $b_{123} x_1 x_2 x_3$ to contribute approximately the same to the model as the quadratic term $b_{ij} x_i x_j$ at the centroid?

2.9. The first six sets of numbers in the following data set were collected at points 1 to 6 of a $\{3, 2\}$ simplex-lattice. Points 7 to 10 were chosen after the fitted second-degree model was obtained and observations were collected at these points to check the fit of the second-degree model.

(a) Fit a second-degree model to the data observed at points 1 to 6.

Design Point	Components Proportions (x_1, x_2, x_3)	Observed Response (y_u)	Check Point	Components Proportions (x_1, x_2, x_3)	Observed Response (y_u)
1	$(1,0,0)$	4.80, 4.90	7	$\left(\frac{1}{3}, \frac{1}{3}, \frac{1}{3}\right)$	2.75
2	$(0,1,0)$	1.38, 1.18	8	$\left(\frac{2}{3}, \frac{1}{6}, \frac{1}{6}\right)$	2.80
3	$(0,0,1)$	2.58, 2.90	9	$\left(\frac{1}{6}, \frac{2}{3}, \frac{1}{6}\right)$	1.73
4	$\left(\frac{1}{2}, \frac{1}{2}, 0\right)$	1.92, 1.78, 1.76	10	$\left(\frac{1}{6}, \frac{1}{6}, \frac{2}{3}\right)$	2.20
5	$\left(0, \frac{1}{2}, \frac{1}{2}\right)$	2.25, 2.43, 2.46			
6	$\left(\frac{1}{2}, 0, \frac{1}{2}\right)$	2.86, 3.16, 3.25			

(b) Predict the response value at the centroid (point 7) using the fitted model of (a) and compare the observed response value at point 7 with the estimate. Are you satisfied with the form of the fitted model? If not, what course of action do you recommend next? *Hint:* $\widehat{\text{Var}}[\hat{y}(7)] = 0.0048$.

2.10. With a fitted model containing p terms, the C_p statistic is defined as

$$C_p = \frac{\text{SSE}_p}{s^2} - (N - 2p)$$

where SSE_p is the residual sum of squares, and s^2 is an estimate of the error variance. If the fitted model has negligible bias, then the expectation of C_p is $E(C_p$ given zero bias$) = p$. Using as an estimate of the error variance $s^2 = 0.05$, refer to the data at the design points 1 to 6 in Question 2.9 and fit a first-degree model and compute the value of C_3, and then compute C_6 with the model 2.9(**a**). From what does the reduction in the value of C_6 compared to C_3 result?

2.11. The following fitted model containing an estimate of the constant term (β_0) was obtained from data collected at the points of a $\{3, 2\}$ simplex-lattice design:

$$\hat{y}(\mathbf{x}) = 12.5 + 7.2x_2 + 2.0x_3 + 6.2x_1x_2 - 10.3x_1x_3 + 3.0x_2x_3$$
$$\quad (0.9) \quad (2.3) \quad (2.3) \quad (4.5) \quad\quad (4.5) \quad\quad (4.5)$$

The numbers in parentheses represent the estimated standard errors of the coefficient estimates, where $s^2 = 1.7$.

(a) Write the model in the equivalent Scheffé-type form.

(b) Calculate the estimated standard errors of the parameter estimates in the model in (a) considering two observations were collected at each design point. (*Hint:* See Eq. (2A.9).) Can you infer from these estimated standard errors which of the quadratic coefficients are nonzero and which of the linear blending coefficients are different from one another?

2.12. Shown are data values at the seven blends of a three-component simplex-centroid design. High values are considered more desirable than low values. Based on the data, indicate the type of blending (linear vs. nonlinear and if nonlinear, synergistic or antagonistic) that is present between: Compute $t = b_{ij}/\text{s.e.}(b_{ij})$ for each (a, b, c, d).

(a) A and B: _____ and is _____ . $t_{cal} =$ _____ .

(b) A and C: _____ and is _____ . $t_{cal} =$ _____ .

(c) B and C: _____ and is _____ . $t_{cal} =$ _____ .

(d) A and B and C: _____ and is _____ . $t_{cal} =$ _____ .

Test $H_0: \beta_{ij} = 0$ vs. $\beta_{ij} \neq 0$ at the 0.05 level. *Hint:* SST $= 63.79$, SSR $= 58.22$, and MSE $= 5.57/6 = 0.93$. From Appendix Table A, $t_{6,0.025} = 2.447$.

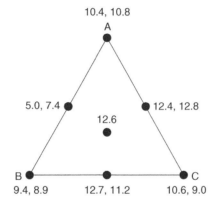

APPENDIX 2A: LEAST-SQUARES ESTIMATION FORMULA FOR THE POLYNOMIAL COEFFICIENTS AND THEIR VARIANCES: MATRIX NOTATION

In Section 7.1 a general review of least squares is presented. In this appendix the variance properties of the parameter estimates and of $\hat{y}(\mathbf{x})$ are reviewed for the case where the second-degree model is fitted to data at the points of the $\{q, 2\}$ simplex-lattice design.

The general form of the mixture model is $\mathbf{y} = \mathbf{X}\boldsymbol{\beta} + \boldsymbol{\varepsilon}$, where $\mathbf{y}$ is an $N \times 1$ vector of observations, $\mathbf{X}$ is an $N \times p$ matrix whose elements are the mixture component proportions and functions (e.g., pairwise products) of the component proportions, $\boldsymbol{\beta}$ is a $p \times 1$ vector of parameters, and $\boldsymbol{\varepsilon}$ is an $N \times 1$ vector of random errors. When the model $\mathbf{y} = \mathbf{X}\boldsymbol{\beta} + \boldsymbol{\varepsilon}$ is of the first degree, then $p = q$ and

$$
\mathbf{y} = \begin{bmatrix} y_1 \\ y_2 \\ \vdots \\ y_N \end{bmatrix}, \quad
\mathbf{X} = \begin{bmatrix} x_{11} & x_{12} & \cdots & x_{1q} \\ x_{21} & x_{22} & \cdots & x_{2q} \\ \vdots & \vdots & \vdots & \vdots \\ x_{N1} & x_{N2} & \cdots & x_{Nq} \end{bmatrix}
$$

$$
\boldsymbol{\beta} = \begin{bmatrix} \beta_1 \\ \beta_2 \\ \vdots \\ \beta_q \end{bmatrix}, \quad
\boldsymbol{\varepsilon} = \begin{bmatrix} \varepsilon_1 \\ \varepsilon_2 \\ \vdots \\ \varepsilon_N \end{bmatrix}
$$

In fitting the model $\mathbf{y} = \mathbf{X}\boldsymbol{\beta} + \boldsymbol{\varepsilon}$ over the N observations, suppose that $r_i \geq 1$ observations are collected at the vertex $x_i = 1$, $x_j = 0$, $j \neq i$, $i = 1, 2, \ldots, q$, and $\sum_{i=1}^{q} r_i = N$. The normal equations (7.4) used for estimating the elements of the parameter vector $\boldsymbol{\beta}$ are

$$
\overset{\mathbf{X'X}}{\begin{bmatrix} r_1 & & & \mathbf{0} \\ & r_2 & & \\ & & \ddots & \\ \mathbf{0} & & & r_q \end{bmatrix}}
\overset{\mathbf{b}}{\begin{bmatrix} b_1 \\ b_2 \\ \vdots \\ b_q \end{bmatrix}}
=
\overset{\mathbf{X'y}}{\begin{bmatrix} \sum x_{u1} y_u \\ \sum x_{u2} y_u \\ \vdots \\ \sum x_{uq} y_u \end{bmatrix}}
\tag{2A.1}
$$

where all of the summations $\sum_{u=1}^{N} x_{ui} y_u$ are over $u = 1, 2, \ldots, N$. The solutions to the normal equations provide the estimates

$$
b_i = \sum_{u=1}^{N} \frac{x_{ui} y_u}{r_i} = \bar{y}_i, \qquad i = 1, 2, \ldots, q
\tag{2A.2}
$$

where $\bar{y}_i$ is the average of the r_i observations collected at the vertex $x_i = 1$, $x_j = 0$, $j \neq i$.

Let us consider the case where a second-degree polynomial is fitted to the observations on a $\{q, 2\}$ simplex-lattice, where r observations are collected at each vertex $x_i = 1$, and r observations are collected at the midpoint $x_i = x_j = \frac{1}{2}$, $x_l = 0, l \neq i, j$ of the edge connecting the vertices corresponding to components i and j. The second-degree polynomial is

$$
y_u = \beta_1 x_{u1} + \beta_2 x_{u2} + \cdots + \beta_q x_{uq} + \beta_{12} x_{u1} x_{u2} + \cdots + \beta_{q-1,q} x_{uq-1} x_{uq} + \varepsilon_u,
$$
$$
u = 1, 2, \ldots, N
\tag{2A.3}
$$

If we set $q = 3$ and rearrange the $x_i x_j$ terms in the model, the normal equations are

$$
\begin{array}{ccc}
\mathbf{X'X} & \mathbf{b} & = & \mathbf{X'y} \\
\begin{bmatrix}
A & B & B & C & D & D \\
B & A & B & D & C & D \\
B & B & A & D & D & C \\
C & D & D & E & F & F \\
D & C & D & F & E & F \\
D & D & C & F & F & E
\end{bmatrix}
&
\begin{bmatrix}
b_1 \\
b_2 \\
b_3 \\
b_{23} \\
b_{13} \\
b_{12}
\end{bmatrix}
& = &
\begin{bmatrix}
\sum x_{u1} y_u \\
\sum x_{u2} y_u \\
\sum x_{u3} y_u \\
\sum x_{u2} x_{u3} y_u \\
\sum x_{u1} x_{u3} y_u \\
\sum x_{u1} x_{u2} y_u
\end{bmatrix}
\end{array}
\qquad (2A.4)
$$

where

$$
A = \sum x_{ui}^2 = \frac{3r}{2}, \qquad B = \sum x_{ui} x_{uj} = \frac{r}{4}, \qquad C = \sum x_{u1} x_{u2} x_{u3} = 0,
$$

$$
D = \sum x_{ui}^2 x_{uj} = \frac{r}{8}, \qquad E = \sum x_{ui}^2 x_{uj}^2 = \frac{r}{16}, \qquad F = \sum x_{ui}^2 x_{uj} x_{uk} = 0,
$$

$$
i, j, k = 1, 2, 3, \text{ and } i \neq j \neq k \qquad\qquad (2A.5)
$$

and all summations are over $u = 1, 2, \ldots, N$. Furthermore, if a $\{3, m\}$ lattice is considered where $m > 2$, then both C and F will be nonzero.

The matrix $\mathbf{X'X}$ in Eq. (2A.4) is of the composite form $\begin{bmatrix} \mathbf{u} & \mathbf{v} \\ \mathbf{v} & \mathbf{w} \end{bmatrix}$ where each partition matrix is of the form $a\mathbf{I} + b\mathbf{J}$, where $\mathbf{I}$ is an identity matrix of order six and $\mathbf{J}$ is a 6×6 matrix of ones. Hence the inverse matrix $(\mathbf{X'X})^{-1}$ will also be of the composite form $\mathbf{X'X}^{-1} = \begin{bmatrix} \mathbf{M} & \mathbf{O} \\ \mathbf{O} & \mathbf{P} \end{bmatrix}$, where

$$
\mathbf{M} = \mathbf{WQ}, \quad \mathbf{O} = -\mathbf{VQ}, \quad \mathbf{P} = \mathbf{UQ}, \quad \mathbf{Q} = (\mathbf{UW} - \mathbf{V}^2)^{-1} \qquad (2A.6)
$$

In particular, from Eq. (2A.5),

$$
\mathbf{U} =
\begin{bmatrix}
\dfrac{6r}{4} & \dfrac{r}{4} & \dfrac{r}{4} \\[2mm]
\dfrac{r}{4} & \dfrac{6r}{4} & \dfrac{r}{4} \\[2mm]
\dfrac{r}{4} & \dfrac{r}{4} & \dfrac{6r}{4}
\end{bmatrix}
= \frac{5r}{4}\mathbf{I} + \frac{r}{4}\mathbf{J}, \quad \mathbf{V} = -\frac{r}{8}\mathbf{I} + \frac{r}{8}\mathbf{J}, \quad \mathbf{W} = \frac{r}{16}\mathbf{I}^{\dagger}
$$

and therefore

$$
\mathbf{Q} = \frac{16}{r^2}\mathbf{I}, \quad \mathbf{M} = \frac{1}{r}\mathbf{I}, \quad \mathbf{O} = \frac{2}{r}\mathbf{I} - \frac{2}{r}\mathbf{J}, \quad \mathbf{P} = \frac{20}{r}\mathbf{I} + \frac{4}{r}\mathbf{J} \qquad (2A.7)
$$

---

†Note that if a matrix $\mathbf{T}$ can be expressed as a sum of the matrices $\mathbf{I}$ and $\mathbf{J}$ each of order q in the form $\mathbf{T} = a\mathbf{I} + b\mathbf{J}$, then $\mathbf{T}^{-1} = c\mathbf{I} + d\mathbf{J}$, where $c = 1/a$ and $d = -b/a(a + bq)$.

On the $\{3, 2\}$ simplex-lattice, the right-hand side $\mathbf{X'y}$ of the normal equations in (2A.4) take on the following values:

$$\sum x_{u1}y_u = r[\bar{y}_1 + \tfrac{1}{2}(\bar{y}_{12} + \bar{y}_{13})], \qquad \sum x_{u2}x_{u3}y_u = r(\bar{y}_{23})/4$$

$$\sum x_{u2}y_u = r[\bar{y}_2 + \tfrac{1}{2}(\bar{y}_{12} + \bar{y}_{23})], \qquad \sum x_{u1}x_{u3}y_u = r(\bar{y}_{13})/4$$

$$\sum x_{u3}y_u = r[\bar{y}_3 + \tfrac{1}{2}(\bar{y}_{13} + \bar{y}_{23})], \qquad \sum x_{u1}x_{u2}y_u = r(\bar{y}_{12})/4$$

and upon substituting the partitions $\mathbf{M}$, $\mathbf{Q}$, and $\mathbf{P}$ in $(\mathbf{X'X})^{-1}$, the estimates of the coefficients, or the solutions to the normal equations (2A.4), become

$$\mathbf{b} = \begin{bmatrix} b_1 \\ b_2 \\ b_3 \\ b_{23} \\ b_{13} \\ b_{12} \end{bmatrix} = (\mathbf{X'X})^{-1}\mathbf{X'y} = \begin{bmatrix} \bar{y}_1 + \tfrac{1}{2}(\bar{y}_{12} + \bar{y}_{13}) - \tfrac{1}{2}(\bar{y}_{12} + \bar{y}_{13}) \\ \bar{y}_2 \\ \bar{y}_3 \\ 4\bar{y}_{23} - 2(\bar{y}_2 + \bar{y}_3) \\ 4\bar{y}_{13} - 2(\bar{y}_1 + \bar{y}_3) \\ 4\bar{y}_{12} - 2(\bar{y}_1 + \bar{y}_2) \end{bmatrix} \qquad (2A.8)$$

Furthermore, if a measure of the error variance σ^2 is available, then the variance-covariance matrix of the coefficient estimates is

$$\text{var} - \text{cov}(\mathbf{b}) = (\mathbf{X'X})^{-1}\sigma^2$$

$$= \left[\begin{array}{c|c} \frac{1}{r}\mathbf{I} & \frac{2}{r}\mathbf{I} - \frac{2}{r}\mathbf{J} \\ \hline \frac{2}{r}\mathbf{I} - \frac{2}{r}\mathbf{J} & \frac{20}{r}\mathbf{I} + \frac{4}{r}\mathbf{J} \end{array} \right] \sigma^2 \qquad (2A.9)$$

The predicted value of the response at a point $\mathbf{x} = (x_1, x_2, \ldots, x_q)'$ in the experimental region is expressed in matrix notation as

$$\hat{y}(\mathbf{x}) = \mathbf{x}'_p\mathbf{b}$$

where $\mathbf{x}'_p$ is a $1 \times p$ vector whose elements correspond to the elements in a row of the matrix $\mathbf{X}$. A measure of the precision of the estimate $\hat{y}(\mathbf{x})$, at the point $\mathbf{x}$, is defined as the variance of $\hat{y}(\mathbf{x})$ and is expressed in matrix notation as

$$\text{var}[\hat{y}(\mathbf{x})] = \text{var}[\mathbf{x}'_p\mathbf{b}]$$

$$= \mathbf{x}'_p\text{var}[\mathbf{b}]\mathbf{x}_p$$

$$= \mathbf{x}'_p(\mathbf{X'X})^{-1}\mathbf{x}_p\sigma^2$$

Since the estimates of the parameters in $\boldsymbol{\beta}$ are expressible as linear functions of the average responses at the lattice points, as shown in Eq. (2A.8), $\hat{y}(\mathbf{x})$ can be written in terms of the averages. Thus the variance of $\hat{y}(\mathbf{x})$ can be written in terms of the variances of the averages as follows.

Quadratic model:

$$\hat{y}(\mathbf{x}) = \sum_{i=1}^{q} a_i \bar{y}_i + \sum_{i<j}^{q} a_{ij} \bar{y}_{ij}$$

where

$$a_i = x_i(2x_i - 1), \qquad a_{ij} = 4x_i x_j$$

and

$$\text{var}[\hat{y}(\mathbf{x})] = \sigma^2 \left[\sum_{i=1}^{q} \frac{a_i^2}{r_i} + \sum_{i<j}^{q} \frac{a_{ij}^2}{r_{ij}} \right]$$

Special cubic model:

$$\hat{y}(\mathbf{x}) = \sum_{i=1}^{q} b_i \bar{y}_i + \sum_{i<j}^{q} b_{ij} \bar{y}_{ij} + \sum_{i<j<k}^{q} b_{ijk} \bar{y}_{ijk}$$

where

$$b_i = \frac{x_i}{2}(6x_i^2 - 2x_i + 1) - 3 \sum_{j \neq 1}^{q} x_j^2$$

$$b_{ij} = 4x_i x_j(3x_i + 3x_j - 2), \qquad b_{ijk} = 27 x_i x_j x_k$$

and

$$\text{var}[\hat{y}(\mathbf{x})] = \sigma^2 \left[\sum_{i=1}^{q} \frac{b_i^2}{r_i} + \sum_{i<j}^{q} \frac{b_{ij}^2}{r_{ij}} + \sum_{i<j<k}^{q} \frac{b_{ijk}^2}{r_{ijk}} \right]$$

Similar expressions corresponding to the full cubic model and quartic model can be found in Gorman and Hinman (1962).

APPENDIX 2B: CUBIC AND QUARTIC POLYNOMIALS AND FORMULAS FOR THE ESTIMATES OF THE COEFFICIENTS

In q components, the *cubic* model would be fitted to the points of a $\{q, 3\}$ simplex-lattice. The model and the expressions for the coefficient estimates in terms of the observed responses at the lattice points are

$$\eta = \sum_{i=1}^{q} \beta_i x_i + \sum_{i<j}^{q} \beta_{ij} x_i x_j + \sum_{i<j}^{q} \gamma_{ij} x_i x_j (x_i - x_j) + \sum_{i<j<k}^{q} \beta_{ijk} x_i x_j x_k$$

$$(2B.1)$$

Denoting the observed responses at the vertices by y_i, at the point $x_i = \frac{2}{3}, x_j = \frac{1}{3}, x_k = 0, k \neq i, j$ by y_{iij}, and at the centroids $x_i = x_j = x_k = \frac{1}{3}, x_l = 0$ of the faces by y_{ijk}, respectively, we have for the coefficient estimates

$$b_i = y_i, \qquad\qquad\qquad i = 1, 2, \ldots, q$$

$$b_{ij} = \tfrac{9}{4}(y_{iij} + y_{ijj} - y_i - y_j), \qquad i < j$$

$$g_{ij} = \tfrac{9}{4}(3y_{iij} - 3y_{ijj} - y_i + y_j), \qquad i < j \qquad\qquad (2B.2)$$

$$b_{ijk} = 27y_{ijk} - \tfrac{27}{4}(y_{iij} + y_{ijj} + y_{iik} + y_{ikk} + y_{jjk} + y_{jkk}) + \tfrac{9}{2}(y_i + y_j + y_k)$$

If replicate observations are collected at some or all of the lattice points, we replace y_i, y_{iij}, and so on, in Eq. (2B.2) with the averages $\bar{y}_i$, $\bar{y}_{iij}$, and so on, of the replicates.

To the points of a $\{q, 4\}$ simplex-lattice, the *quartic* model to be fitted is

$$\eta = \sum_{i=1}^{q} \beta_i x_i + \sum_{i<j}^{q} \beta_{ij} x_i x_j + \sum_{i<j}^{q} \gamma_{ij} x_i x_j (x_i - x_j) + \sum_{i<j}^{q} \delta_{ij} x_i x_j (x_i - x_j)^2$$

$$+ \sum_{i<j<k}^{q} \beta_{iijk} x_i^2 x_j x_k + \sum_{i<j<k}^{q} \beta_{ijjk} x_i x_j^2 x_k + \sum_{i<j<k}^{q} \beta_{ijkk} x_i x_j x_k^2$$

$$+ \sum_{i<j<k<l}^{q} \beta_{ijkl} x_i x_j x_k x_l \qquad\qquad (2B.3)$$

The formulas for the estimates of the coefficients, in terms of y_i, y_{ij}, and so on, are

$$b_i = y_i$$

$$b_{ij} = 4y_{ij} - 2(y_i + y_j)$$

$$g_{ij} = \tfrac{8}{3}(2y_{iiij} - 2y_{ijjj} - y_i + y_j)$$

$$d_{ij} = \tfrac{8}{3}(4y_{iiij} - 6y_{ij} + 4y_{ijjj} - y_i - y_j)$$

$$b_{iijk} = 32(3y_{iijk} - y_{ijjk} - y_{ijkk}) + \tfrac{8}{3}(6y_i - y_j - y_k) - 16(y_{ij} + y_{ik})$$

$$- \tfrac{16}{3}(5y_{iiij} + 5y_{iiik} - 3y_{ijjj} - 3y_{ikkk} - y_{jjjk} - y_{jkkk})$$

$$b_{ijjk} = 32(3y_{ijjk} - y_{iijk} - y_{ijkk}) + \tfrac{8}{3}(6y_j - y_i - y_k) - 16(y_{ij} + y_{jk}) \qquad \text{(2B.4)}$$

$$- \tfrac{16}{3}(5y_{ijjj} + 5y_{jjjk} - 3y_{iiij} - 3y_{jkkk} - y_{iiik} - y_{ikkk})$$

$$b_{ijkk} = 32(3y_{ijkk} - y_{iijk} - y_{ijjk}) + \tfrac{8}{3}(6y_k - y_i - y_j) - 16(y_{ik} + y_{jk})$$

$$- \tfrac{16}{3}(5y_{ikkk} + 5y_{jkkk} - 3y_{iiik} - 3y_{jjjk} - y_{iiij} - y_{ijjj})$$

$$b_{ijkl} = 256y_{ijkl} - 32(y_{iijk} + y_{iijl} + y_{iikl} + y_{ijjk} + y_{ijjl} + y_{jjkl} + y_{ijkk} + y_{ikkl}$$

$$+ y_{jkkl} + y_{ijll} + y_{jkll} + y_{ikll}) + \tfrac{32}{3}(y_{iiij} + y_{iiik} + y_{iiil} + y_{ijjj} + y_{jjjk}$$

$$+ y_{jjjl} + y_{ikkk} + y_{jkkk} + y_{kkkl} + y_{illl} + y_{jlll} + y_{klll})$$

Notice that the formula (2B.3) for the quartic model does not contain any of the third-degree terms $\beta_{ijk}x_ix_jx_k$ that appear in the cubic model (2B.1). Also none of the face centroids $x_i = x_j = x_k = \tfrac{1}{3}, x_l = 0, l \neq i, j, k$ are included as points of the $\{q, 4\}$ simplex-lattice. Therefore the parameters β_{ijk} are not estimable if included in the model (2B.3) unless additional points are added to the design.

APPENDIX 2C: THE PARTITIONING OF THE SOURCES IN THE ANALYSIS OF VARIANCE TABLE WHEN FITTING THE SCHEFFÉ MIXTURE MODELS

In Section 2.7 the partitioning of the total variation in the 15 yarn elongation values of Table 2.3 was described as separating the variation among the six blends in Eq. (2.26) from the variation among the replicate observations within the blends. The sums of squares formulas and corresponding degrees of freedom for regression, the residual, and the total are listed in Table 2.5.

From the standpoint of testing hypotheses about the shape of the response surface based on fitting a candidate model, typically the form of the candidate model is

$$E(y) = \beta_0 + \sum_{j=1}^{p} \beta_j z_j \qquad \text{(2C.1)}$$

where each $z_j, j = 1, 2, \ldots, p$, may be a linear, quadratic, or other function of the component proportions $x_i, i = 1, 2, \ldots, q$. The usual null hypothesis that says the response does not depend on the z_j is

$$H_0 : \beta_j = 0 \qquad \text{all } j = 1, 2, \ldots, p \qquad \text{(2C.2)}$$

When Eq. (2C.2) is true, the model (2C.1) is

$$E(y) = \beta_0 \tag{2C.3}$$

The least-squares estimate of the parameter β_0 is $\hat{\beta}_0 = \bar{y}$, where $\bar{y}$ is the average of the N observations. Note that the null hypothesis model (2C.3) is a strict subset of the candidate model (2C.1).

Since the null hypothesis model (2C.3) is estimated by a single linear combination of the N observations, the "total" sum of squares, denoted by SST $= \sum_{u=1}^{N} (y_u - \bar{y})^2$, is the residual sum of squares remaining after the null hypothesis model has been fitted. Thus the total sum of squares has $N - 1$ degrees of freedom (d.f.). Next, the regression sum of squares, SSR $= \sum_{u=1}^{N} (\hat{y}_u - \bar{y})^2$, is the contribution explained by fitting the candidate model (2C.1) after the null hypothesis model has been fitted. The regression d.f. is the number of additional *independent* parameters estimated for the candidate model after the null hypothesis model has been fitted. If the z_j, $j = 1, 2, \ldots, p$, in model (2C.1) are independent, then the regression sum of squares has p d.f.

Consider Scheffé's canonical form of polynomial in q components x_i, $i = 1, 2, \ldots, q$. The first-degree linear blending model is

$$E(y) = \sum_{i=1}^{q} \beta_i x_i \tag{2C.4}$$

owing to the constraint $\sum_{i=1}^{q} x_i = 1$, which forces the β_0 term to be confounded with the β_i. Letting the $z_j = x_i$, $i = j = 1, 2, \ldots, q = p$, the appropriate null hypothesis that states the response does not depend on the mixture components (i.e., change from blend to blend) is

$$H_0 : \beta_1 = \beta_2 = \cdots = \beta_q = \beta_0 \tag{2C.5}$$

The resulting null hypothesis model is once again (2C.3), since

$$E(y) = \sum_{i=1}^{q} \beta_0 x_i = \beta_0 \sum_{i=1}^{q} x_i = \beta_0$$

and the least-squares estimate of β_0 is $\bar{y}$.

The candidate model (2C.4) contains $q - 1$ independent parameters and thus SSR associated with fitting (2C.4) has $q - 1$ degrees of freedom. Similarly, if the candidate model is the Scheffé quadratic model

$$E(y) = \sum_{i=1}^{q} \beta_i x_i + \sum \sum_{i<j} \beta_{ij} x_i x_j \tag{2C.6}$$

then the appropriate null hypothesis that states the response does not depend on the mixture components is

$$H_0 : \beta_1 = \beta_2 = \cdots = \beta_q = \beta_0 \quad \text{and} \quad \beta_{ij} = 0, \quad i < j = 2, 3, \ldots, q \qquad (2C.7)$$

Once again, the SSR associated with fitting (2C.6) has $q - 1$ or $(q + 2)(q - 1)/2$ degrees of freedom. Hence, whether one is testing (2C.2) or (2C.7) through the fitting of (2C.1) or (2C.6), the appropriate analysis of variance table is Table 2.5.

CHAPTER 3

Multiple Constraints on the Component Proportions

In the last chapter we focused on the construction of lattice designs and the fitting of model equations in the interest of exploring all, or almost all, of the simplex region. There are occasions, however, when one is not completely at freedom to explore the entire simplex because of certain restrictions that are placed on the component proportions over and above $0 \le x_i \le 1$, $i = 1, 2, \ldots q$ and $x_1 + x_2 + \cdots + x_q = 1$. For example, in the yarn manufacturing experiment of Section 2.6, one might wish to learn the properties of the yarn spun only from mixtures in which the fractional proportions of polystyrene and polypropylene are greater than or equal to $x_2 \ge 0.20$ and $x_3 \ge 0.35$, respectively. These lower-bound restrictions placed on x_2 and x_3 would limit the desired mixtures to a subregion of the simplex.

Upper bounds on some of the component proportions can also limit the experimentation to some subregion of the simplex. But the case that occurs frequently is where both lower and upper bounds are placed on some or all of the component proportions. In any of these situations where some subset of the simplex is the region one is confined to, if one is able to isolate the design and model the subregion, then a decrease in experimentation cost and time, as well as an increase in precision of the model estimates, should result. We now discuss the methodology necessary to design and model blending systems where lower bounds only are placed on the component proportions.

3.1 LOWER-BOUND RESTRICTIONS ON SOME OR ALL OF THE COMPONENT PROPORTIONS

We begin our study of how to design an experiment when the component proportions are restricted by lower bounds by assuming, for the sake of simplicity, that there are

A Primer on Experiments with Mixtures, By John A. Cornell
Copyright © 2011 John Wiley & Sons, Inc. Published by John Wiley & Sons, Inc.

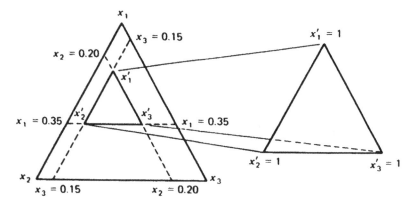

Figure 3.1 A subregion of the original simplex redefined as a simplex in the L-pseudocomponents x'_i, $i = 1, 2, 3$.

only three components in the system. We let the current product be manufactured within the following blending limits:

$$x_1 = 0.45 \pm 0.01, \quad x_2 = 0.30 \pm 0.02, \quad x_3 = 0.25 \pm 0.02$$

However, we would like to study the shape of the surface over the region of the simplex that is defined by the placing of the lower bounds:

$$x_1 \geq 0.35, \quad x_2 \geq 0.20, \quad x_3 \geq 0.15 \tag{3.1}$$

on the respective component proportions. In such a region at least a proportion $x_1 = 0.35$ of component 1 is required to be present in each blend, combined with at least a proportion $x_2 = 0.20$ of component 2 combined with at least $x_3 = 0.15$ of component 3. Of course, not all three components can assume these lower bounds simultaneously, since in this case the sum $x_1 + x_2 + x_3 = 0.70$ is less than unity and thus would not form a valid mixture.

Corresponding to the restrictions on the x_i listed in Eq. (3.1), the factor space of feasible mixtures, involving components 1, 2, and 3, is represented by the interior triangle shown in Figure 3.1 The placing of lower bounds only on the component proportions does not distort the shape of the subregion; it retains the shape of a regular simplex or equilateral triangle in three components. If the sizes of the lower bounds are equal, then the centroid of the subregion is the centroid $(1/q, 1/q, \ldots, 1/q)$ of the simplex in the original components. If upper bounds are placed on the component proportions, however, or if both upper and lower bounds are placed on the component proportions, the resulting region usually will not assume the shape of a simplex. We will see how these other constraints affect the shape of the region later in Sections 3.6 through 3.11.

3.2 INTRODUCING *L*-PSEUDOCOMPONENTS

Since the placing of lower bounds on the x_i limits one's attention to a simplex-shaped subregion of the original x-simplex, it seems only natural to redefine the coordinates of the subregion in terms of "pseudo" components. (This is similar to introducing coded variables that are scaled to take values $-1, 0$, and $+1$, for example, instead of using the original variables such as temperature, time, etc., in a standard independent factor [variable] experiment). The "pseudo" components are defined as combinations of the original components and the primary reason for introducing the pseudocomponents is that usually both the construction of designs and the fitting of models are much easier when done in the pseudocomponent system than when done in the original components system. One word of caution, however, is that one must remember that the pseudocomponents are "pseudo" and if one wishes to make inferences concerning the components that actually comprise the blending system one must either fit a model in the original components or make the inverse transformation from the pseudocomponent back to the original components in order to produce a fitted model in the original components.

To show how the L-pseudocomponents are defined in terms of the original components and their lower bounds, let us talk in general terms by saying the system consists of q components and let $L_i \geq 0$ denote the lower bound for component i, $i = 1, 2, \ldots, q$. The lower-bound constraints in Eq. (3.1) are expressed in the more general form

$$0 \leq L_i \leq x_i \qquad \text{for} \quad i = 1, 2, \ldots, q \tag{3.2}$$

where some of the L_i might be equal to zero. If we subtract the lower bound L_i from x_i and divide the difference by $1 - $ (sum of the L_i), then the *L-pseudocomponent* x_i' is defined, using the linear transformation, as

$$x_i' = \frac{x_i - L_i}{1 - L} \tag{3.3}$$

where $L = \sum_{i=1}^{q} L_i < 1$. This transformation to L-pseudocomponents was introduced by Kurotori (1966). To illustrate, let us refer to the earlier restrictions in Eq. (3.1), where $L = 0.35 + 0.20 + 0.15 = 0.70$, so that the L-pseudocomponents are

$$x_1' = \frac{x_1 - 0.35}{0.30}, \quad x_2' = \frac{x_2 - 0.20}{0.30}, \quad x_3' = \frac{x_3 - 0.15}{0.30} \tag{3.4}$$

The factor space shown in Figure 3.1 is a regular two-dimensional simplex in the L-pseudocomponents x_i' since $x_1' + x_2' + x_3' = 1$. The range of each x_i' in terms of a range of unity for each x_i as specified by the original simplex is $1-L$; that is, the height of the L-pseudocomponent triangle in Figure 3.1 is $1-0.70 = 0.30$.

The orientation of the L-pseudocomponents simplex is the same as the orientation of the original components simplex. Closest to the vertex $x_i = 1$ of the original

Table 3.1 Original Component Settings and L-Pseudocomponent Settings

L-Pseudocomponent Setting			Original Component Setting			
x_1'	x_2'	x_3'	x_1	x_2	x_3	Data Value
1	0	0	0.65	0.20	0.15	28.6
0.5	0.5	0	0.50	0.35	0.15	42.4
0	1	0	0.35	0.50	0.15	20.0
0	0.5	0.5	0.35	0.35	0.30	12.5
0	0	1	0.35	0.20	0.45	15.3
0.5	0	0.5	0.50	0.20	0.30	32.7

simplex is the vertex $x_i' = 1$ of the L-pseudocomponents simplex. The coordinates x_i' $= 1, x_j' = 0, j \neq i$, of the L-pseudocomponents vertices correspond to the coordinates $x_i = L_i + 1 - L, x_j = L_j, j \neq i$, in the original components; that is,

$$
\begin{aligned}
(x_1', x_2', x_3') = (1, 0, 0) = (x_1, x_2, x_3) &= (1 - L_2 - L_3, L_2, L_3) \\
= (0, 1, 0) = &= (L_1, 1 - L_1 - L_3, L_3) \\
= (0, 0, 1) = &= (L_1, L_2, 1 - L_1 - L_2)
\end{aligned}
$$

The ease in constructing designs in the L-pseudocomponent system can be illustrated as follows. For simplicity, let us choose a second-degree polynomial to model the surface over the region in the x_i', and let us choose a $\{3, 2\}$ lattice in the x_i' at which to observe the values of the response. The design settings $x_i' = 0, \frac{1}{2}, 1$, are shown on the left side of Table 3.1. The settings in the original x_i components, corresponding to the lattice settings in the x_i', are obtained by reversing the equations in Eq. 3.4 and solving. In other words, the settings in the original components are obtained using

$$ x_i = L_i + (1 - L)x_i' \tag{3.5} $$

so that for $i = 1, L_1 = 0.35$ and $L = 0.70$, we have for the value of x_1 corresponding to $x_1' = 1.0, x_1 = 0.35 + (0.30)1.0 = 0.65$. Similarly, corresponding to $x_1' = 0.5$, we have $x_1 = 0.35 + (0.30)(0.5) = 0.5$. The remaining settings in the original components are listed in Table 3.1. Note that the range of values for each x_i is $1 - L$; that is, x_i goes from L_i to $L_i + (1 - L)$ as x_i' goes from zero to one.

Once the mixture blends in the original system are defined from the L-pseudocomponent settings, the next step is to collect observed values of the response at the design settings so that a model either in terms of the L-pseudocomponents or in terms of the original components can be obtained. A second-degree polynomial in the L-pseudocomponents is

$$ \eta = \gamma_1 x_1' + \gamma_2 x_2' + \gamma_3 x_3' + \gamma_{12} x_1' x_2' + \gamma_{13} x_1' x_3' + \gamma_{23} x_2' x_3' \tag{3.6} $$

while the corresponding model in the original components would be of the form

$$\eta = \gamma_1 \frac{(x_1 - L_1)}{1 - L} + \gamma_2 \frac{(x_2 - L_2)}{1 - L} + \gamma_3 \frac{(x_3 - L_3)}{1 - L} + \gamma_{12} \frac{(x_1 - L_1)(x_2 - L_2)}{(1 - L)^2}$$
$$+ \gamma_{13} \frac{(x_1 - L_1)(x_3 - L_3)}{(1 - L)^2} + \gamma_{23} \frac{(x_2 - L_2)(x_3 - L_3)}{(1 - L)^2}$$

or

$$\eta = \beta_1 x_1 + \beta_2 x_2 + \beta_3 x_3 + \beta_{12} x_1 x_2 + \beta_{13} x_1 x_3 + \beta_{23} x_2 x_3 \qquad (3.7)$$

where the β's can be expressed in terms of the γ's as

$$\beta_1 = \frac{\gamma_{12} L_2 (L_1 - 1) + \gamma_{13} L_3 (L_1 - 1) + \gamma_{23} L_2 L_3}{(1 - L)^2} + \frac{\gamma_1 - \sum_{i=1}^{3} \gamma_i L_i}{1 - L}$$

$$\beta_2 = \frac{\gamma_{12} L_1 (L_2 - 1) + \gamma_{13} L_1 L_3 + \gamma_{23} L_3 (L_2 - 1)}{(1 - L)^2} + \frac{\gamma_2 - \sum_{i=1}^{3} \gamma_i L_i}{1 - L} \qquad (3.8)$$

$$\beta_3 = \frac{\gamma_{12} L_1 L_2 + \gamma_{13} L_1 (L_3 - 1) + \gamma_{23} L_2 (L_3 - 1)}{(1 - L)^2} + \frac{\gamma_3 - \sum_{i=1}^{3} \gamma_i L_i}{1 - L}$$

$$\beta_{ij} = \gamma_{ij}/(1 - L)^2, \quad i, j = 1, 2, 3, \text{ and } i < j$$

For the L-pseudocomponent model in Eq. (3.6), the interpretation of the parameters γ_i and γ_{ij}, $i, j = 1, 2,$ and 3, $i < j$, in the L-pseudocomponent system is the same as was previously described in the earlier chapters when discussing the original component models. The γ_i, $i = 1, 2,$ and 3, represent the heights of the surface above the triangular subregion at the vertices $x_i' = 1$, $i = 1, 2,$ and 3, respectively, in the L-pseudocomponent system and the γ_{ij}, $i < j$, represent deviations of the surface from the planar surface owing to the nonlinear blending of the $x_i' x_j'$ binaries. The β_i and β_{ij}, $i = 1, 2,$ and 3, $i < j$, in Eq. (3.7), on the other hand, are functions of the γ_i and γ_{ij}, where β_i represents the height of the surface extrapolated to the vertex $x_i = 1$, $x_j = 0$, $j \neq i$, of the simplex region in the original components and β_{ij} represents the deviation of the extrapolated surface from the extrapolated plane $\eta = \beta_1 x_1 + \beta_2 x_2 + \beta_3 x_3$ taken along the $x_i - x_j$ edge. This is illustrated now using the data values from Table 3.1.

3.3   A NUMERICAL EXAMPLE OF FITTING AN
L-PSEUDOCOMPONENT MODEL

With the L-pseudocomponent model in Eq. (3.6), the estimates g_i and g_{ij} of the model parameters γ_i and γ_{ij}, respectively, are found using the same formulas as in Eqs. (2.14) in Section 2.4, that is, $g_i = y_i$, $g_{ij} = 4 y_{ij} - 2(y_i + y_j)$, $i < j$. From the data values of Table 3.1, the estimates are

$$g_1 = 28.6, \quad g_2 = 20.0, \quad g_3 = 15.3$$
$$g_{12} = 4(42.4) - 2(28.6 + 20.0) = 72.4, \quad g_{13} = 43.0, \quad g_{23} = -20.6$$

Substituting these estimates into the model form of Eq. (3.6) obtains the prediction equation for the response in the L-pseudocomponent system as

$$\hat{y}(\mathbf{x}') = 28.6x_1' + 20.0x_2' + 15.3x_3' + 72.4x_1'x_2' + 43.0x_1'x_3' - 20.6x_2'x_3' \qquad (3.9)$$

The corresponding set of parameter estimates for the model in the original components as expressed by Eq. (3.7) is

$$b_1 = \frac{72.4(0.20)(0.35 - 1.0) + 43.0(0.15)(0.35 - 1.0) + (-20.6)(0.20)(0.15)}{(1 - 0.70)^2}$$
$$+ \frac{28.6 - 28.6(0.35) - 20.0(0.20) - 15.3(0.15)}{1 - 0.70}$$
$$= -117.04$$

$$b_2 = -160.38, \quad b_3 = -50.27, \quad b_{12} = \frac{72.4}{(1 - 0.70)^2} = 804.44,$$
$$b_{13} = 477.78, \quad b_{23} = -228.89$$

and the prediction equation for the response in the original components, rounding off the estimates to tenths, is

$$\hat{y}(\mathbf{x}) = -177.0x_1 - 160.4x_2 - 50.3x_3 + 804.4x_1x_2 + 477.8x_1x_3 - 228.9x_2x_3 \qquad (3.10)$$

It is important to keep in mind, especially when attempting to interpret the coefficient estimates, that model (3.10) is a description of the surface only over the region where $x_1 \geq 0.35$, $x_2 \geq 0.20$, $x_3 \geq 0.15$.

A contour plot of the estimated surface can be drawn using either the L-pseudocomponent model Eq. (3.9) or the model in the original components, Eq. (3.10). In Figure 3.2 is presented the contour plot of the estimated surface using the L-pseudocomponent model (3.9).

In summary, when one or more of the component proportions is restricted by lower bounds and the sum of the lower bounds is less than unity, $\sum_{i=1}^{q} L_i < 1$, the resulting subspace or subregion is a simple one-to-one transformation of the original simplex space. The lattice designs may be applied directly to the subspace and the surface can be modeled by a model in the L-pseudocomponents or by a model in the original restricted components. Also it is important to keep in mind that software, such as DESIGN-EXPERT and Minitab, exists today to relieve us of having to estimate the parameters using Eqs. (3.8).

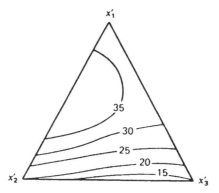

Figure 3.2 Contours of the predicted surface in the L-pseudocomponent system.

3.4 UPPER-BOUND RESTRICTIONS ON SOME OR ALL COMPONENT PROPORTIONS

When one or more of the component proportions is restricted by upper bounds $x_i \leq U_i$, the simplest modification to the simplex-lattice design consists of replacing the restricted components with mixtures consisting of combinations of the restricted components and predetermined proportions of the unrestricted components. These mixtures are then used to obtain observations from which estimates of the parameters in the standard mixture polynomials can be calculated.

Let us assume, for simplicity, that only one component is restricted, say, $x_1 \leq U_1$, where the system consists of the four components whose proportions are denoted by x_1, x_2, x_3, and x_4. We assume also that the following second-degree polynomial is to be fitted over the restricted region:

$$y = \sum_{i=1}^{4} \beta_i x_i + \sum_{i<j}^{3} \sum^{4} \beta_{ij} x_i x_j + \varepsilon \tag{3.11}$$

The feasible experimental region consists of the lower frustrum $EFGDCB$ of the tetrahedron $ABCD$ shown in Figure 3.3.

Since the design points in the simplex-lattice arrangement where $x_1 > U_1$ cannot be used (i.e., the design space is now the frustrum of the simplex satisfying $x_1 \leq U_1$), we may consider replacing the usual lattice points, where $x_1 > U_1$ with component combinations consisting of the proportion U_1 of x_1 and the proportion x_j ($2 \leq j \leq 4$) of the remaining three components such that

$$\sum_{j=2}^{4} x_j = 1 - U_1 \tag{3.12}$$

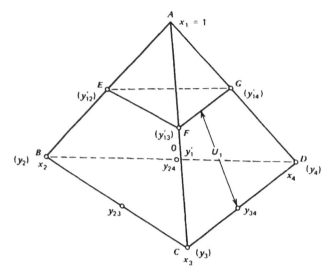

Figure 3.3 Lower frustrum of the tetrahedron defined as $x_1 \le U_1$ with values of response observed at the design points.

For each blend, if two or more components other than x_1 are present, then they are equally represented, so that, with the standard notation of binary, ternary, and quartenary blends, some of the combinations that might be used are

$$
\begin{aligned}
x_1 x_i : \quad & x_1 = U_1, \quad x_i = 1 - U_1 \\
x_1 x_i x_j : \quad & x_1 = U_1, \quad x_i = x_j = \frac{1 - U_1}{2} \\
x_1 x_i x_j x_k : \quad & x_1 = \frac{U_1}{2}, \quad x_i = x_j = x_k = \frac{1 - U_1/2}{3}
\end{aligned}
\tag{3.13}
$$

where the subscripts i, j, and $k = 2, 3$, and $4, i \ne j \ne k$. The blends $x_1 x_i$ are represented by the points E, F, and G in Figure 3.3, whereas the blends $x_1 x_i x_j$ are located at the midpoints of the edges joining the points E, F, and G. The point $x_1 x_2 x_3 x_4$ that is defined in Eq. (3.13) is located on the x_1 axis midway between the triangle EFG and the bottom triangle BCD and is denoted by 0.

In addition to the design points above, we may also consider taking the standard $\{q - 1, 2\}$ lattice points where $x_1 = 0$ and the other components are set at the six combinations $x_i = 1, x_j = x_k = 0$ and $x_i = 0, x_j = x_k = \frac{1}{2}, i, j, k = 2, 3, 4, i \ne j \ne k$. These are the three vertices associated with pure components $2, 3$, and 4 and the midpoints of the three edges (i.e., $x_i = x_j = \frac{1}{2}, i < j, i, j, k \ne 1$) connecting the vertices associated with the components $2, 3$, and 4. If these latter six points plus the four points $x_1 = U_1, x_j = 1 - U_1, j = 2, 3$, and 4 and $x_1 = U_1/2, x_2 = x_3 = x_4 = (1 - U_1/2)/3$ from Eq. (3.13) make up the design, the estimates of the parameters β_i and $\beta_{ij} (j > i > 1)$ in Eq. (3.11) can be obtained as follows: let the observed response at $x_i = 1, x_1 = x_j = x_k = 0 (i \ne j \ne k)$ be denoted by y_i, and let the observed response at $x_i = x_j = \frac{1}{2}, x_1 = x_k = 0$ be denoted by y_{ij}. Upon substituting these values of x_1,

x_i, x_j, x_k, y_i and y_{ij} into Eq. (3.11), we obtain the estimates

$$b_i = y_i, \quad i = 2, 3, 4$$
$$b_{ij} = 4y_{ij} - 2y_i - 2y_j, \quad j > i > 1 \tag{3.14}$$

These formulas are identical to the formulas presented in Section 2.4, as we know they must be because we are fitting the parameters associated with the components 2, 3, and 4 to the {3, 2} simplex-lattice that defines the base of the frustrum in Figure 3.3.

Still to be determined are the estimates of β_1 and β_{1j} ($j > 1$) in Eq. (3.11). If the observed response taken at the combination $x_1 = U_1$, $x_j = 1-U_1$ ($j = 2, 3, 4$) is denoted by y'_{1j} (see Figure 3.3), and the observed response taken at the combination $x_1 = U_1/2$, $x_2 = x_3 = x_4 = (1-U_1/2)/3$ is denoted by y'_1, then the estimates b_1 and b_{1j} can be found by solving the following equations:

$$U_1 b_1 + U_1(1 - U_1)b_{1j} = y'_{1j} - (1 - U_1)b_j, \quad j = 2, 3, 4 \tag{3.15}$$

$$\frac{U_1}{2} b_1 + \frac{U_1}{2} \frac{(1 - U_1/2)}{3} \sum_{j=2}^{4} b_{1j} = y'_1 - \frac{(1 - U_1/2)}{3} \sum_{j=2}^{4} b_j$$

$$- \frac{(1 - U_1/2)^2}{9} \sum_{i}^{3} \sum_{<j}^{4} b_{ij} \tag{3.16}$$

We now illustrate the use of the estimating formulas in Eqs. (3.14) through (3.16) for modeling the shape characteristics associated with the flavor surface of a tropical beverage.

3.5 AN EXAMPLE OF THE PLACING OF AN UPPER BOUND ON A SINGLE COMPONENT: THE FORMULATION OF A TROPICAL BEVERAGE

A tropical beverage was formulated by combining juices of watermelon (x_1), orange (x_2), pineapple (x_3), and grapefruit (x_4). It was decided to restrict the percentage of watermelon in all blends to at most 80%. However, the feasibility of a punch consisting of 80% watermelon was of interest because watermelon is so much less expensive than the other fruits. Thus several combinations of $x_1 = 0.80$ with $x_2 + x_3 + x_4 = 1.0 - 0.80 = 0.20$ were studied. The response of interest for this example is the average flavor score (based on a scale of 1 to 9) where the flavor scores in Table 3.2 represent average values taken over 40 samples of each blend.

In Table 3.2 the first 4 blends, designated as points 1 to 4, respectively, are, from Eq. (3.13),

$$x_1 x_i, \quad \text{where } x_1 = 0.80, \ x_i = 0.20, \ i = 2, 3, \ \text{and } 4$$
$$x_1 x_2 x_3 x_4, \quad \text{where } x_1 = 0.40, \ x_2 = x_3 = x_4 = 0.60/3 = 0.20$$

Table 3.2 Tropical Beverage Data

Design Point	Watermelon (x_1)	Orange (x_2)	Pineapple (x_3)	Grapefruit (x_4)	Average Flavor Score (y)	Response Designation in Figure 3.4
1	0.80	0.20	0	0	6.50	y'_{12}
2	0.80	0	0.20	0	6.96	y'_{13}
3	0.80	0	0	0.20	6.00	y'_{14}
4	0.40	0.20	0.20	0.20	6.82	y'_1
5	0	1.00	0	0	5.80	y_2
6	0	0	1.00	0	5.65	y_3
7	0	0.50	0.50	0	5.93	y_{23}
8	0	0	0	1.00	5.05	y_4
9	0	0.50	0	0.50	5.36	y_{24}
10	0	0	0.50	0.50	5.72	y_{34}
11	0.80	0.10	0.10	0	7.25	
12	0.80	0	0.10	0.10	6.20	
13	0.80	0.10	0	0.10	6.47	
14	0.40	0.30	0.30	0	7.21	
15	0.40	0.30	0	0.30	6.53	
16	0.40	0	0.30	0.30	6.88	

The six additional blends designated as points 5–10 in Table 3.2 are

$$x_i x_j, \quad \text{where } x_i = x_j = 0.50, i, j = 2, 3, \text{ and } 4, i \neq j$$
$$x_i = 1, \text{ where } i = 2, 3, \text{ and } 4$$

The average flavor scores for the first 10 blends, as well as for 6 additional blends to be mentioned shortly are presented in Figure 3.4.

Fitted to the first 10 observations is the second-degree model

$$y = \sum_{i=1}^{4} \beta_i x_i + \sum_{i<j}^{3} \sum_{}^{4} \beta_{ij} x_i x_j + \varepsilon \tag{3.17}$$

The estimates of the parameters are computed using Eqs. (3.14)–(3.16). From Eq. (3.14) we have

$$b_2 = y_2 = 5.80, \quad b_3 = y_3 = 5.65, \quad b_4 = y_4 = 5.05$$
$$b_{23} = 4(5.93) - 2(5.80 + 5.65) = 0.82$$
$$b_{24} = 4(5.36) - 2(5.80 + 5.05) = -0.26$$
$$b_{34} = 4(5.72) - 2(5.65 + 5.05) = 1.48$$

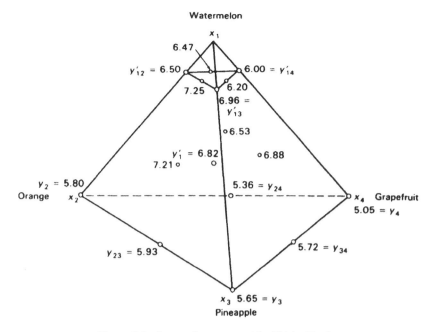

Figure 3.4 Average flavor scores at the 16 juice blends.

From Eqs. (3.15) and (3.16), the four equations to solve for b_1 and b_{1j} are

$$0.80b_1 + 0.80(1.0 - 0.80)b_{12} = 6.50 - (1.0 - 0.80)b_2$$
$$0.80b_1 + 0.80(1.0 - 0.80)b_{13} = 6.96 - (1.0 - 0.80)b_3$$
$$0.80b_1 + 0.80(1.0 - 0.80)b_{14} = 6.00 - (1.0 - 0.80)b_4$$

$$0.40b_1 + 0.40\frac{(1.0 - 0.40)}{3}[b_{12} + b_{13} + b_{14}]$$

$$= 6.82 - \frac{(1.0 - 0.40)}{3}[b_2 + b_3 + b_4] - \frac{(1.0 - 0.40)^2}{9}[b_{23} + b_{24} + b_{34}]$$

or in the case of this last equation

$$0.40b_1 + 0.40\frac{(0.60)}{3}[b_{12} + b_{13} + b_{14}] = 6.82 - (0.20)[16.50] - 0.04[2.04]$$
$$= 3.4384$$

and therefore

$$b_1 = 5.80, \quad b_{12} = 4.37, \quad b_{13} = 7.43, \quad b_{14} = 2.18$$

The 10-term prediction equation for flavor score is

$$\hat{y}(\mathbf{x}) = 5.80x_1 + 5.80x_2 + 5.65x_3 + 5.05x_4 + 4.37x_1x_2 + 7.43x_1x_3$$
$$\phantom{\hat{y}(\mathbf{x}) = }(0.60)\quad(0.35)\quad(0.35)\quad(0.35)\quad(3.01)\quad\quad(3.01)$$
$$+\ 2.18x_1x_4 + 0.82x_2x_3 - 0.26x_2x_4 + 1.48x_3x_4$$
$$(3.01)\quad\quad(1.73)\quad\quad(1.73)\quad\quad(1.73)$$

$$(3.18)$$

where the quantities in parentheses below the coefficient estimates are the estimated standard errors of the coefficient estimates. The estimate of the variance of each average response value is $s_{\bar{y}}^2 = 0.125$, which was calculated using the pooled within-blend variance $s_y^2 = 5.0/40$, where the number 40 represents the sample size used for calculating each y_u value in Table 3.2.

Prior to using Eq. (3.18) to predict the flavor score of blends other than the 10 blends used in the experiment, the fit of the model in Eq. (3.18) should be verified at other blends. To do so would require that data be collected on additional blends in the factor space and the model refitted. This is because Eq. (3.18) contains 10 terms and was fitted to data collected at exactly 10 blends. If data are collected at other points in the composition space and the fitted model does not perform well (see Section 2.10) at these points, this would alert us to the "lack of fit" or the inadequacy of the fitted model. Testing for model lack of fit was discussed previously in Sections 2.10 and 2.11.

Six additional blends, listed as design points 11 to 16 in Table 3.2, were chosen because of their importance and the second-degree equation (3.17) was fitted to the data at all 16 points, resulting in

$$\hat{y}(\mathbf{x}) = 5.87x_1 + 5.79x_2 + 5.65x_3 + 5.05x_4 + 5.29x_1x_2 + 7.13x_1x_3$$
$$\phantom{\hat{y}(\mathbf{x}) = }(0.37)\quad(0.36)\quad(0.36)\quad(0.36)\quad(1.95)\quad\quad(1.95)$$
$$+\ 1.90x_1x_4 + 0.68x_2x_3 - 0.16x_2x_4 + 1.53x_3x_4$$
$$(1.95)\quad\quad(1.73)\quad\quad(1.73)\quad\quad(1.73)$$

$$(3.19)$$

The estimated standard errors in parentheses were calculated using the new pooled error estimate $s_y^2 = [5.0(390) + 5.23(234)]/624(40) = 0.13$, where the second within-blend estimate $s_y^2 = 5.23$ was obtained from the 40 responses to each of the six new blends.

The d.f., SS Regression and SS Residual in the top two lines of Table 3.3 were calculated from the 16 average flavor scores in Table 3.2. The pure-error sum of squares in Table 3.3, 79.35 was computed by calculating the sum of squares among the 40 samples within each blend and pooling the 16 sums of squares to produce $16 \times (40 - 1) = 624$ d.f. An F-test comparing the mean square for residual to mean square for pure error was valued at $F = 0.058/0.13 < 1$. Since there was no reason to suspect that the residual mean square contained any source of variation other than error variation, the model was thought to be adequate.

Table 3.3 Analysis of Variance for the Tropical
Beverage Data

Source of Variation	d.f.	Sums of Squares	Mean Square
Regression	9	6.22	0.69
Residual	6	0.35	0.058
Pure error from replicates	624	79.35	0.13

3.6 INTRODUCING U-PSEUDOCOMPONENTS

When two or more of the component proportions are restricted by upper bounds, $x_i \leq U_i$, Crosier (1984) suggested defining the U-pseudocomponents,

$$u_i = \frac{U_i - x_i}{U - 1}, \quad i = 1, 2, \ldots, q \tag{3.20}$$

where $U = \sum_{i=1}^{q} U_i > 1$. Crosier denoted the U-pseudocomponents by z_i but we will use u_i so as not to confuse the U-pseudocomponents with the independent variables, z_i, $i = 1$ and 2 whose nine-term interaction model is Eq. (3.66). The region of the U-pseudocomponents, u_i, $i = 1, 2, \ldots, q$, is an inverted simplex, that is, $\sum_{i=1}^{q} u_i = 1$, which we will call the U-simplex. For example, let $q = 3$, and suppose we have the upper bounds

$$x_1 \leq 0.70 = U_1, \quad x_2 \leq 0.60 = U_2, \quad x_3 \leq 0.80 = U_3 \tag{3.21}$$

Then the U-simplex defined by (3.21) is the inverted triangle shown in Figure 3.5.

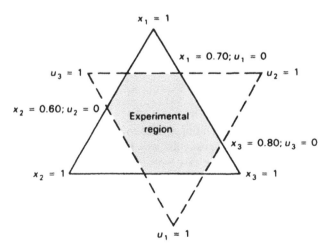

Figure 3.5 Inverted U-pseudocomponent simplex.

As seen in Figure 3.5, the vertices, $u_i = 1$, $i = 1, 2, 3, \ldots, q$, of the U-simplex may extend beyond the boundaries of the original simplex. When this happens, the experimental region of feasible mixtures is the region common to (defined as the intersection of) the original and the inverted simplexes. This region is not a simplex. The *U-simplex is the experimental region* only when it lies entirely inside the original simplex, and this happens if and only if

$$\sum_{i=1}^{q} U_i - U_{\min} \leq 1 \qquad (3.22)$$

where $U_{\min}$ is the minimum of the q upper bounds.

When the U-simplex lies entirely inside the original simplex, the denominator, $U - 1$, in the formula (3.20) for u_i is the experimental range of the values of u_i (relative to the range of unity for x_i). For convenience, let us define the linear size of the U-simplex as

$$R_U = \sum_{i=1}^{q} U_i - 1 \qquad (3.23)$$

where $0 < R_U \leq q - 1$. Of course, when the U-simplex lies inside the original simplex, we have $R_U \leq \frac{1}{2}$.

The placing of upper bounds on some or all of the x_i creates *implied lower bounds*, L_i^*, for all of the x_i. For x_i, the implied lower bound is

$$L_i^* = U_i - R_U \qquad (3.24)$$

Moreover, when a vertex of the U-simplex lies outside the boundaries of the original simplex, the resulting implied lower bound for x_i, whose vertex $x_i = 1$ is opposite the outlying vertex of the U-simplex, is negative, that is, $L_i^* < 0$. For example, the implied lower bounds for x_1, x_2, and x_3 of (3.21) are, respectively,

$$L_1^* = 0.70 - 1.10 = -0.40, \quad L_2^* = 0.60 - 1.10 = -0.50,$$
$$L_3^* = 0.80 - 1.10 = -0.30$$

where $R_U = (0.70 + 0.60 + 0.80) - 1 = 2.1 - 1 = 1.10$. Thus for each L_i^* that is negative (and unattainable), this signals to us that a vertex of the U-simplex (specifically the $u_i = 1$ vertex opposite the $x_i = 1$ vertex) lies outside the boundaries of the original simplex. Such is the case with all three vertices in Figure 3.5.

Implied lower pounds (3.24) as well as *implied upper bounds* on the x_i,

$$U_i^* = L_i + R_L \qquad (3.25)$$

where $R_L = 1 - \sum_{i=1}^{q} L_i$, are used to determine whether the bounds are *consistent* (Piepel, 1983a; Crosier, 1984). As we will see in Section 3.8, consistent bounds are necessary for determining the shape (number of vertices, edges, faces, etc.) and size of the experimental region of feasible mixture proportions when both upper and lower bounds are present on the component proportions. An upper bound U_i that is *greater* than U_i^* is not attainable, and a lower bound L_i that is *less* than L_i^* is not attainable. When one or more of the original bounds L_i or U_i are not attainable, the set of bounds on the x_i is said to be *inconsistent*. Inconsistent bounds are not useful, and in order for the bounds to be useful, they must be made to be consistent by adding quantities to them or subtracting quantities from them.

To set up designs when upper bounds only are stated for the x_i, and the *U-simplex lies inside the original simplex* (see Eq. 3.22), we use the *U*-pseudocomponents, u_i, in Eq. (3.20). Any of the standard lattice designs in the u_i maps into a lattice design in the x_i by applying the inverse transformation

$$x_i = U_i - R_U u_i \tag{3.26}$$

Data that are collected at the points of the design can be fitted either with a model in the x_i or with a model in the u_i. However, since the orientation of the *U*-simplex is opposite that of the original simplex, this reverse orientation of the simplexes must be kept in mind when interpreting the coefficients in the fitted model in the u_i in order to make inferences about the nature of the surface (or to describe the type of blending that occurs) in the original *x*-component system.

Let us illustrate the use of the *U*-pseudocomponent transformation by working through the following three-component example, where

$$x_1 \leq 0.4, \quad x_2 \leq 0.6, \quad x_3 \leq 0.3 \tag{3.27}$$

The constraints (3.27) imply that feasible mixture blends consist of *at most* 40% of component 1, 60% of component 2, and 30% of component 3, respectively. Since $(U_1 + U_2 + U_3 - U_{\min}) = (0.4 + 0.6 + 0.3 - 0.3) = 1.0 \leq 1.0$, the *U*-simplex is entirely inside the boundaries of the original x_i triangle; see Figure 3.6.

For the *U*-pseudocomponents we use Eq. (3.20) with the denominator $R_U = U - 1 = 0.3$:

$$u_1 = \frac{0.4 - x_1}{0.3}, \quad u_2 = \frac{0.6 - x_2}{0.3}, \quad u_3 = \frac{0.3 - x_3}{0.3} \tag{3.28}$$

Now, suppose that a seven-point simplex-centroid design is to be set up in u_1, u_2, and u_3. The settings in x_1, x_2, and x_3, corresponding to the seven design point settings in u_1, u_2, and u_3, are listed in Table 3.4. The values of x_1, x_2, and x_3 in Table 3.4 were obtained using Eq. (3.26), $x_i = U_i - R_U u_i$; that is,

$$x_1 = 0.4 - 0.3u_1, \quad x_2 = 0.6 - 0.3u_2, \quad x_3 = 0.3 - 0.3u_3$$

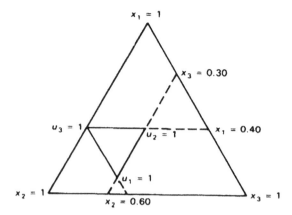

Figure 3.6   The U-simplex defined by $x_1 \leq 0.40$, $x_2 \leq 0.60$, and $x_3 \leq 0.30$.

Fitted to the data values in Table 3.4 are the several models in the u_i and in the x_i listed in Table 3.5, along with the corresponding summary statistics R^2, R_A^2, and s_e^2 for each pair of equivalent models.

Owing to the orientations of the U-simplex and the original simplex being opposite one another, the corresponding coefficients in the u_i and x_i models have interpretations that are reversed. The coefficient estimates in the u_i models are measures of the shape of the surface directly above the U-simplex. The coefficient estimates in the x_i models, however, are extrapolations of the surface above the U-simplex to the boundaries of the original simplex. Owing to the U-simplex comprising only approximately one-tenth the area of the original simplex [see Crosier (1984) for a discussion of the content of a subregion], the sizes of the coefficients in the x_i models, particularly the quadratic or second-degree model and the special cubic model, are of several orders of magnitude larger than the coefficients in the corresponding models in the u_i.

Extrapolation of the surface above the U-simplex, in an attempt to make inferences about the shape of the surface in the x-component system, is often risky. For example, let us refer to the first-degree models of Table 3.5 in the u_i and x_i. Each model

Table 3.4   Mixture Settings in the Constrained Region Defined by the Simplex-Centroid Design in the U-Pseudocomponents

U-Pseudocomponents			Original Components			Response Value
u_1	u_2	u_3	x_1	x_2	x_3	(Y)
1	0	0	0.1	0.6	0.3	4.2, 4.5
0	1	0	0.4	0.3	0.3	8.6, 8.1
0	0	1	0.4	0.6	0	12.7, 13.2
1/2	1/2	0	0.25	0.45	0.3	9.2, 9.5
1/2	0	1/2	0.25	0.6	0.15	10.3, 10.4
0	1/2	1/2	0.4	0.45	0.15	7.8, 8.6
1/3	1/3	1/3	0.3	0.5	0.2	12.3, 11.9, 13.0

Table 3.5 Fitted Models and Summary Statistics in the U-Pseudocomponents and in the Original Constrained Components

Fitted Model	R^2	R_A^2	s_e^2 (d.f.)
First degree			
$\hat{y} = 6.70u_1 + 9.04u_2 + 13.12u_3$ $\quad\quad (1.27)\quad\ (1.27)$	0.4770	0.3898	4.82 (12)
$\hat{y} = 17.80x_1 + 10.00x_2 - 3.60x_3$ $\quad\quad (4.13)\quad\ (2.91)\quad\ (4.62)$			
Second degree			
$\hat{y} = 4.11u_1 + 8.11u_2 + 12.71u_3 + 16.89u_1u_2 + 11.69u_1u_3 - 4.91u_2u_3$ $\quad\quad (0.87)\ (0.87)\ (0.87)\quad\quad (3.94)\quad\quad (3.94)\quad\quad (3.94)$	0.8751	0.8050	1.54 (9)
$\hat{y} = -80.04x_1 - 0.51x_2 + 23.65x_3 + 187.62x_1x_2 + 129.84x_1x_3 - 54.60x_2x_3$ $\quad\quad (18.76)\ (9.78)\quad (23.69)\quad\quad (43.74)\quad\quad (43.74)\quad\quad (43.74)$			
Special cubic			
$\hat{y} = 4.35u_1 + 8.35u_2 + 12.95u_3 + 12.00u_1u_2 + 6.80u_1u_3 - 9.80u_2u_3 + 76.95u_1u_2u_3$ $\quad\quad (0.28)\ (0.28)\ (0.28)\quad\quad (0.28)\quad\quad (1.39)\quad\quad (1.39)\quad\quad (8.71)$	0.9884	0.9797	0.16 (8)
$\hat{y} = -368.05x_1 - 128.38x_2 - 429.94x_3 + 988.33x_1x_2 + 1785.56x_1x_3 + 1031.11x_2x_3 - 2850.00x_1x_2x_3$ $\quad\quad (33.17)\quad (14.82)\quad (51.92)\quad\quad (91.75)\quad\quad (187.98)\quad\quad (123.73)\quad\quad (322.66)$			

Note: The quantities in parentheses are the estimated standard errors of the coefficient estimates.

111

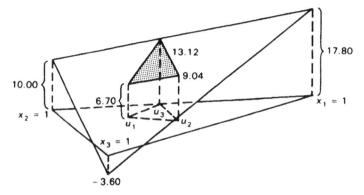

Figure 3.7 Estimated planar surface above the U-simplex extrapolated to the three vertices of the original simplex.

describes a tilted planar surface but the x_i model depicts a plane in which most but not all of it lies above the original simplex as seen in Figure 3.7. The lowest height of the plane above the $u_1 = 1$ vertex projects to the greatest height of the extrapolated plane at $x_1 = 1$, yet because of the tilt of the plane the projection to the $x_3 = 1$ vertex actually dips below the plane of the original simplex, causing the estimate of β_3 to be negative.

In the actual experiment from which the data in Table 3.4 were taken, a negative response to any pure component or any blend of the components is considered to be impossible. Hence the negative coefficient estimates of the linear terms in each of the x_i models serve as a warning to us that extrapolations are risky and that these models are to be used for purposes of interpretation and prediction only over the subregion of the simplex defined by the constraints (3.27).

As to the precision of the coefficient estimates in the fitted models with the u_i and with the x_i in Table 3.5, one notice that the precision of the estimates of the coefficients of the terms of a particular degree in each of the u_i models is constant (i.e., they have equal standard errors) but this is not the case with the estimates of the coefficients in the x_i models. The reason for this is that the design in the u_i is a symmetrical arrangement over the U-simplex but the design is not symmetrical with respect to the orientation of the original simplex. If the upper bounds U_1, U_2, U_3 are set at 0.5, then centroids of the U-simplex and the original simplex are the same, and the symmetrical design in the u_i is also symmetrical in the x_i. In this latter case the estimated standard errors of the coefficients of the terms of a particular degree in each of the x_i models are the same.

3.7 THE PLACING OF BOTH UPPER AND LOWER BOUNDS ON THE COMPONENT PROPORTIONS

In Section 3.2, L-pseudocomponents were introduced for the primary purpose of simplifying the design problem, as well as for modeling the surface over the interior region, which was defined from the placing of lower bounds on the component proportions. In Sections 3.4–3.6, we discussed the estimation of the model coefficients (as well as the selection of certain component blends) for the case where upper bounds are placed on some or all of the component proportions. Quite often, in practice, one is faced with both situations occurring at the same time because some of the component proportions are restricted in value by upper and lower bounds. Such situations arise, for example, when to form a valid blend we require or need at least L_i, but no more than U_i, of component i, and similar bounds are specified for the other component proportions as well.

To give an example of placing constraints on the component proportions, let us recall the fruit punch formulations discussed in Section 3.5. Suppose that now we want the punch to be at least 40% but not more than 80% watermelon (x_1), we want at least 10% orange juice (x_2), and we insist that pineapple (x_3) and grapefruit (x_4) juices contribute at least 5% each but not more than 30% each. We can write these restrictions as

$$
\begin{array}{llcl}
40\% \le \text{Watermelon} \le 80\% & & & 0.40 \le x_1 \le 0.80 \\
10\% \le \text{Orange} & & & 0.10 \le x_2 \\
5\% \;\le \text{Pineapple} \;\;\le 30\% & \text{or} & & 0.05 \le x_3 \le 0.30 \\
5\% \;\le \text{Grapefruit} \;\;\le 30\% & & & 0.05 \le x_4 \le 0.30
\end{array}
$$

Although the definition of the formulations desired did not state an upper bound for the percentage of orange, an upper bound of 50% is *implied* by the presence of the lower bounds of 40%, 5%, and 5% for watermelon, pineapple, and grapefruit, respectively. Therefore the restrictions on the x_i are more correctly written as

$$
\begin{aligned}
0.40 &\le x_1 \le 0.80 \\
0.10 &\le x_2 \le 0.50 \\
0.05 &\le x_3 \le 0.30 \\
0.05 &\le x_4 \le 0.30
\end{aligned}
\tag{3.29}
$$

With q components, the multiple constraints are written as

$$
0 \le L_i \le x_i \le U_i \le 1, \quad i = 1, 2, \dots, q
\tag{3.30}
$$

When only one or two of the component proportions are restricted in value, the shape of the resulting factor space is not so difficult to envision. However, if nearly all of the component proportions are constrained above and below, then the resulting factor space takes the form of a hyperpolyhedron that is convex and that will often be considerably more complicated in shape than the simplex.

As for the boundaries of the constrained region (3.30) that are to be used for design-point locations, the particular boundaries that are chosen depend on the form or degree of the equation that is to be used to model the surface over the region. Very often we shall require for our design-point locations at least some of the extreme vertices of the region as well as midpoints and centroids of some of the edges and two-dimensional faces, respectively, of the region. For example, suppose that we wish to fit a second-degree model to data collected at the various combinations of the x_i $i = 1, 2, \ldots, q$, over the region (3.30) where the form of the model is

$$\eta = \sum_{i=1}^{q} \beta_i x_i + \sum_{i<j} \sum^{q} \beta_{ij} x_i x_j \qquad (3.31)$$

Then a minimum of $q + q(q - 1)/2$ distinct points are needed at which to collect observations, since this is the number of parameters (or unknown coefficients) in Eq. (3.31) to be estimated. In general, the set of design points would consist of at least q extreme vertices, the midpoints of at least $q(q - 1)/2$ edges, and a subset of the face centroids. The edges and faces of the polyhedron are found by taking convex combinations of the coordinates of the extreme vertices. The coordinates of the centroids of the edges and faces can be calculated using the procedure described in Piepel (1983b). Note that the face centroid points are used primarily for testing lack of fit of the second-degree model (3.31).

Before we discuss several algorithms that can be used to determine the coordinates of the extreme vertices of the constrained regions, we will present formulas for calculating the number of vertices and higher dimensional boundaries of the region. The formulas, however, require that the upper- and lower-bound constraints on the x_i be *consistent*.

3.7.1 Detecting Inconsistent Constraints

Piepel (1983a) presents formulas for checking the consistency of the constraints

$$0 \leq L_i \leq x_i \leq U_i \leq 1, \quad i = 1, 2, \ldots, q \quad \text{and} \quad \sum_{i=1}^{q} x_i = 1 \qquad (3.32)$$

The constraints (3.32) are said to be *consistent* when, upon listing the feasible combinations of the x_i, $i = 1, 2, \ldots, q$ for the region (3.32), each and every component proportion (not necessarily all simultaneously) attains its lower bound ($x_i = L_i$) and each and every component proportion attains its upper bound ($x_i = U_i$). For example, the set of constraints

$$0 \leq x_1 \leq 0.1$$

$$0.1 \leq x_2 \leq 0.2 \qquad (3.33)$$

$$0.6 \leq x_3 \leq 0.8$$

is not consistent, or is *inconsistent*, because x_3 cannot be as low as 0.6 (i.e., there are no blends for which $x_3 = L_3 = 0.6$). The minimum value of x_3 is

$x_3 = (1 - U_1 - U_2) = 0.7$. As another example, the set of constraints

$$0.3 \leq x_1 \leq 0.8$$
$$0.1 \leq x_2 \leq 0.5 \qquad (3.34)$$
$$0.2 \leq x_3 \leq 0.6$$

is *inconsistent* because the upper bound, $U_1 = 0.8$, for x_1 *is unattainable*. If U_1 is set at 0.7, then the set of constraints is consistent.

To check on the consistency of the constraints (3.32) or what amounts to the same, to detect any inconsistencies in the constraints (3.32), first we calculate the range of each x_i,

$$R_i = U_i - L_i, \quad i = 1, 2, \ldots, q$$

Then

$$R'_i = \frac{R_i}{1 - \sum_{i=1}^{q} L_i}, \quad i = 1, 2, \ldots, q \qquad (3.35)$$

represents the *range* of L-pseudocomponent x'_i. Now, if for any i, $R'_i > 1$, that is,

$$R_i > 1 - \sum_{i=1}^{q} L_i \quad \text{or} \quad U_i + \sum_{j \neq i}^{q} L_j > 1 \qquad (3.36)$$

then U_i is *unattainable*. Such was the case of U_1 in (3.34) since $R'_i = 0.5/0.4 = 1.25 > 1$. Also, the lower bound, L_i, is unattainable if, for any i,

$$\sum_{j \neq i}^{q} R'_j < 1 \quad \text{or} \quad L_i + \sum_{j \neq i}^{q} U_j < 1 \qquad (3.37)$$

As an example, L_3 of (3.33) is unattainable because $R'_1 + R'_2 = (0.1/0.3 + 0.1/0.3) = 0.67 < 1$.

Another way of detecting inconsistent constraints caused by an unattainable L_i is by using the U-pseudocomponents, u_i, $i = 1, 2, \ldots, q$. In Section 3.6, the linear size of the U-simplex was defined as

$$R_U = \sum_{i=1}^{q} U_i - 1$$

Now, if for any i, $R_i > R_U$, then L_i is unattainable (Crosier, 1984, p. 211).

When the constraints are not consistent and adjustments are to be made in the bounds in order to make the constraints consistent, the adjustments are made either in the bounds that are unattainable (i.e., implied bounds are defined), or in some of the bounds on the other x_i. Consistent constraints are necessary in Section 3.8 when enumerating the d-dimensional boundaries of the constrained region as well as in Section 3.9.5 when using U-pseudocomponents in order to compute the coordinates of the extreme vertices of the constrained region.

3.7.2 Adjusting Inconsistent Constraints to Make Them Consistent

The set of inconsistent constraints can be adjusted to form a consistent set but this will often alter the shape and size of the experimental region. To illustrate, let us consider the following constraints on x_1, x_2, and x_3:

$$0.20 \leq x_1 \leq 0.40 \quad 0.20 \leq x_2 \leq 0.60, \quad 0.18 \leq x_3 \leq 0.70 \tag{3.38}$$

The component ranges are $R_1 = 0.4 - 0.2 = 0.2$, $R_2 = 0.4$, and $R_3 = 0.52$. Now,

$$R_L = 1 - \sum_{i=1}^{3} L_i = 1 - 0.58 = 0.42, \quad R_U = \sum_{i=1}^{3} U_i - 1.7 - 1 = 0.7$$

and since $R_3 = 0.52 > R_L$, this means that the upper bound, $U_3 = 0.7$, is not attainable and thus the constraints (3.38) are inconsistent. If U_3 is replaced by the implied upper bound (see 3.25), $U_3^* = L_3 + R_L = 0.18 + 0.42 = 0.6$, then the constraints are consistent and $R_U = 0.6$.

We can also force the constraints to be consistent by altering the lower bounds. For example, suppose that we decide to keep $U_3 = 0.7$ as the upper bound for x_3. Then to alter the constraints (3.38) to make them consistent would require lowering $L_1 = 0.2$ to $L_1 = 0.1$ or lowering $L_2 = 0.2$ to $L_2 = 0.1$ or lowering both L_1 and L_2 by an amount equal to 0.05, say, to give $L_1 = 0.15$ and $L_2 = 0.15$. We will check each of these cases to see how altering the constraints affects the shape and size of the experimental region.

Let us replace the upper bound $U_3 = 0.7$ with the implied $U_3^* = 0.6$ and keep the constraints on x_1 and x_2 as shown in (3.38). The experimental region is shown in Figure 3.8. The constrained region has five vertices whose coordinates (x_1, x_2, x_3) are (0.40, 0.42, 0.18) for vertex 1, (0.40, 0.20, 0.40) for vertex 2, (0.20, 0.20, 0.60) for vertex 3, (0.20, 0.60, 0.20) for vertex 4, and (0.22, 0.60, 0.18) for vertex 5.

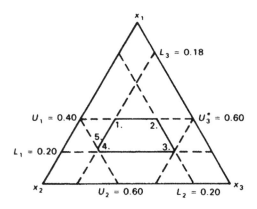

Figure 3.8 Region defined by $0.2 \leq x_1 \leq 0.4, 0.2 \leq x_2 \leq 0.6, 0.18 \leq x_3 \leq 0.6 = U_3^*$.

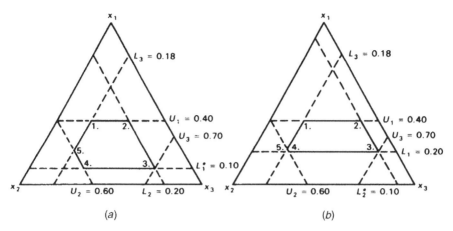

(a)　　　　　　　　　　　　　(b)

Figure 3.9 (a) Region $L_1^* = 0.10 \le x_1 \le 0.40, 0.2 \le x_2 \le 0.6, 0.18 \le x_3 \le U_3 = 0.7$. (b) Region $0.2 \le x_1 \le 0.4, L_2^* = 0.1 \le x_2 \le 0.6, 0.18 \le x_3 \le U_3 = 0.7$.

As an alternative to the constrained region in Figure 3.8, let us define two different regions that result from keeping the upper bound $U_3 = 0.7$ and lower L_1 and/or L_2. For the first case, we reduce the lower bound on x_1 from $L_1 = 0.2$ to $L_1 = 0.1$ while keeping $L_2 = 0.2$, and in the second case, we reduce the lower bound on x_2 to $L_2 = 0.1$ and keep $L_1 = 0.2$. This action produces the experimental regions shown in Figures 3.9a,b, respectively. The five vertices of the regions in Figures 3.9a,b, respectively, are as follows:

Vertex	Figure 3.9a (x_1, x_2, x_3)	Figure 3.9b (x_1, x_2, x_3)
1	(0.40, 0.42, 0.18)	(0.40, 0.42, 0.18)
2	(0.40, 0.20, 0.40)	(0.40, 0.10, 0.50)
3	(0.10, 0.20, 0.70)	(0.20, 0.10, 0.70)
4	(0.10, 0.60, 0.30)	(0.20, 0.60, 0.20)
5	(0.22, 0.60, 0.18)	(0.22, 0.60, 0.18)

Moreover, if both L_1 and L_2 are reduced to 0.15, the shape of the experimental region is the same as in Figures 3.9a,b, but the coordinates of the five vertices become $1 = (0.40, 0.42, 0.18)$, vertex $2 = (0.40, 0.15, 0.45)$, vertex $3 = (0.15, 0.15, 0.70)$, vertex $4 = (0.15, 0.60, 0.25)$, and vertex $5 = (0.22, 0.60, 0.18)$.

The four regions that we have defined by altering the upper and lower bounds on the component proportions ever so slightly are of the same shape in terms of the number of vertices (five) and edges (five). The *areas* of the regions are different, however. What this means is that if data are collected at the five vertices of each of the four regions for the purpose of fitting the first-degree model

$$y = \beta_1 x_1 + \beta_2 x_2 + \beta_3 x_3 + \varepsilon$$

the estimated variances and covariances, $\widehat{\mathrm{var}}(\hat{\beta}) = (\mathbf{X'X})^{-1}\hat{\sigma}^2$, of the parameter estimates in the four fitted models will be different with the four designs.

To illustrate this variance property of the four designs, let us define the design matrix for a particular design to be the 5×3 matrix

$$\mathbf{X} = \begin{bmatrix} \text{coordinates of point 1} \\ \text{coordinates of point 2} \\ \text{coordinates of point 3} \\ \text{coordinates of point 4} \\ \text{coordinates of point 5} \end{bmatrix}$$

Then the sum of the variances of the parameter estimates in the first-degree fitted model, $\mathrm{var}(b_1) + \mathrm{var}(b_2) + \mathrm{var}(b_3)$, is equal to the trace, or sum of the diagonal elements, of the matrix $(\mathbf{X'X})^{-1}\sigma^2$. The trace, when divided by σ^2, along with the sum of the ranges of the x_i corresponding to each of the four design is as follows:

Design	Trace/σ^2	$\sum_{i=1}^{3} R_i$
Figure 3.8	19.56	1.02
Figure 3.9a	10.95	1.22
Figure 3.9b	18.07	1.22
$L_1 = L_2 = 0.15$	13.79	1.22

Based on the trace criterion, we would say the design in Figure 3.9a is the preferred design among the four listed because its value, 10.95, is lowest among the four designs. The trace criterion (or generalized variance criterion) is just one of several criteria that are used often when choosing among two or more competing designs; see Snee (1975).

Altering the constraints can also affect the shape of the experimental region. In Figure 3.8 and Figure 3.9b, we see that if the lower bound $L_3 = 0.18$ is raised to $L_3 = 0.20$, the number of vertices of the region is reduced from five to four. In cases dealing with only three or four components, any attempt at reducing the number of vertices of the constrained region by altering the upper and lower bounds of some of the x_i may not be worth the trouble. However, in problems involving a large number of components, for example, when there are five or more ($q \geq 5$) components, adjusting the constraints in an attempt to reduce the number of extreme vertices can be highly desirable. This is illustrated by Crosier (1984), who, in working with a five-component octane-blending example taken from Snee and Marquardt (1974), showed that a 28-vertices region could be modified to a region with only 15 vertices by adding 0.01 to the lower and upper bounds of each of the five components. There seems little argument to the contention that a candidate list of 15 vertices is easier to work with than is a list of 28 candidate points when selecting a design to fit either a first- or second-degree model.

Knowing the actual number of vertices, edges, faces, and so on, of the constrained region is very helpful when trying to make up the list of candidate points from which to choose a design for fitting any of the Scheffé-type mixture models. As an example, when fitting the first-degree model

$$y = \sum_{i=1}^{q} \beta_i x_i + \varepsilon$$

it is useful to know how many extreme vertices the constrained region has. This is because extreme-vertices designs are among the class of minimum-variance designs for estimating the parameters in the first-degree model as well as for producing a prediction equation that also has the minimum-variance property. When the number of extreme vertices is large (e.g., $\geq 2q$), generally one would select only a subset of the vertices along with certain convex combinations of some of the vertices if that is all that is needed in order to fit the model. For fitting the second-degree model

$$y = \sum_{i=1}^{q} \beta_i x_i + \sum_{i<j}\sum^{q} \beta_{ij} x_i x_j + \varepsilon$$

knowing the number of extreme vertices and the number of edges of the region is important. Again, the vertices are most useful for estimating the coefficients of the linear blending terms while the midpoints of the edges are useful design points for estimating the binary blending parameters, β_{ij}. Finally, when fitting the special cubic model

$$y = \sum_{i=1}^{q} \beta_i x_i + \sum_{i<j}\sum^{q} \beta_{ij} x_i x_j + \sum_{i<j<k}\sum\sum^{q} \beta_{ijk} x_i x_j x_k + \varepsilon$$

it is important to know how many vertices, edges, and faces the region has. The centroids of the faces are desirable points at which to collect data for estimating the ternary blending parameters, β_{ijk}.

In summary, knowing the numbers and types of boundaries the constrained region possesses permits us to be selective in choosing only a subset of them when setting up a design. Without this information, we are helpless and often totally dependent on some available computer algorithm to generate and select a set of design points for us. Moreover, without knowing how many boundaries the region has, we can be left without any assurance that the total number of points in the candidate list is correct.

3.8 FORMULAS FOR ENUMERATING THE NUMBER OF EXTREME VERTICES, EDGES, AND TWO-DIMENSIONAL FACES OF THE CONSTRAINED REGION

A formula that can be used to calculate the number of d-dimensional boundaries $(d = 0,1,\ldots,q-2)$ of a region, defined by the set of consistent constraints, is given in

$$x_1 + x_2 + \cdots + x_q = 1, \quad 0 \leq L_i \leq x_i \leq U_i \leq 1 \tag{3.39}$$

Crosier (1986). To understand the terms used in the formula, we present the following steps. First, compute the ranges, $R_i = U_i - L_i$, $i = 1, 2, \ldots, q$, of each of the q components and calculate $R_L = 1 - \sum_{i=1}^{q} L_i$ and $R_U = \sum_{i=1}^{q} U_i - 1$. Let R_p be the minimum of R_L and R_U. [Note that the number of combinations of the q ranges taken r at a time is $C(q, r) = q!/r!(q-r)!$] Second, divide the $C(q,r)$ combinations of ranges into three mutually exclusive and exhaustive subsets by letting, for $r = 1, 2, \ldots, q$:

$L(r)$ = number of combinations of component ranges that sum to a number that is lower than R_p.

$E(r)$ = number of combinations of component ranges that sum to R_p.

$G(r)$ = number of combinations of component ranges that sum to a number that is higher than R_p.

As a check on the counts, $C(q, r) = L(r) + E(r) + G(r)$. Now, the formula used to calculate the number of d-dimensional boundaries of the region defined by (3.39) is

$$N_d = C(q, q-d-1) + \sum_{r=1}^{q-d-1} L(r)C(q-r, q-r-d-1)$$

$$- \sum_{r=d+1}^{q} [L(r) + E(r)]C(r, r-d-1) \tag{3.40}$$

where $d = 1, 2, \ldots, q-2$. For $d = 0$ (the vertices), Eq. (3.40) is simplified to

$$N_0 = q + \sum_{r=1}^{q} [L(r)(q-2r) - E(r)(r-1)] \tag{3.41}$$

To illustrate the use of the formulas (3.40) and (3.41), we calculate the number of vertices, edges, and faces of the four-component experimental region defined by the set of consistent constraints

$$0.80 \leq x_1 \leq 0.90, \quad 0.05 \leq x_2 \leq 0.15, \quad 0.02 \leq x_3 \leq 0.10, \, 0.03 \leq x_4 \leq 0.05 \tag{3.42}$$

The ranges are $R_1 = 0.10$, $R_2 = 0.10$, $R_3 = 0.08$, and $R_4 = 0.02$. The values of R_L and R_U are 0.10 and 0.20, respectively, so that $R_p = \min(0.10, 0.20) = 0.10$. For $r = 1, 2, 3$, and 4, the component ranges with a value lower than $R_p = 0.10$ are R_3 and R_4, so that for $r = 1$, $L(1) = 2$ and for $r > 1$, $L(2) = L(3) = L(4) = 0$. Similarly, since R_1 and R_2 equal R_p and $R_3 + R_4 = R_p$, then $E(1) = 2$ and $E(2) = 1$, $E(3) = E(4) = 0$.

Finally, $G(1) = 0$, $G(2) = 5$, $G(3) = 4$, and $G(4) = 1$. As a check,

$$C(4, 1) = 4 = L(1) + E(1) + G(1) = 2 + 2 + 0$$
$$C(4, 2) = 6 = L(2) + E(2) + G(2) = 0 + 1 + 5$$
$$C(4, 3) = 4 = L(3) + E(3) + G(3) = 0 + 0 + 4$$
$$C(4, 4) = 1 = L(4) + E(4) + G(4) = 0 + 0 + 1$$

The number of vertices ($d = 0$) of the region (3.42) is, from (3.41),

$$N_0 = 4 + \{L(1)(2) - E(1)(0)\} + \{L(2)(0) - E(2)(1)\}$$
$$+\{L(3)(-2) - E(3)(2)\} + \{L(4)(-4) - E(4)(3)\}$$
$$= 4 + \{4\} + \{-1\} + \{0\} + \{0\}$$
$$= 7$$

From (3.40), the number of edges ($d = 1$) is

$$N_1 = C(4, 2) + L(1)C(3, 1) + L(2)C(2, 0) - [L(2) + E(2)]C(2, 0)$$
$$-[L(3) + E(3)]C(3, 1) - [L(4) + E(4)]C(4, 2)$$
$$= 6 + 2(3) + 0 - [0 + 1] - [0]3 - [0]6$$
$$= 11$$

and the number of faces ($d = 2$) is

$$N_2 = C(4, 3) + L(1)C(3, 0) - [L(3) + E(3)]C(3, 0)$$
$$-[L(4) + E(4)]C(4, 1)$$
$$= 4 + 2 - 0 - 0$$
$$= 6$$

The constrained region (3.42) consisting of 7 vertices, 11 edges, and 6 faces is shown in Figure 3.10.

Once we have determined the number of vertices, edges, faces, and so on, of the constrained region using formulas (3.40) and (3.41), the next order of business is to define the coordinates of the vertices and the coordinates of the centroids of the edges, faces, and so on. These coordinates comprise the set of candidate design points from which a subset may be chosen for fitting any of the Scheffé-type or other mixture model forms. In cases involving only three or four components, the candidate set of points is relatively easy to determine. For example, with three components, one only needs to draw the constrained region on triangular coordinate paper and then determine where the bounds on x_1, x_2, and x_3 intersect to locate the coordinates of the extreme vertices. With four components, one can either draw the region and determine the coordinates from the figure, or one can compute the coordinates of the vertices by listing the $2^4 = 16$ possible combinations of lower and upper bounds and

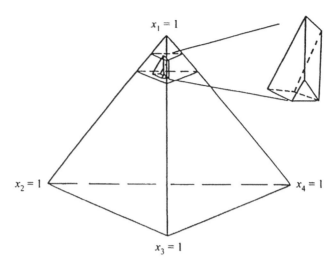

Figure 3.10 Four-component constrained region defined by the constraints $0.80 \leq x_1 \leq 0.90$, $0.05 \leq x_2 \leq 0.15$, $0.02 \leq x_3 \leq 0.10$, and $0.03 \leq x_4 \leq 0.05$.

then select those that are admissible and adjust those that are not admissible but can be made admissible, to form the vertices. With five or more components, the procedures above are not so simple, and algorithms have been developed for generating the coordinates of the vertices. A few of the more popular algorithms are discussed in Section 3.9.

3.8.1 Choosing a Particular Set of Pseudocomponents

Up to this point in this chapter, we have discussed the problems associated with selecting a design when constraints are present on the component proportions. Moreover both L-pseudocomponents and U-pseudocomponents were introduced as alternatives to using the original constrained components, where lower- and upper-bound constraints are stated for the x_i. To summarize what we have learned thus far, we now review the two types of pseudocomponents and also make some suggestions as to when each type of pseudocomponent is the appropriate one to use.

When some of all of the mixture component proportions are restricted in value by placing lower bounds (L_i) on the x_i, the experimental region, or constrained region, lies inside the original simplex and is also a simplex whose vertices are aligned with the vertices of the original simplex. The L-pseudocomponent transformation is the appropriate one where

$$x_i' = \frac{x_i - L_i}{1 - \sum_{i=1}^{q} L_i}$$

Typically a design is set up in the L-pseudocomponents and the original component proportions can be calculated using $x_i = L_i + (1 - \sum_{i=1}^{q} L_i)x_i'$. A model in the L-pseudocomponents is then fitted to data that are collected at the design points. All of the standard summary calculations and statistics such as the ANOVA table and the values of R^2 and R_A^2 are easily computed for the L-pseudocomponent model in the usual manner.

When some or all of the component proportions are restricted in value by placing upper bounds (U_i) on the x_i, the U-pseudocomponents are defined as

$$u_i = \frac{U_i - x_i}{\sum_{i=1}^{q} U_i - 1}$$

The U-simplex has an orientation that is opposite, or the reverse of, that of the orientation of the simplex region in the x_i, hereafter referred to as the x-simplex. The vertices of the U-simplex may extend beyond the boundaries of the x-simplex. However, if $\sum_{i=1}^{q} U_i - U_{min} < 1$, where U_{min} is the minimum of the q U_i, then the U-simplex lies entirely inside the x-simplex and the U-simplex is the experimental region. In this case, the quantity $R_U = \sum_{i=1}^{q} U_i - 1$ is the length of the axes of the U-pseudocomponents.

When some or all of the component proportions are restricted by placing lower (L_i) and upper (U_i) bounds on the x_i, the first step in determining whether to use the L-pseudocomponents or the U-pseudocomponents is to check on the consistency of the upper and lower bounds. This is done by comparing each range, $R_i = U_i - L_i$, against $R_L = 1 - \sum_{i=1}^{q} L_i$ and $R_U = \sum_{i=1}^{q} U_i - 1$. If any $R_i > R_L$, then this indicates an inconsistent upper bound, U_i, and if any $R_i > R_U$ this indicates an inconsistent lower bound L_i. Inconsistent bounds should be replaced by their respective implied bounds, $U_i^* = L_i + R_L$ and $L_i^* = U_i - R_U$, in order to make the set of constraints consistent.

The choice of working with the L-pseudocomponents, x_i', or the U-pseudocomponents, u_i, depends on the shape of the experimental region. If $R_L < R_U$, then the L-pseudocomponent simplex is smaller in size than the U-pseudocomponent simplex, and also, if the L-simplex is entirely inside the U-simplex, then the x_i' are chosen for use; otherwise, the u_i are chosen. If $R_U < R_L$, the U-simplex is smaller in size than the L-simplex, and furthermore if the U-simplex lies entirely inside the L-simplex, then the u_i are chosen for use. If neither simplex is inside the other, or if $R_L = R_U$ so that the experimental region is not a simplex, then the u_i are chosen for the purpose of constructing designs.

3.9 McLEAN AND ANDERSON'S ALGORITHM FOR CALCULATING THE COORDINATES OF THE EXTREME VERTICES OF A CONSTRAINED REGION

Generating the coordinates of the extreme vertices of the constrained region

$$0 \leq L_i \leq x_i \leq U_i \leq 1 \quad \text{and} \quad \sum_{i=1}^{q} x_i = 1 \tag{3.43}$$

can be accomplished by working with the original constrained components or by working with the U-pseudocomponents, u_i, $i = 1, 2, \ldots, q$.

In this section we briefly describe McLean and Anderson's (1966) extreme-vertices (EV) algorithm and simply mention two other algorithms: Snee and Marquardt's (1974) XVERT algorithm, and Nigam, Gupta, and Gupta's (1983) XVERT1 algorithm. Each of these algorithms works with the original constrained components and their upper and lower bounds. XVERT and XVERT1 are discussed briefly in Cornell (2002, ch. 4.)

The coordinates of some of the extreme vertices of the polyhedron—specifically, those vertices that are defined as combinations of the lower- and upper-bound constraints (3.43)—can be obtained directly by using the upper and lower bounds of $q - 1$ components. McLean and Anderson (1966) suggested the following two-step procedure:

> **Step 1.** List all possible combinations (as in a two-level factorial arrangement) of the values of L_i and U_i for $q - 1$ components and leave the value of the remaining component blank. This procedure produces 2^{q-1} combinations. With three components, for example, where L_1, L_2, L_3 are the lower bounds and U_1, U_2, and U_3 are the upper bounds, the component level combinations are $L_1 L_2$ __, $L_1 U_2$ __, $U_1 L_2$ __, $U_1 U_2$__, where the proportion for component 3 is left blank. The procedure is repeated q times, allowing each component to be the one whose level is left blank and is therefore to be computed. This list will consist of $q \times (2^{q-1})$ possible combinations.
>
> **Step 2.** Go through all possible combinations generated in step 1 and fill in those blanks that are admissible; that is, fill in the level (necessarily falling within the constraints of the missing factor) that will make the total of the levels for that treatment combination sum to unity. Each admissible combination of the levels of all q components defines an extreme vertex.

After the procedure above is used to define the extreme vertices of the convex polyhedron, the next step is to define a variety of centroids or points centrally positioned on the faces and the edges and so on, which are referred to as the centroids. There is one centroid point located in each bounding two-dimensional face, three-dimensional face, $\ldots$, d-dimensional face where $d \leq q - 2$, as well as the "overall" centroid of the polyhedron. This latter point is the component combination obtained by averaging all of the factor levels that define the existing extreme vertices, and because it may or may not coincide with the true centroid of the polyhedron, quotation marks are used. The centroids of the two-dimensional faces are found by isolating all vertices that have identical $q - 3$ factor levels and by averaging the factor levels of each of the three remaining factors. All remaining centroids are found in similar fashion using all vertices that have identical $q - r - 1$ factor levels for an r-dimensional face when $r \leq d \leq q - 2$. It should be noted that d may be less than $q - 2$, since the constraints listed and applied to the proportions may reduce the dimensionality of the hyperpolyhedron to less than $q - 1$.

To illustrate the procedure for locating the extremities of a region that is defined from the placing of constraints on the x_i, let us refer to the tropical beverage example of Section 3.7, where, in addition to insisting that between 40% and 80% of the beverage be watermelon juice, we require at least 10%, 5%, and 5% of orange, pineapple, and grapefruit, respectively, be present, but at most 30% of each of the latter two fruit juices. The constraints on the proportions were listed previously in Eqs. (3.29) as

$$0.40 \le x_1 \le 0.80$$
$$0.10 \le x_2 \le 0.50$$
$$0.05 \le x_3 \le 0.30$$
$$0.05 \le x_4 \le 0.30$$

To generate the $q \times (2^{q-1}) = 4 \times (2^{4-1}) = 32$ possible component combinations that may or may not be admissible, we begin our listing as follows:

Vertex	x_1	x_2	x_3	x_4	Vertex	x_1	x_2	x_3	x_4
	0.40	0.10	0.05	___	3	0.40	0.50	0.05	0.05
1	0.40	0.10	0.30	0.20	4	0.40	0.25	0.05	0.30
3	0.40	0.50	0.05	0.05	5	0.40	0.25	0.30	0.05
	0.40	0.50	0.30	___		0.40	___	0.30	0.30
2	0.80	0.10	0.05	0.05	2	0.80	0.10	0.05	0.05
	0.80	0.10	0.30	___		0.80	___	0.05	0.30
	0.80	0.50	0.05	___		0.80	___	0.30	0.05
	0.80	0.50	0.30	___		0.80	___	0.30	0.30
	0.40	0.10	___	0.05	2	0.80	0.10	0.05	0.05
6	0.40	0.10	0.20	0.30	7	0.55	0.10	0.05	0.30
3	0.40	0.50	0.05	0.05	8	0.55	0.10	0.30	0.05
	0.40	0.50	___	0.30		___	0.10	0.30	0.50
2	0.80	0.10	0.05	0.05	3	0.40	0.50	0.05	0.05
	0.80	0.10	___	0.30		___	0.50	0.05	0.30
	0.80	0.50	___	0.05		___	0.50	0.30	0.05
	0.80	0.50	___	0.30		___	0.50	0.30	0.30

Eight admissible vertices are present and are numbered as 1, 2, ..., 8. The two-dimensional faces are found by grouping the vertices of the polyhedron into groups of three or more vertices, where each vertex has the same value x_i for one of the components. For example, the vertices 2, 3, 4, and 7 each have $x_3 = 0.05$, and thus these four vertices define a face. The polyhedron contains six two-dimensional faces and the coordinates of the centroids of the six faces are listed as the design points 9 to 14 in Table 3.6. The overall centroid, whose coordinates are defined as the average of the coordinates of the eight vertices, is listed as point 15. The resulting factor space and the design-point designations are presented in Figure 3.11.

Table 3.6 Coordinates of the 15 Design Points for the Constrained Beverage Example

Design Point Designation	Type of Boundary	Mixture Component Proportions (Coordinates)				Combination of Vertices
		x_1	x_2	x_3	x_4	
1	Vertex	0.40	0.10	0.30	0.20	
2	Vertex	0.80	0.10	0.05	0.05	
3	Vertex	0.40	0.50	0.05	0.05	
4	Vertex	0.40	0.25	0.05	0.30	
5	Vertex	0.40	0.25	0.30	0.05	
6	Vertex	0.40	0.10	0.20	0.30	
7	Vertex	0.55	0.10	0.05	0.30	
8	Vertex	0.55	0.10	0.30	0.05	
9	Face centroid	0.58	0.24	0.05	0.13	2, 3, 4, 7
10	Face centroid	0.54	0.24	0.17	0.05	2, 3, 5, 8
11	Face centroid	0.40	0.24	0.18	0.18	1, 3, 4, 5, 6
12	Face centroid	0.54	0.10	0.18	0.18	1, 2, 6, 7, 8
13	Face centroid	0.45	0.15	0.10	0.30	4, 6, 7
14	Face centroid	0.45	0.15	0.30	0.10	1, 5, 8
15	Overall centroid	0.49	0.19	0.16	0.16	1, 2, ..., 8

Once the values of the response are collected at the design points, the second-degree model of the form in Eq. (3.31) can be fitted. The centroids of the faces, points 9 to 14, were obtained by McLean and Anderson because centroid points are useful for determining whether the surface is nonlinear. It is interesting that McLean and Anderson did not choose to use midedge points, which in general are more useful

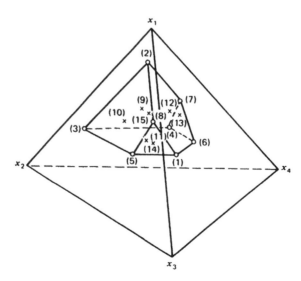

Figure 3.11 Constrained factor space inside the tetrahedron.

than the centroids of the faces for fitting cross-product terms such as $\beta_{ij}x_ix_j$ in the model (3.31).

The model in Eq. (3.31) contains 10 terms, and yet the number of vertices and centroids used is 15, five more than the number of coefficients to be estimated. We might ask ourselves whether some of the centroids and some of the vertices could have been removed from consideration prior to fitting the model without sacrificing the precision of the parameter estimates in terms of their variances, as well as the precision of the predictor $\hat{y}(\mathbf{x})$. Then again, by using all 15 points, one can obtain an estimate of the error variance or use the extra five degrees of freedom to test for the adequacy of the fitted model if an independent estimate of the error variance is available. Although we would probably use all 15 points in small examples like the previous one, the choice of which design points to use and not use may not always be straightforward. A method for choosing a subset of the points of an extreme-vertices design to use when fitting a first-degree model over the constrained region inside a simplex is presented in Snee and Marquardt (1974). In a later paper Snee (1975) discusses designs for fitting quadratic models in constrained mixture regions. We mention Snee's suggestions now.

3.9.1 Some Comments on the Design Strategy for Fitting the Scheffé Quadratic Model over a Constrained Region

An algorithm called CONAEV written by Gregory F. Piepel (1988), goes one step further than the McLean and Anderson algorithm for generating coordinates of the extreme vertices of a constrained region. The Piepel algorithm uses the coordinates of the extreme vertices to generate the midpoints of the edges and the centroids of the two-dimensional faces, three-dimensional flats and so on, up to the overall centroid of the region itself. Once the entire group of extreme vertices and centroids have been generated, the next course of action is to decide whether to use all of the points that are generated or to use only a subset of them. Earlier Snee and Marquardt (1974) had shed some light on the problem of selecting only a subset of the total number of extreme vertices for fitting a first-degree mixture model. They compared the performance of XVERT with that of another algorithm called CADEX. The CADEX algorithm was developed by Kennard and Stone (1969).

To obtain a design for fitting the Scheffé quadratic model, Snee (1975) pointed out that other authors (Farrell, Keifer, and Walbran, 1967) have shown that the support points of an optimum design on the hypercube consist of points belonging to each of three classes:

1. The vertices.
2. The edge centroids, or two-dimensional face centroids (if $q \geq 7$).
3. The c-dimensional centroids ($2 \leq c \leq q - 1$ if $3 \leq q \leq 6$, and $3 \leq c \leq q - 1$ if $q \leq 7$).

In this regard the centroid of a c-dimensional face is the average of all the vertices that lie on the same constraint plane formed by imposing $q - c - 1$ constraints. For example, an edge centroid is the average of the two vertices that lie on the same constraint plane generated by imposing $q - 2$ constraints. A constraint plane centroid is the average of all the vertices that lie on the same constraint plane. For a regular simplex, the constraint plane centroids are the q blends that consist of $q - 1$ components present in equal proportions, $1/(q - 1)$.

Using these principles, Snee arrived at the following design recommendations for fitting a quadratic model. The points of an efficient design [G-efficiency[†] $\geq$ 50% and/or a maximum prediction variance, var $\hat{y}(\mathbf{x})$, of $\leq 1.0\sigma^2$] consist of the following:

1. Select a subset of the extreme vertices, edge centroids, constraint plane centroids, and the overall centroid.
2. For $q = 3$, use the extreme vertices and the overall centroid. If the number, N_0, of extreme vertices is less than $q(q + 1)/2$, then add the centroids of the longest edges to the design.
3. For $q = 4$, use the extreme vertices, constraint planes centroids, the overall centroid, and the centroids of the longest edges.
4. For $q = 5$, select a subset of the extreme vertices, the constraint planes centroids, and the overall centroid. For $q \geq 6$, add the edge centroids to the list for $q = 5$.

In point 4, to select the subset, Snee suggests two algorithms: Wynn (1970) and the exchange algorithm of Wheeler (1972). Other algorithms that are described in the literature are DUPLEX (Snee, 1977) and DETMAX (Mitchell, 1974). Still another algorithm that adequately meets the challenge of design selection and for which the source code is available on magnetic tape from the author, William J. Welch, is ACED (Welch, 1982, 1983, 1984).

3.10 MULTICOMPONENT CONSTRAINTS

In addition to upper and lower bounds on the component proportions, $L_i \leq x_i \leq U_i$, $i = 1, 2, \ldots, q$, there often exists linear constraints (e.g., h in number) among the x_i of the form

$$C_j \leq A_{1j}x_1 + A_{2j}x_2 + \cdots + A_{qj}x_q \leq D_j, \qquad j = 1, 2, \ldots, h \qquad (3.44)$$

where the A_{ij} are scalar constants (some A_{ij} may be zero). For example, in the production of iron, Koons (1989) lists the following constraints that are present

---

[†] G-efficiency is defined as $(p/nd) \times 100\%$, where p is the number of terms in the model, n is the number of points in the design, and d is the maximum value of var$[\hat{y}(\mathbf{x})]/\sigma^2$ over all candidate points.

among the proportions of eight components, x_i, i = 1, 2, ..., 8:

1. Ores: $x_1 \leq 0.45$, $x_2 \leq 0.90$ and $x_1/(x_1 + x_2) \leq 1/3$ or, $0 \leq -x_1 + 0.5x_2$.
2. Reverts: $x_3 \leq 0.35$, $x_4 \leq 0.20$, $x_5 \leq 0.30$ and $x_3 + x_4 + x_5 \leq 0.35$.
3. Ore-revert: $0 \leq x_1 + x_2 - x_3 - x_4 - x_5$.
4. Total iron: $0.46 \leq 0.6x_1 + 0.5x_2 + 0.35x_3 + 0.2x_4 + 0.7x_5$. $\qquad$ (3.45)
5. Flux: $0.04 \leq x_6 \leq 0.08$, $0.06 \leq x_7 \leq 0.12$.
6. Fuel: $0.029 \leq x_8 \leq 0.072$ and $0.043 \leq 0.17x_3 + 0.85x_8 \leq 0.085$.

Stated in more precise terms, the constraints (3.45) are as follows:

1. Earthy hematite ore (x_1) cannot comprise more than one-third of the total ore present while specular hematite ore (x_2) cannot make up more than two-thirds of the total ore. With the constraint $x_1/(x_1 + x_2) \leq \frac{1}{3}$, the implied upper bounds for x_1 and x_2, respectively, are $x_1 \leq \frac{1}{3}$ and $x_2 \leq \frac{2}{3}$.
2. The total steelmaking revert consisting of flue dust (x_3) and/or BOF slag (x_4) and mill scale (x_5) cannot exceed 35% of the total blend.
3. The total steelmaking revert cannot exceed the amount of raw ore used.
4. All blends must contain at least 46% iron. The ores x_1 and x_2 are known to be 60% iron while flue dust (x_3), BOF slag (x_4), and mill scale (x_5) are known to have iron contents of 35%, 20%, and 70%, respectively.
5. The sum of the revert (x_3), which is known to be 17% fuel, plus 85% of pure fuel (x_8), which is known to be present in a percentage between 2.9% and 7.2%, must together comprise at least 4.3% of the total blend but cannot exceed 8.5% of the total blend.

The presence of the multicomponent constraints can change the shape (the number of extreme vertices, edges, faces, etc.) of the region that was defined previously by the placing only of lower and upper bounds on the x_i. To see this, let us work through a smaller three-component example, which is discussed in Snee (1979), and derive the resulting experimental region. The upper- and lower-bound constraints on x_1, x_2, and x_3 are

$$0.1 \leq x_1 \leq 0.5$$
$$0.1 \leq x_2 \leq 0.7 \qquad (3.46)$$
$$x_3 \leq 0.7$$

In addition there are the two multicomponent constraints

$$90 \leq 85x_1 + 90x_2 + 100x_3 \leq 95$$
$$0.4 \leq 0.7x_1 + x_3 \qquad (3.47)$$

The constraints (3.46) and (3.47) can be written in the form of the constraints (3.44) or as the modified constraints

$$0 \leq A_{0j} + \sum_{i=1}^{q} A_{ij}x_i, \qquad j = 1, 2, \ldots, \text{up to } 2h \qquad (3.48)$$

To illustrate, consider:

Number	(3.46) and (3.47)	(3.44) or (3.48)
1	$0.1 \leq x_1$	$0.1 \leq (1)x_1 + (0)x_2 + (0)x_3$
		or $0 \leq -0.1 + x_1$
2	$x_1 \leq 0.5$	$(1)x_1 + (0)x_2 + (0)x_2 \leq 0.5$
		or $0 \leq 0.5 - x_1$
3	$0.1 \leq x_2$	$0.1 \leq (0)x_1 + (1)x_2 + (0)x_3$
		or $0 \leq -0.1 + x_2$
4	$x_2 \leq 0.7$	$(0)x_1 + (1)x_2 + (0)x_3 \leq 0.7$
		or $0 \leq 0.7 - x_2$
5	$x_3 \leq 0.7$	$(0)\,x_1 + (0)x_2 + (1)x_3 \leq 0.7$
		or $0 \leq 0.7\text{--}x_3$
6	$90 \leq 85x_1 + 90x_2 + 100x_3$	$0.9 \leq 0.85x_1 + 0.90x_2 + x_3$
		or $0 \leq -0.9 + 0.85x_1 + 0.9x_2 + x_3$
7	$85x_1 + 90x_2 + 100x_3 \leq 95$	$0.85x_1 + 0.9x_2 + x_3 \leq 0.95$
		or $0 \leq 0.95\text{--}0.85x_1\text{--}0.9x_2\text{--}x_3$
8	$0.4 \leq 0.7x_1 + x_3$	$0.4 \leq 0.7x_1 + (0)x_2 + x_3$
		or $0 \leq -0.4 + 0.7\,x_1 + x_3$

The constrained region corresponding to the lower- and upper-bound constraints (3.46) only, or to the modified constraints numbered 1 to 5, is drawn, using solid lines, in Figure 3.12. This region has six vertices. The multicomponent constraints (3.47), or modified constraints numbered 6 to 8, are indicated by the dashed lines in Figure 3.12. The additional multicomponent constraints (3.47) pare off portions of the constrained region defined by the constraints (3.46), resulting in a smaller region than before but still containing six vertices.

An algorithm, called CONSIM (CONstrained SIMplex), was introduced by Snee (1979), for generating the extreme vertices of a constrained region when constraints of the form (3.48) are imposed on the mixture component proportions. We shall not work through the steps in the CONSIM algorithm, since a newer program, called CONVRT, appeared later in the literature (Piepel, 1988), and CONVRT performs the same operations as CONSIM. The program listings for CONVRT and CONAEV were written by Piepel and appeared in the April 1988 issue of the *Journal of Quality Technology*. Incidentally, for the set of constraints (3.45) listed in the iron-production example at the beginning of this section, the CONVRT program generated the coordinates of 184 extreme vertices.

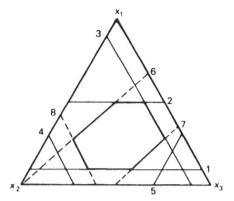

Figure 3.12 Doubly constrained region. Upper- and lower-bound constraints are indicated by the solid lines numbered 1 to 5, while the multicomponent constraints are indicated by the dashed lines numbered 6, 7, and 8.

3.11 SOME EXAMPLES OF DESIGNS FOR CONSTRAINED MIXTURE REGIONS: CONVRT AND CONAEV PROGRAMS

In this section are presented four examples of constrained mixture problems. The number of components in the first three examples is at most five, so that the coordinates of the extreme vertices of the constrained regions are listed. The fourth example is of an eight-component system and only the numbers of extreme vertices, edges, and faces of the constrained region are given. The purpose in presenting these examples is to illustrate different types of constrained mixture problems that arise in industrial settings as well as to offer examples that enable the users of these techniques an opportunity to compare the results they found using their software packages against the results obtained with the CONVRT and CONAEV programs of Piepel (1988), which generated the points listed in Tables 3.7, 3.8, and 3.9.

Piepel's CONVRT and CONAEV algorithms calculate the extreme vertices and centroids, respectively, of an experimental region that is defined by constraints of the general form

$$A_1 x_1 + A_2 x_2 + \cdots + A_q x_q + A_0 \geq 0$$

where $x_1, x_2, \ldots, x_q$ are the mixture component proportions and/or the settings of other nonmixture variables. The A_i, $i = 0, 1, 2, \ldots, q$, are constants. For example, the less-than-or-equal constraint $2x_1 - 3x_2 \leq 1$ can be rewritten as $-2x_1 + 3x_2 + 1 \geq 0$, where $A_1 = -2$, $A_2 = 3$, and $A_0 = 1$. The ratio constraint $2x_1/x_2 \geq 1$ can be written as $2x_1 - x_2 \geq 0$, where $A_1 = 2$, $A_2 = -1$, and $A_0 = 0$.

Source code listings of CONVRT and CONAEV are given in of Cornell (2002, app. 10A and 10B). The listings are the same as given in Piepel (1988) with a couple of minor corrections provided later by Piepel. More recently Piepel has developed the MIXSOFT software package for the design and analysis of mixture experiments. MIXSOFT contains updated versions of CONVRT and CONAEV (under different names) as well as many other capabilities. (For information on MIXSOFT, contact Dr. Gregory F. Piepel, 649 Cherrywood Loop, Richland, Washington 99352 or by email at MIXSOFT@aol.com.)

Example 1

A group of experiments is to be set up in four ingredients (P, E, F, and C), where the sum of the four individual amounts is fixed. Each ingredient is to be varied from the setting that is currently being used to see what kinds of changes in the response result. The current settings in grams and the ranges of the four ingredients are

	P	E	F	C
Current setting =	1.2	0.5	0.3	0.4
Range =	0.8–1.6	0.3–0.7	0.1–0.5	0.2–0.6

Expressed in terms of proportions of the fixed total amount, $P + E + F + C = 2.4$ g, the component proportions and the intervals of interest for the individual components are

$$0.3333 \le x_1 = \frac{P}{2.4} \le 0.6667, \quad 0.0417 \le x_3 = \frac{F}{2.4} \le 0.2083,$$

$$0.1250 \le x_2 = \frac{E}{2.4} \le 0.2917, \quad 0.0833 \le x_4 = \frac{C}{2.4} \le 0.2500 \tag{3.49}$$

The constrained region defined by Eq. (3.49) contains 12 vertices, 18 edges, and 8 faces as determined by Eqs. (3.40) and (3.41). The coordinates of the extreme vertices, midpoints of the edges, and centroids of the faces, as determined by CONVRT, are listed in Table 3.7. Since the candidate list of points numbers $39(= 12 + 18 + 8 + 1)$, to fit the quadratic model

$$y = \sum_{i=1}^{4} \beta_i x_i + \sum \sum_{i<j}^{4} \beta_{ij} x_i x_j + \varepsilon$$

we need only a subset of the extreme vertices and midpoints of the edges as previously mentioned in Section 3.9.2. Additional data collected from a subset of the face centroids along with the overall centroid could than be used for estimating the error variance or for testing the adequacy of the fitted quadratic model if an estimate of the error variance were available either from previous experience or from replicate observations.

Table 3.7 Candidate List of Design Points for Fitting the Four-Component Quadratic Model Over the Constrained Region Defined in (3.49)

Extreme Vertices				Midpoints of Edges				Face Centroids			
x_1	x_2	x_3	x_4	x_1	x_2	x_3	x_4	x_1	x_2	x_3	x_4
0.6667	0.1250	0.0417	0.1666	0.3333	0.2917	0.1666	0.2084	0.3333	0.2639	0.1805	0.2223
0.3333	0.2917	0.2083	0.1667	0.3333	0.2500	0.2083	0.2084	0.6667	0.1528	0.0695	0.1110
0.6667	0.1250	0.1250	0.0833	0.3333	0.2501	0.1666	0.2500	0.5834	0.1250	0.1250	0.1666
0.5834	0.1250	0.2083	0.0833	0.6667	0.1250	0.0833	0.1250	0.4166	0.2917	0.1250	0.1667
0.6667	0.2083	0.0417	0.0833	0.6667	0.1666	0.0417	0.1250	0.5833	0.2083	0.0417	0.1667
0.5833	0.2917	0.0417	0.0833	0.6667	0.1666	0.0834	0.0833	0.4167	0.2084	0.2083	0.1666
0.4167	0.2917	0.2083	0.0833	0.6250	0.1250	0.0417	0.2083	0.5834	0.2083	0.1250	0.0833
0.5833	0.1250	0.0417	0.2500	0.5001	0.1250	0.2083	0.1666	0.4166	0.2084	0.1250	0.2500
0.3333	0.2917	0.1250	0.2500	0.6251	0.1250	0.1666	0.0833	0.5000	0.2084	0.1250	0.1666[a]
0.4166	0.2917	0.0417	0.2500	0.5000	0.1250	0.1250	0.2500				
0.3333	0.2084	0.2083	0.2500	0.4999	0.2917	0.0417	0.1667				
0.4167	0.1250	0.2083	0.2500	0.3750	0.2917	0.2083	0.1250				
				0.5000	0.2917	0.1250	0.0833				
				0.3749	0.2917	0.0834	0.2500				
				0.6250	0.2500	0.0417	0.0833				
				0.4999	0.2084	0.0417	0.2500				
				0.5000	0.2084	0.2083	0.0833				
				0.3750	0.1667	0.2083	0.2500				

[a]Overall centroid obtained by averaging the 12 extreme vertices.

Table 3.8 Coordinates of the Extreme Vertices, Midpoints of the Edges, and Centroids of the Faces for the Constrained Regions of Example 2. A, Region Defined by Eq. (3.50); B, Region Defined by (3.50) and $0.25 \leq x_3 + x_4 \leq 0.45$; C, Region Defined by Eqs. (3.50) and (3.51)

	Extreme Vertices					Midpoints of Edges					Face Centroids			
	Point	x_1	x_2	x_3	x_4	Point	x_1	x_2	x_3	x_4	Point	x_1	x_2	x_3	x_4
A	1	0.650	0.100	0.100	0.150	9	0.200	0.450	0.100	0.250	21	0.200	0.400	0.150	0.250
	2	0.200	0.550	0.100	0.150	10	0.200	0.350	0.200	0.250	22	0.500	0.100	0.150	0.250
	3	0.550	0.100	0.200	0.150	11	0.200	0.500	0.150	0.150	23	0.375	0.275	0.100	0.250
	4	0.200	0.450	0.200	0.150	12	0.200	0.300	0.150	0.350	24	0.325	0.325	0.200	0.250
	5	0.450	0.100	0.100	0.350	13	0.550	0.100	0.100	0.250	25	0.400	0.300	0.150	0.150
	6	0.200	0.350	0.100	0.350	14	0.450	0.100	0.200	0.250	26	0.300	0.200	0.150	0.350
	7	0.350	0.100	0.200	0.350	15	0.600	0.100	0.150	0.150					
	8	0.200	0.250	0.200	0.350	16	0.400	0.100	0.150	0.350		0.350^a	0.250	0.150	0.250
						17	0.425	0.325	0.100	0.150					
						18	0.325	0.225	0.100	0.350					
						19	0.375	0.275	0.200	0.150					
						20	0.275	0.175	0.200	0.350					

B

Vertices 1–6 in panel A

0.450	0.100	0.200	0.250
0.200	0.350	0.200	0.250

Points 9, 11, 13, 14, 17–19 in A

27	0.200	0.400	0.200	0.200
28	0.200	0.350	0.150	0.300
29	0.500	0.100	0.200	0.200
30	0.450	0.100	0.150	0.300
31	0.325	0.225	0.200	0.250

Point 23, 25 in A

32	0.200	0.425	0.150	0.225
33	0.525	0.100	0.150	0.225
34	0.350	0.250	0.200	0.200
35	0.325	0.225	0.150	0.300
	0.3625^a	0.2625	0.150	0.225

C

Vertices 3–6 in panel A

0.600	0.100	0.150	0.150
0.600	0.100	0.100	0.200
0.200	0.500	0.150	0.150
0.200	0.500	0.100	0.200
0.450	0.100	0.200	0.250
0.200	0.350	0.200	0.250

Points 19, 27–30

0.200	0.425	0.100	0.275
0.200	0.475	0.175	0.150
0.200	0.500	0.125	0.175
0.525	0.100	0.100	0.275
0.575	0.100	0.175	0.150
0.600	0.100	0.125	0.175
0.400	0.300	0.100	0.200
0.325	0.225	0.100	0.350
0.325	0.225	0.200	0.250
0.400	0.300	0.150	0.150

Points 34, 35

0.200	0.430	0.150	0.220
0.530	0.100	0.150	0.220
0.3625	0.2625	0.100	0.275
0.3875	0.2875	0.175	0.150
0.400	0.300	0.125	0.175
0.365^a	0.265	0.150	0.220

^{a}Overall centroid coordinates.

Example 2

Four components (A, G, M, and N) were restricted by lower and upper bounds of the form, where $A = x_1$, $G = x_2$, $M = x_3$, and $N = x_4$:

$$0.20 \leq x_1 \leq 0.65, \quad 0.10 \leq x_2 \leq 0.55,$$
$$0.10 \leq x_3 \leq 0.20, \quad 0.15 \leq x_4 \leq 0.35 \tag{3.50}$$

In addition, two multicomponent constraints involving components 1 and 3, and 3 and 4 were specified. These constraints are

$$0.25 \leq x_1 + x_3 \leq 0.70, \quad 0.25 \leq x_3 + x_4 \leq 0.45 \tag{3.51}$$

The constrained region defined only by the constraints in Eq. (3.50) contains 8 extreme vertices, 12 edges, and 6 faces. When the two-component constraint $0.25 \leq x_3 + x_4 \leq 0.45$ is imposed, the shape of the region does not change. The effect of imposing the additional constraint is to alter the coordinates of two of the extreme vertices, the coordinates of the midpoints of five of the edges, the coordinates of the centroids of four of the faces, and of course the coordinates of the overall centroid. The resulting designs are listed in Table 3.8.

When both of the two-component constraints in Eq. (3.51) are imposed, the resulting constrained region contains 10 extreme vertices 15 edges, and 7 faces.

Example 3

In an experiment designed to study the light stability (resistance to fading) of a coating material for photographic films, the relative proportions of two couplers (C_1 and C_2), one coupler solvent (CS), and two stabilizers (S_1 and S_2) are to be varied. The individual component proportions and the combined proportions, respectively, are

$$0.05 \leq C_1 \leq 0.50, \quad 0.15 \leq C_2 \leq 0.20, \quad \text{and} \quad 0.20 \leq C_1 + C_2 \leq 0.65$$
$$0.20 \leq CS \leq 0.40, \quad 0.10 \leq S_1 \leq 0.30, \tag{3.52}$$
$$0.05 \leq S_2 \leq 0.35, \quad \text{and} \quad 0.15 \leq S_1 + S_2 \leq 0.50$$

The presence of the nonzero lower bounds for all five components implies that all blends must contain both couplers, the coupler solvent, and both stabilizers. Further discussion on the blending of groups of components is contained in Sections 3.12 and 3.13.

The constraints (3.52) were put in to the CONVRT and CONAEV algorithms in order to generate the candidate list of extreme vertices, midpoints of the edges, and centroids of the two-dimensional faces of the constrained region. The constrained region contains 23 extreme vertices, 47 edges, 34 two-dimensional faces, and 10

Table 3.9 Listing of the Extreme Vertices of the Constrained Region of Example 3

Vertex	Component Proportions				
	C_1	C_2	CS	S_1	S_2
1	0.50	0.15	0.20	0.10	0.05
2	0.05	0.15	0.40	0.10	0.30
3	0.05	0.20	0.40	0.10	0.25
4	0.05	0.15	0.40	0.30	0.05
5	0.05	0.20	0.40	0.30	0.05
6	0.45	0.20	0.20	0.10	0.05
7	0.30	0.15	0.40	0.10	0.05
8	0.25	0.20	0.40	0.10	0.05
9	0.30	0.15	0.20	0.30	0.05
10	0.25	0.20	0.20	0.30	0.05
11	0.10	0.15	0.40	0.30	0.05
12	0.20	0.15	0.20	0.10	0.35
13	0.05	0.15	0.35	0.10	0.35
14	0.05	0.20	0.30	0.10	0.35
15	0.15	0.20	0.20	0.10	0.35
16	0.05	0.15	0.30	0.30	0.20
17	0.15	0.15	0.20	0.30	0.20
18	0.05	0.20	0.25	0.30	0.20
19	0.10	0.20	0.20	0.30	0.20
20	0.15	0.15	0.20	0.15	0.35
21	0.05	0.15	0.30	0.15	0.35
22	0.05	0.20	0.25	0.15	0.35
23	0.10	0.20	0.20	0.15	0.35

three-dimensional constraint planes. Table 3.9 contains a listing of the 23 extreme vertices.

Example 3A

As a follow-up to the generation of the design in Example 3, it was decided to include a second coupler solvent and another stabilizer. This strategy resulted in altering the constraints on the individual components as well as altering the constraints on the combined component proportions. The new constraints were

$$0.05 \le C_1 \le 0.30, \quad 0.10 \le C_2 \le 0.40, \quad \text{and} \quad 0.15 \le C_1 + C_2 \le 0.60$$

$$0 \le CS_1 \le 0.20, \quad 0 \le CS_2 \le 0.25, \quad \text{and} \quad 0.05 \le CS_1 + CS_2 \le 0.40$$

$$0 \le S_1 \le 0.30, \quad 0.05 \le S_2 \le 0.20, \quad 0 \le S_3 \le 0.25,$$

$$\text{and} \quad 0.05 \le S_1 + S_2 + S_3 \le 0.45 \tag{3.53}$$

For the region defined by the constraints (3.53), the CONVRT and CONAEV algorithms generated 172 extreme vertices, the midpoints of 567 edges, and the centroids of 685 two-dimensional faces.

Example 4

In producing vinyl for automobile seat covers, blends may consist of up to eight components. The following represents a listing of eight components with their respective lower- and upper-bound constraints,

$$0.50 \leq P_1 \leq 0.748, \quad 0.20 \leq P_2 \leq 0.40, \quad 0.015 \leq AP \leq 0.045,$$
$$0.007 \leq CR \leq 0.021, \quad 0.002 \leq Cat\ 1 \leq 0.006, \quad 0.001 \leq Cat\ 2 \leq 0.002,$$
$$0.008 \leq Cat\ 3 \leq 0.012, \quad 0.019 \leq Wa \leq 0.025$$

When the eight components and their constraints were input to the CONVRT and CONAEV algorithms, the resulting region was found to possess 156 extreme vertices, 555 edges, and 833 two-dimensional faces. These numbers of boundaries agree with the results obtained using Eqs. (3.40) and (3.41).

We next discuss mixture experiments in which the components are categorized. The categories are classified as major components, and each category consists of some number of member components that are referred to as minor components. The designs used for studying the blending characteristics of all of the components are known as multiple-lattice designs.

3.12 MULTIPLE LATTICES FOR MAJOR AND MINOR COMPONENT CLASSIFICATIONS

Some mixture experiments involve two or more distinct classes or categories of components. For example, we might have a major category of chemical salt types to be blended, where each salt type consists of one or more subset salts that differ only slightly from one another. The salt types are called *major* components and the subset salts are called *minor* components. Another example, taken from Smith and Cornell (1993), is shown in Figure 3.13 and displays the makeup of a photographic coating where the emulsion layer consists of an oil phase that has been dispensed in a solution of water, gelatin, and silver halide. The oil phase is comprised of five components—a dye precursor (called a coupler), two coupler solvents (SOLV1, SOLV2), and two stabilizers (STAB1, STAB2). The five components are classified as belonging to three major components—(coupler, coupler solvent, and stabilizer)—having one, two, and two minor components, respectively. In practice, only oil-phase blends consisting of the coupler with one or both coupler solvents and one or both stabilizers are of interest.

One of the measured responses of interest is the color stability to light of the coated photograph. Color stability to light is known to be influenced by the relative

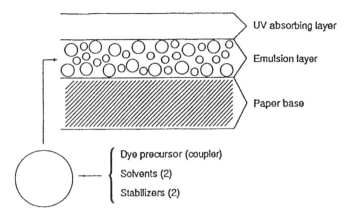

Figure 3.13 Single-layer coating of a photographic paper. The emulsion layer contains the mixture components. Reprinted with permission from *Technometrics*. Copyright 1993 by the American Statistical Association. All rights reserved.

proportions of coupler, coupler solvent, and stabilizer in the oil phase. Varying the proportions of the three major components and modeling the changes in the recorded stability measurement is what is called by Cornell and Ramsey (1998) as measuring the intercategory blending of the major components. Furthermore, for fixed proportions of coupler, coupler solvent, and stabilizer, color stability is expected to be influenced also by the relative proportions of the SOLV1 and SOLV2 components comprising the solvent as well as the relative proportions of STAB1 and STAB2 making up the stabilizer. Measuring the changes in color stability while varying only the relative proportions of SOLV1 and SOLV2 as well as the relative proportions of STAB1 and STAB2 is likewise called measuring the intracategory blending among the minor components within the coupler solvent and stabilizer major components. Thus, to study the behavior of the color stability values across blends of the five components, we need to model the inter- and intracategory blending of the major and minor components. We will designate the major components as M-components and the minor components as m-components.

3.12.1 Fixed Values for the Major Component Proportions

We confine our discussion here to only M-components consisting of at least two m-components, where there are p M-components present in the mixture system under study. Let c_i be the proportion of the total mixture contributed by the ith M-component,

$$0 < c_i < 1 \quad \text{for } i = 1, 2, \ldots, p \tag{3.54}$$

All of the M-components are blended together so that

$$c_1 + c_2 + \cdots + c_p = 1 \tag{3.55}$$

Next let $n_i \geq 2$ be the number of m-components belonging to M-component i so that $\sum_{i=1}^{p} n_i = q$. Each M-component consisting of at least two m-components is itself a mixture so that the number of m-components in each blend of the p M-components varies from p to q.

The constraints on the M- and m-component proportions are as follows. Let us denote by X_{ij} the proportion of the jth m-component belonging to the ith M-component. Then

$$0 \leq X_{ij} \leq c_i \qquad \text{for } i = 1, 2, \ldots, p; \qquad j = 1, 2, \ldots, n_i$$

$$\sum_{j=1}^{n_i} X_{ij} = c_i, \qquad i = 1, 2, \ldots, p \tag{3.56}$$

and

$$\sum_{i=1}^{p} \sum_{j=1}^{n_i} X_{ij} = 1 \tag{3.57}$$

For fixed values of $c_1, c_2, \ldots, c_p$, the equality constraint Eq. (3.56) reduces the dimensionality of the factor space of the q components from $q - 1$ to $\sum_{i=1}^{p} (n_i - 1) = q - p$. Furthermore, because each M-component consisting of $n_i \geq 2$ m-components is a mixture of n_i m-components, and we scale the m-component proportions by defining $x_{ij} = X_{ij}/c_i$, Eqs. (3.56) and (3.57) become

$$\sum_{j=1}^{n_i} x_{ij} = 1 \tag{3.58}$$

and

$$\sum_{i=1}^{p} \sum_{j=1}^{n_i} x_{ij} = p \tag{3.59}$$

To explore the factor space of feasible mixtures of all q components, a multiple-lattice configuration was proposed by Lambrakis (1968, 1969). The configuration is constructed by defining the points of the multiple-lattice as combinations of the points of the separate simplex-lattices associated with the individual M-components. To see this, first let us consider the simple case where $p = 2$ and let the respective M-component proportions be equal, that is, $c_1 = \frac{1}{2}$ and $c_2 = \frac{1}{2}$. If we fit a Scheffé-type model of degree m_1 over the simplex-lattice associated with M-component 1, and a Scheffé-type model of degree m_2 over the simplex-lattice associated with M-component 2, then the proportions x_{ij} in each M-component will take the values

$$x_{ij} = 0, \frac{1}{m_i}, \frac{2}{m_i}, \ldots, 1, \qquad i = 1, 2$$

As an example, if $m_1 = 2$ and $m_2 = 3$, then

$$x_{1j} = 0, \tfrac{1}{2}, 1, \qquad j = 1, 2$$
$$x_{2j} = 0, \tfrac{1}{3}, \tfrac{2}{3}, 1, \qquad j = 1, 2, 3$$

Notice that the m_i $i = 1, 2, \ldots, p$, for the individual M-components do not have to be equal; in fact they probably will not be. So we should consider fitting polynomials of different degrees in the m-components associated with the separate M-components.

The point arrangement of the multiple-lattice configuration is constructed by crossing the design points from each of the individual M-component lattices. The dimensionality of the multiple-lattice is $q - p$. For example, let us consider the two M-components case described above, and for simplicity let $n_1 = 2$ and $n_2 = 2$. As before, suppose that we elect to fit a $m_1 = 2$ or second-degree Scheffé model over the lattice of M-component 1, and this we can do with a $\{2, 2\}$ lattice. Suppose that we then elect to fit a $m_2 = 3$ or third-degree Scheffé model to the points of $\{2, 3\}$ lattice of M-component 2. The points of the multiple-lattice will be the result of crossing each of the three points of the $\{n_1, m_1\} = \{2, 2\}$ simplex-lattice with each of the four points of the $\{n_2, m_2\} = \{2, 3\}$ simplex-lattice to produce the 12 points of a $\{n_1, n_2; m_1, m_2\} = \{2, 2; 2, 3\}$ multiple- or double-lattice in this case. And since each $\{2, m_i\}$ simplex-lattice is of dimensionality 1, the $\{2, 2; 2, 3\}$ double-lattice will be of dimensionality $1 + 1 = 2$. The number of points in the multiple-lattice of p M-components will be

$$\prod_{i=1}^{p} \binom{n_i + m_i - 1}{m_i} = \prod_{i=1}^{p} \frac{(n_i + m_i - 1)!}{m_i!(n_i - 1)!}$$

To illustrate the construction of a double-lattice, let us define the proportions of the individual m-components from M-component 1 to be X_{11} and X_{12} and, similarly, the m-components in M-component 2 to be X_{21} and X_{22}. The possible mixture blends formed by combining the m-components from each of the two M-components are

$$
\begin{aligned}
&X_{1j}X_{2j} \\
&X_{11}X_{12}X_{2j} \\
&X_{1j}X_{21}X_{22} \\
&X_{11}X_{12}X_{21}X_{22}
\end{aligned} \qquad j = 1, 2
$$

where $X_{1j}X_{2j}$ represents a blend consisting of one m-component from each M-component.

The individual m-component proportions of Eq. (3.56) that are present in blends are determined by recalling that $X_{11} + X_{12} = \tfrac{1}{2}$ and $X_{21} + X_{22} = \tfrac{1}{2}$. Therefore the mixture $X_{1j}X_{2j}$, which represents a single m-component from each M-component, requires that each m-component proportion X_{ij} ($i = 1,2$) be equal to $\tfrac{1}{2}$. Each of the four combinations $X_{11}X_{21}$, $X_{12}X_{21}$, $X_{11}X_{22}$, and $X_{12}X_{22}$ thus consists of a single m-component from each M-component and each m-component makes up one-half

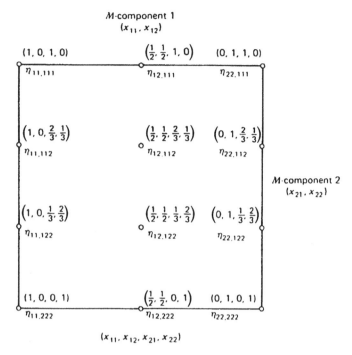

Figure 3.14 Minor component proportions and the expected responses at the points of the $\{2, 2; 2, 3\}$ double-lattice.

the mixture. The following combinations comprise the remainder of the $\{2, 2; 2, 3\}$ double-lattice:

$$X_{11}X_{12}X_{2j} : X_{11} = \tfrac{1}{4}, X_{12} = \tfrac{1}{4}, X_{2j} = \tfrac{1}{2}, \quad j = 1, 2$$
$$X_{1j}X_{21}X_{22} : X_{1j} = \tfrac{1}{2}, X_{21} = \tfrac{1}{6}, X_{22} = \tfrac{2}{6}, \quad j = 1, 2$$
$$X_{1j} = \tfrac{1}{2}, X_{21} = \tfrac{2}{6}, X_{22} = \tfrac{1}{6}, \quad j = 1, 2$$
$$X_{1j}X_{21}X_{22} : X_{11} = X_{12} = \tfrac{1}{4}, X_{21} = \tfrac{1}{6}, X_{22} = \tfrac{2}{6}$$
$$X_{11} = X_{12} = \tfrac{1}{4}, X_{21} = \tfrac{2}{6}, X_{22} = \tfrac{1}{6}$$

The coordinates of the 12 mixture compositions or blends are displayed in the form of the $\{2,2; 2,3\}$ double-lattice in Figure 3.14.

A regression function that can be used to model the response over the multiple-lattice is formed by multiplying the terms of the individual models that could be fitted separately to the points of the simplex-lattices of each of the M-components. In other words, the regression function that would be fitted to the points of the double-lattice is formed by multiplying the terms in each of the two polynomials that could be used over the two single-lattices. For example, let us refer to our previous two M-component example where with M-component 1 a $\{2, 2\}$ lattice was set up to support the second-degree polynomial

$$\eta_{s_1} = d_1 x_{11} + d_2 x_{12} + d_{12} x_1 x_2, \quad S_1 = 11, \ 12, \ 22$$

whereas to the points of the $\{2, 3\}$ lattice for M-component 2, we consider fitting the cubic polynomial

$$\eta_{s_2} = e_1 x_{21} + e_2 x_{22} + e_{12} x_{21} x_{22} + e_{112} x_{21} x_{22} (x_{21} - x_{22}), \quad S_2 = 111, \ 112, \ 122, \ 222$$

The meanings of the subscripts S_1 and S_2 are to be provided shortly. Setting up cross-products between each of the terms in η_{s_1} and each of the terms in η_{s_2}, that is, $\eta_{s_1} \eta_{s_2} = (\eta_{s_1})(\eta_{s_2})$, we have the resulting 12-term regression function that is used to represent the response over the $\{2, 2; 2, 3\}$ double-lattice:

$$\eta_{s_1} \eta_{s_2} = \sum_{j=1}^{2} \sum_{l=1}^{2} \gamma_{jl} x_{1j} x_{2l} + \sum_{j=1}^{2} \gamma_{j12} x_{1j} x_{21} x_{22}$$

$$+ \sum_{j=1}^{2} \gamma_{j112} x_{1j} x_{21} x_{22} (x_{21} - x_{22}) + \sum_{l=1}^{2} \gamma_{12l} x_{11} x_{12} x_{2l}$$

$$+ \gamma_{1212} x_{11} x_{12} x_{21} x_{22} + \gamma_{12112} x_{11} x_{12} x_{21} x_{22} (x_{21} - x_{22}) \qquad (3.60)$$

where the subscripts S_1 and S_2 are defined in the next paragraph. The polynomial equation (3.60) of degree $m_1 + m_2 = 2 + 3 = 5$ contains the same number of terms as there are points on the double-lattice of Figure 3.14.

The estimates of the parameters in the double-lattice model in Eq. (3.60) are calculated using simple linear combinations of the observations collected at the points of the double-lattice. To see this, let us refer to Figure 3.14, where each response is denoted by suffixed letters η_{s_1, s_2}. Each of the suffixes S_1 and S_2 consists of a sequence of numbers. In $S_j, j = 1, 2$, the number l ($l = 1$ or 2) appears $m_j x_{jl}$ times, indicating that the lth m-component is present with proportions x_{jl}. For example, the response $\eta_{11,111}$ is from the blend $X_{11} = \frac{1}{2}$, $X_{21} = \frac{1}{2}$, which in the m-components is $x_{11} = 1$, $x_{21} = 1$, because $x_{jl} = x_{jl}/c_j$ and $c_j = \frac{1}{2}$, and therefore we have the suffices $S_1 = 2x_{11}$ $= 11$ and $S_2 = 111 = 3x_{21}$; the response $\eta_{12,111}$ belongs to blend $X_{11} = \frac{1}{4}, X_{12} = \frac{1}{4}$, $X_{21} = \frac{1}{2}$, which in the m-components is $X_{11} = \frac{1}{2}$, $x_{12} = \frac{1}{2}$, $x_{21} = 1$; the response $\eta_{12,112}$ corresponds to the blend $X_{11} = \frac{1}{4}, X_{12} = \frac{1}{4}, X_{21} = \frac{2}{6}, X_{22} = \frac{1}{6}$, which in the m-components is $x_{11} = \frac{1}{2}$, $x_{12} = \frac{1}{2}$, $x_{21} = \frac{2}{3}$, $x_{22} = \frac{1}{3}$, and so on.

If the averages $\bar{y}_{s_1 s_2}$ of the responses at the lattice points are substituted along with the corresponding proportions x_{1j} into Eq. (3.60), we have 12 equations that when solved simultaneously produce the formulas for calculating the estimates of the 12 unknown parameters. The coefficients of the averages that are used for estimating the parameters in Eq. (3.60) are listed in Table 3.10.

To set up an equation for predicting the response over the double-lattice, the estimates $g_{ij}, \ldots, g_{12112}$ are substituted into Eq. (3.60) or one may use the average

Table 3.10 Coefficients of the Average Response Values at the Lattice Points Used to Estimate the Parameters in the Double-Scheffé Model

g_{11}	g_{12}	g_{21}	g_{22}	g_{112}	g_{212}	g_{121}	g_{122}	g_{1112}	g_{2112}	g_{1212}	g_{12112}	Mean Response at $(x_{11},x_{12};x_{21},x_{22})$
1				$-\frac{9}{4}$			-2	$-\frac{9}{4}$		$\frac{9}{2}$	$\frac{9}{2}$	$\bar{y}_{11,111}$ $(1,0:1,0)$
				$\frac{9}{4}$				$\frac{27}{4}$		$-\frac{9}{2}$	$\frac{27}{2}$	$\bar{y}_{11,112}$ $(1,0:\frac{2}{3},\frac{1}{3})$
				$\frac{9}{4}$				$-\frac{27}{4}$		$-\frac{9}{2}$	$-\frac{9}{2}$	$\bar{y}_{11,122}$ $(1,0:\frac{1}{3},\frac{2}{3})$
	1			$-\frac{9}{4}$			-2	$\frac{9}{4}$		$\frac{9}{2}$	$-\frac{27}{2}$	$\bar{y}_{11,222}$ $(1,0:0,1)$
						4				-9	-9	$\bar{y}_{12,111}$ $(\frac{1}{2},\frac{1}{2}:1,0)$
										9	-27	$\bar{y}_{12,112}$ $(\frac{1}{2},\frac{1}{2}:\frac{2}{3},\frac{1}{3})$
										-9	9	$\bar{y}_{12,122}$ $(\frac{1}{2},\frac{1}{2}:\frac{1}{3},\frac{2}{3})$
							4			-9	27	$\bar{y}_{12,222}$ $(\frac{1}{2},\frac{1}{2}:0,1)$
		1			$-\frac{9}{4}$		-2		$-\frac{9}{4}$	$\frac{9}{2}$	$\frac{9}{2}$	$\bar{y}_{22,111}$ $(0,1:1,0)$
					$\frac{9}{4}$				$\frac{27}{4}$	$-\frac{9}{2}$	$\frac{27}{2}$	$\bar{y}_{22,112}$ $(0,1:\frac{2}{3},\frac{1}{3})$
					$\frac{9}{4}$				$-\frac{27}{4}$	$-\frac{9}{2}$	$-\frac{9}{2}$	$\bar{y}_{22,122}$ $(0,1:\frac{1}{3},\frac{2}{3})$
			1		$-\frac{9}{4}$		-2		$\frac{9}{4}$	$\frac{9}{2}$	$-\frac{27}{2}$	$\bar{y}_{22,222}$ $(0,1:0,1)$

responses directly. In this latter case

$$\hat{y}(\mathbf{x}) = \sum_{j=1}^{2}\sum_{l=1}^{2} C_{jj,lll}\,\bar{y}_{jj,lll} + \sum_{j=1}^{2} C_{jj,112}(\bar{y}_{jj,112} + \bar{y}_{jj,122}) + \sum_{l=1}^{2} C_{12,lll}\,\bar{y}_{12,lll}$$

$$+ C_{12,112}\bar{y}_{12,112} + C_{12,122}\bar{y}_{12,122} \tag{3.61}$$

where the coefficients of the $\bar{y}_{s_1,s_2}$ are

$$
\begin{aligned}
C_{11,111} &= x_{11}x_{21}(1 - 2x_{12})[1 - \tfrac{9}{4}x_{22}(1 + x_{12} - x_{22})] \\
&= [x_{11}(2x_{11} - 1)][\tfrac{1}{2}x_{21}(3x_{21} - 2)(3x_{21} - 1)] \\
&= a_{11}b_{111} \\
C_{jj,lll} &= [x_{ij}(2x_{1j} - 1)][\tfrac{1}{2}x_{2l}(3x_{2l} - 2)(3x_{2l} - 1)] = a_{jj}b_{lll}, \quad j, l = 1, 2 \\
C_{jj,112} &= \tfrac{9}{2}x_{1j}x_{21}x_{22}(2x_{1j} - 1)(3x_{21} - 1) = a_{jj}b_{112}, \quad j = 1, 2 \\
C_{jj,122} &= \tfrac{9}{2}x_{1j}x_{21}x_{22}(2x_{1j} - 1)(3x_{22} - 1) = a_{jj}b_{122}, \quad j = 1, 2 \\
C_{12,lll} &= 2x_{11}x_{12}x_{21}(3x_{2i} - 2)(3x_{2l} - 1) = a_{12}b_{lll}, \quad l = 1, 2 \\
C_{12,112} &= 4x_{11}x_{12}[\tfrac{9}{2}x_{21}x_{22}(3x_{21} - 1)] = a_{12}b_{112} \\
C_{12,122} &= 4x_{11}x_{12}[\tfrac{9}{2}x_{21}x_{22}(3x_{22} - 1)] = a_{12}b_{122}
\end{aligned}
\tag{3.62}
$$

The constants C_{s_1,s_2} are products of the orthogonal polynomials a_{s_1} and b_{s_2} as shown in (3.62) and are discussed in Lambrakis (1968).

Rather than continue the discussion with using the $C_{S1,S2}$ in (3.61), there is an easier approach that was used by Cornell and Ramsey (1998) to estimate the coefficients in (3.60). This approach is discussed in the following section.

3.12.2 Allowing the Major Component Proportions to Vary: Mixtures of Mixtures

Suppose that the c_i, $i = 1, 2, \ldots, p$, which are the proportions of the total mixture contributed by the individual M-components, are allowed to vary as well. As an example, shown in Figure 3.15 are three planes corresponding to the values $(c_1, c_2) = (0.75, 0.25)$, $(0.50, 0.50)$, and $(0.25, 0.75)$ for the simple case of $p = 2$ major components where each M-component consists of two m-components. The three planes represent three blends of the two M-components, and this example was taken from Cornell and Ramsey (1998). The two M-components are two types of resins, R_1 and R_2, to be used to form the base resin of a photoresist formulation and each resin type consists of two minor resins possessing different dissolution rates (slow and fast). In Figure 3.15 the R_1 minor resins are denoted by X_{11} and X_{12} and the R_2 minor resins by X_{21} and X_{22}, respectively.

When the M-component proportions or c_i's, $i = 1, 2, \ldots, p$, are allowed to vary, it becomes possible to measure the blending properties of the major components as well as the blending properties of the minor components within each major component. It is also possible to measure the impact, if any, that the blending among the major components has on the blending properties of the minor components. Stated another way, to support the fitting of data collected from the three planes in Figure 3.15 a

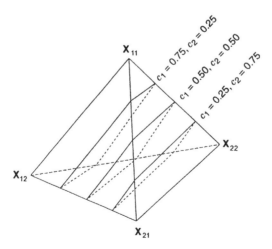

Figure 3.15 Three planes $X_{11} + X_{12} = c_1$, $X_{21} + X_{22} = c_2$ corresponding to the values $(c_1, c_2) = (0.75, 0.25)$, $(0.50, 0.50)$, and $(0.25, 0.75)$.

combined model in the major and minor component proportions can be set up to measure the following:

1. Blending among the p major components, or the intercategory blending of the M-components.
2. Blending among the $n_i \geq 2$ minor components, or the intracategory blending, among the m-components within each M-component.
3. Blending between the major and minor components, or the impact of the blending in (1) on the blending in (2).

For the case where $p = 2$ and $n_1 = n_2 = 2$, Cornell and Ramsey (1998) proposed modeling the above blending properties by modeling the shape characteristics of the unknown response surfaces above each of the three planes in Figure 3.15. The shapes of the response surfaces directly above the three planes were said to represent the intraresin blending properties (linear and nonlinear) of the minor resins of each resin type while the differences in the shapes of the surfaces above the three planes were assumed to be influenced by the resin composition and therefore the interresin blending properties of resin types R_1 and R_2.

To better understand the modeling of intra- and intercategory blending of m-components and M-components, respectively, let us recall the experimental scenario involving the two resin types and two minor resins of each type presented by Cornell and Ramsey (1998). Over each of the three planes in Figure 3.15 the response studied was the dissolution rate of blends consisting of at most four components. For the purposes of constructing designs as well as when interpreting the model coefficient estimates, it is convenient with this four-component case to rescale the X_{ij} in Eqs. (3.56) and (3.57) once again only this time using

$$x_1 = \frac{X_{11}}{c_1}, \quad x_2 = \frac{X_{12}}{c_1}, \qquad x_3 = \frac{X_{21}}{c_2}, \quad x_4 = \frac{X_{22}}{c_2} \tag{3.63}$$

so that $x_1 + x_2 = 1$ and $x_3 + x_4 = 1$. Equation (3.63) allows us to write each resin type as a mixture of the slow and fast dissolving minor resins of that type.

Selecting the blends of the four minor resins, or the points on each plane of Figure 3.15 is the next order of business. At the four corners of each plane in Figure 3.15 are the blends of the scaled minor components whose coordinates are $(x_1, x_2, x_3, x_4) = (1,0,1,0)$, $(1,0,0,1)$, $(0,1,1,0)$, and $(0,1,0,1)$. These blends consist of only a single minor resin of each type. Since one of the objectives of the experiment is to study the within or intraresin blending of the minor resins, and since the blending could be nonlinear, additional blends with $x_1 > 0$, $x_2 > 0$, and with $x_3 > 0$, $x_4 > 0$, are needed on each plane.

The response of interest with the resin experiment is the dissolution rate (DR) of the mixture formulation. The higher the value of DR, the faster the formulation dissolves. Within each resin type, mixing the slow and fast dissolving minor resins is expected to affect the dissolution rate of the formulation in a nonlinear manner. What this means is the DR value of the blend of x_1 with x_2, in the presence of x_3, say, could be different (higher or lower) from the weighted average of the DR values of the two binary blends

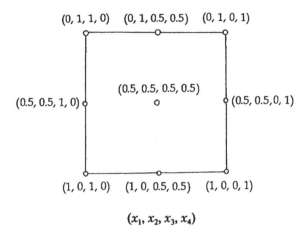

$$(x_1, x_2, x_3, x_4)$$

Figure 3.16 Minor resin (x_1, x_2, x_3, x_4) coordinates at the points of the double-lattice.

of x_1 with x_3 and x_2 with x_3. When this happens so that nonlinear blending (NLB) is said to exist with the minor resins x_1 and x_2, or x_3 with x_4, and the DR of the blend is higher (lower) than the linear blending value, this is called positive (negative) NLB of x_1 and x_2 in the presence of x_3. To model this intraresin nonlinear blending of DR, a second-degree Scheffé model is chosen for each resin type of the form

$$\text{DR}_{R_1} = \alpha_1 x_1 + \alpha_2 x_2 + \alpha_{12} x_1 x_2 + \varepsilon$$
$$\text{DR}_{R_2} = \beta_1 x_3 + \beta_2 x_4 + \beta_{12} x_3 x_4 + \varepsilon \tag{3.64}$$

where DR_{R_1}, and DR_{R_2} represent the dissolution rates for the resin types R_1 and R_2, respectively.

To support the fit of the separate models in Eq. (3.64), at least three distinct minor resin blends of each type are needed. For each resin type, suppose that we elect to measure the DR of each minor resin as well as the DR of the 50% : 50% blend of the two. For each set of (c_1, c_2) values in Figure 3.15 this corresponds to selecting the minor resin proportions for the type R_1 resin to be $(x_1, x_2) = (1, 0)$, $(\frac{1}{2}, \frac{1}{2})$, $(0, 1)$, and for type R_2, resin $(x_3, x_4) = (1, 0)$, $(\frac{1}{2}, \frac{1}{2})$, $(0, 1)$. Then on each plane of Figure 3.15 the coordinates of the four minor resin blends are found by multiplying the coordinates of each resin R_1 blend by the coordinates of each resin R_2 blend. The resulting nine blends of the four minor resins is a $\{2, 2; 2, 2\}$ double-lattice arrangement having the coordinate settings (x_1, x_2, x_3, x_4) as shown in Figure 3.16. Furthermore, if the same nine blends of the double-lattice are performed at each of the three planes in Figure 3.15 this results in a total of 27 distinct blends.

A model for studying the intraresin blending of the four minor resins results from multiplying each term of the DR_{R_1} model in Eq. (3.64) by each term of the DR_{R_2} model in Eq. (3.64), to produce the nine-term double-Scheffé model

$$\text{DR}_{R_1 R_2} = \gamma_{13} x_1 x_3 + \gamma_{14} x_1 x_4 + \gamma_{23} x_2 x_3 + \gamma_{24} x_2 x_4 + \gamma_{123} x_1 x_2 x_3 + \gamma_{124} x_1 x_2 x_4$$
$$+ \gamma_{134} x_1 x_3 x_4 + \gamma_{234} x_2 x_3 x_4 + \gamma_{1234} x_1 x_2 x_3 x_4 + \varepsilon \tag{3.65}$$

The coefficients $\gamma_{13}, \gamma_{123}, \ldots, \gamma_{1234}$ of the terms in Eq. (3.65) are simple functions of the DR values that are collected at the nine blends of Figure 3.16.

A second modeling strategy, and one that is certainly more straightforward than fitting the model of Eq. (3.65) is that of fitting a model in the two independent variables z_1 and z_2, each of which is expressible as a difference between the minor resins of each type, that is, $z_1 = x_2 - x_1$ and $z_2 = x_4 - x_3$. Only two independent variables are needed, since the dimensionality of each plane in Figure 3.15 is two. Such a model is the nine-term interaction (see Cornell and Montgomery, 1996) model

$$
\begin{aligned}
\text{DR} = {} & \delta_0 + \delta_1 z_1 + \delta_2 z_2 + \delta_{12} z_1 z_2 + \delta_{11} z_1^2 + \delta_{22} z_2^2 + \delta_{112} z_1^2 z_2 \\
& + \delta_{122} z_1 z_2^2 + \delta_{1122} z_1^2 z_2^2 + \varepsilon
\end{aligned}
\tag{3.66}
$$

In Appendix 3A the equivalence of the nine-term double–Scheffé model of Eq. (3.65) and the nine-term interaction model of Eq. (3.66) is shown.

The effect of resin composition (interresin blending) of the $R_1 : R_2$ blends on the intraresin blending properties in Eqs. (3.65) and (3.66) can be modeled with any of several types of combined model forms. One form in particular that comes to mind is the following. First let us define the coded resin composition variable

$$
c = \frac{c_1 - 0.50}{0.25}
$$

where $c_1 = 0.25, 0.50$, and 0.75 corresponding to the bottom, middle, and top plane, respectively, and then write the combined inter- and intraresin blending model as

$$
\begin{aligned}
\text{DR} = {} & \{\text{Eq. 3.65 or 3.66}\} + \{\text{Eq. (3.65) or (3.66)}\}c + \{\text{Eq. (3.65) or (3.66)}\}c^2 + \varepsilon \\
= {} & \text{Initial portion} \quad + \quad \text{Middle portion} \quad + \quad \text{Final portion} \quad + \varepsilon
\end{aligned}
\tag{3.67}
$$

If c^2 is coded as 1, 0, and 1 when $c_1 = 0.75, 0.50$, and 0.25, respectively, then in Eq. (3.67) the initial portion represents the DR surface over the middle plane at $c = 0$ or at $(c_1, c_2) = (0.50, 0.50)$, while the middle portion represents the differences between the shapes of the surfaces (in terms of the blending properties of x_1 through x_4) at the two extreme planes, that is, at $(c_1, c_2) = (0.75, 0.25)$ or $c = 1$ and at $(c_1, c_2) = (0.25, 0.75)$ or $c = -1$. The final portion of Eq. (3.67) represents the differences between the shape of the DR surface at $(c_1, c_2) = (0.50, 0.50)$ and the average shape of the DR surfaces at $(c_1, c_2) = (0.75, 0.25)$ and $(c_1, c_2) = (0.25, 0.75)$. Coefficient estimates of terms in the final portion of Eq. (3.67) that are significantly different from zero represent the effects of interresin nonlinear blending of the resin types on the intraresin blending of the minor resins.

We now look deeper into the resin experiment presented by Cornell and Ramsey (1998). A fitted intraresin model and a fitted combined inter- and intraresin model are used to illustrate the types of information that are obtainable from categorized mixtures of mixtures experiments.

3.12.3 A Numerical Example of Interresin and Intraresin Blending

Dissolution rates ($\mathring{A}$/s) were measured for each of the 27 blends of the minor resins as defined by the double-lattice configuration of Figure 3.16 on each of the three planes of Figure 3.15. The five 3- and 4- resin blends located at the midpoints of the edges and the center of each of the planes were replicated a second time for a total of 42 DR values. The coordinates of the minor components X_{11}, X_{12}, X_{21}, and X_{22}, the scaled minor or mixture components x_1, x_2, x_3, and x_4, the values of z_1 and z_2, and the DR values are listed in Table 3.11.

As we mentioned earlier with the intraresin model of Eq. (3.65), the coefficient estimates of the terms $\gamma_{13}, \gamma_{123}, \ldots, \gamma_{1234}$ in Eq. (3.65) are simple functions of the DR values collected at the nine blends of Figure 3.16. To be more specific, let us take the upper plane in Figure 3.15 and the DR values listed in Table 3.11 in the column marked (0.75, 0.25) for the blend of resin types R_1 and R_2. Shown in Figure 3.17 are the individual and average DR values at the nine points of the double-lattice.

The formulas for the estimates $g_{13}, g_{123}, \ldots, g_{1234}$ of the coefficients $\gamma_{13}, \gamma_{123}, \ldots, \gamma_{1234}$ of the terms in the model

$$DR_{R_1 R_2} = \gamma_{13} x_1 x_3 + \gamma_{14} x_1 x_4 + \gamma_{23} x_2 x_3 + \gamma_{24} x_2 x_4 + \gamma_{123} x_1 x_2 x_3 + \gamma_{124} x_1 x_2 x_4$$
$$+ \gamma_{134} x_1 x_3 x_4 + \gamma_{234} x_2 x_3 x_4 + \gamma_{1234} x_1 x_2 x_3 x_4 + \varepsilon$$

when expressed in terms of the average response values at the nine blends are

$$g_{13} = y_{13} = 14.256, \quad g_{14} = y_{14} = 19.464,$$
$$g_{23} = y_{23} = 32.922, \quad g_{24} = y_{24} = 41.868$$
$$g_{123} = 4(\bar{y}_{123}) - 2(y_{13} + y_{23}) = 4(21.888) - 2(14.256 + 32.922) = -6.804$$
$$g_{124} = 4(\bar{y}_{124}) - 2(y_{14} + y_{24}) = 4(28.842) - 2(19.464 + 41.868) = -7.296$$
$$g_{134} = 4(\bar{y}_{134}) - 2(y_{13} + y_{14}) = 4(18.324) - 2(14.256 + 19.464) = 5.856$$
$$g_{234} = 4(\bar{y}_{234}) - 2(y_{23} + y_{24}) = 4(40.524) - 2(32.922 + 41.868) = 12.516$$
$$g_{1234} = 16(\bar{y}_{1234}) - 8(\bar{y}_{123} + \bar{y}_{124} + \bar{y}_{134} + \bar{y}_{224}) + 4(y_{13} + y_{14} + y_{23} + y_{24})$$
$$= 16(25.386) - 8(21.888 + 28.842 + 18.324 + 40.524)$$
$$+ 4(14.256 + 19.464 + 32.922 + 41.868)$$
$$= -36.408$$

The fitted double–Scheffé model for measuring the intraresin blending over the top plane in Figure 3.17 is

$$\widehat{DR}_{R_1 R_2} = 14.256 x_1 x_3 + 19.464 x_1 x_4 + 32.922 x_2 x_3 + 41.868 x_2 x_4 - 6.804 x_1 x_2 x_3$$
$$\quad (0.446) \qquad (0.446) \qquad (0.446) \qquad (0.446) \qquad (1.786)$$
$$- 7.296 x_1 x_2 x_4 + 5.856 x_1 x_3 x_4 + 12.516 x_2 x_3 x_4 - 36.408 x_1 x_2 x_3 x_4$$
$$\quad (1.786) \qquad\quad (1.786) \qquad\quad (1.786) \qquad\quad (7.986)$$

$$(3.68)$$

Table 3.11 Response Values at the Nine Blends of Minor Resins and at the Three Blends of Resin Types R_1 and R_2

Minor Resin Value[a]				Scaled Mixture Component						Blend (c_1, c_2) or Resin Types R_1 and R_2		
x_{11}	x_{12}	x_{21}	x_{22}	x_1	x_2	x_3	x_4	z_1	z_2	$(0.75, 0.25)$	$(0.50, 0.50)$	$(0.25, 0.75)$
c_1	0	c_2	0	1	0	1	0	−1	−1	14.256	14.004	15.546
c_1	0	$c_2/2$	$c_2/2$	1	0	0.5	0.5	−1	0	18.084, 18.564	19.680, 19.380	25.032, 25.560
c_1	0	0	c_2	1	0	0	1	−1	1	19.464	26.250	38.586
$c_1/2$	$c_1/2$	c_2	0	0.5	0.5	1	0	0	−1	21.492, 22.284	18.990, 19.446	18.816, 18.420
$c_1/2$	$c_1/2$	$c_2/2$	$c_2/2$	0.5	0.5	0.5	0.5	0	0	25.638, 25.134	25.410, 25.938	28.572, 29.268
$c_1/2$	$c_1/2$	0	c_2	0.5	0.5	0	1	0	1	29.226, 28.458	34.296, 34.116	46.176, 46.836
0	c_1	c_2	0	0	1	1	0	1	−1	32.922	24.300	20.658
0	c_1	$c_2/2$	$c_2/2$	0	1	0.5	0.5	1	0	40.254, 40.794	33.420, 32.604	32.886, 33.258
0	c_1	0	c_2	0	1	0	1	1	1	41.868	43.440	50.358

[a]Measured response values for individual minor resins are $x_{11} = 23.250$, $x_{12} = 67.704$, $x_{21} = 17.406$, $x_{22} = 63.456$.

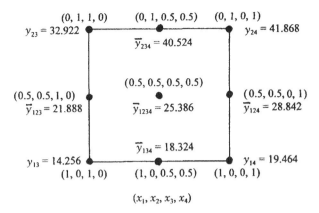

Figure 3.17 Individual (y) and average ($\bar{y}$) DR values at the nine points of the double-lattice of Figure 3.16.

with an $R^2 = 0.9991$, $R_A^2 = 0.9975$, and MSE $= 0.199$. The quantities in parentheses below the coefficient estimates are the estimated standard errors of the coefficient estimates. Listed in Table 3.12 are the fitted double–Scheffé and interaction models along with their R_A^2 and MSE values for estimating the dissolution rate surfaces over the three planes of Figure 3.17.

Several pieces of the overall blending puzzle are uncovered by the results of tests performed on the NLB terms in the double-Scheffé models in Table 3.12. As anticipated, intraresin NLB is present in the shapes of the estimated DR surfaces

Table 3.12 Coefficient Estimates in the Double–Scheffé and Interaction Models Fitted to the Data over the Three Planes of Figure 3.15

	Double–Scheffé Model (3.65)				Interaction Model (3.66)		
Term	(0.75, 0.25)	(0.50, 0.50)	(0.25, 0.75)	Term	(0.75, 0.25)	(0.50, 0.50)	(0.25, 0.75)
x_1x_3	14.256	14.004	15.545	Intercept	25.386	25.674	28.920
x_1x_4	19.465	26.251	38.585	z_1	11.100^b	6.741^b	3.888^b
x_2x_3	32.922	24.300	20.657	z_2	3.477^b	7.494^b	13.944^b
x_2x_4	41.868	43.441	50.357	z_1z_2	0.935^b	1.724^b	1.665^b
$x_1x_2x_3$	-6.804^a	0.263	2.063	z_1^2	4.038^b	0.597	0.264
$x_1x_2x_4$	-7.296^b	-2.556	8.136^b	z_2^2	-0.021	1.038^a	3.642^b
$x_1x_3x_4$	5.857^a	-2.387	-7.081^b	$z_1^2z_2$	0.062	0.353	-0.759^a
$x_2x_3x_4$	12.517^b	-3.431	-9.745^b	$z_1z_2^2$	-0.833^a	0.131	0.333
$x_1x_2x_3x_4$	-36.407^b	-4.968	-24.624^a	$z_1^2z_2^2$	-2.276^b	-0.311	-1.539^a
R_A^2	0.9975	0.9981	0.9988		0.9975	0.9981	0.9988
MSE	0.199	0.127	0.149		0.199	0.127	0.149

[a] Significant at 0.05.
[b] Significant at 0.01.

above both the upper and lower planes. Above the upper plane ($c_1 = 0.75$, $c_2 = 0.25$), *negative intraresin NLB is present with the R_1 minor resins in the presence of each R_2 minor resin, while positive intraresin NLB is present with the R_2 minor resins in the presence of each R_1 minor resin.* Above the lower plane ($c_1 = 0.25$, $c_2 = 0.75$), *negative intraresin NLB is present with the R_2 minor resins in the presence of each R_1 minor resin while positive intraresin NLB is present with the R_1 minor resins in the presence of only the x_4 (the fast dissolving) minor resin.* Above the middle plane ($c_1 = 0.50$, $c_2 = 0.50$), there is no evidence of intraresin NLB among the minor resins.

So what about the effects of interresin blending of the resin types R_1 and R_2 on the intraresin blending of the minor components? When fitted to the total of 42 DR values taken across the three planes, the 27-term combined model in Eq. (3.67), with the double-Scheffé model of Eq. (3.65) substituted in the brackets, is

$$
\begin{aligned}
\widehat{DR} = {} & 14.00x_1x_3 + 26.25x_1x_4 + 24.30x_2x_3 + 43.44x_2x_4 \\
& + 0.26x_1x_2x_3 - 2.56x_1x_2x_4 - 2.39x_1x_3x_4 \\
& \quad (1.59) \qquad\quad (1.59) \qquad\quad (1.59) \\
& - 3.43x_2x_3x_4 - 4.97x_1x_2x_3x_4 - \{0.65x_1x_3 + 9.56x_1x_4 - 6.13x_2x_3 \\
& \quad (1.59) \qquad\quad (7.13) \qquad\qquad (0.28) \qquad\quad (0.28) \qquad\quad (0.28) \\
& + 4.25x_2x_4 + 4.43x_1x_2x_3 \\
& \quad (0.28) \qquad\quad (1.13) \\
& + 7.72x_1x_2x_4 - 6.47x_1x_3x_4 - 11.13x_2x_3x_4 + 5.89x_1x_2x_3x_4\}c \\
& \quad (1.13) \qquad\qquad (1.13) \qquad\qquad (1.13) \qquad\qquad (5.04) \\
& + \{0.90x_1x_3 + 2.78x_1x_4 \\
& \quad (0.49) \qquad\quad (0.49) \\
& + 2.49x_2x_3 + 2.67x_2x_4 - 2.63x_1x_2x_3 + 2.98x_1x_2x_4 + 1.78x_1x_3x_4 \\
& \quad (0.49) \qquad\quad (0.49) \qquad\quad (1.95) \qquad\qquad (1.95) \qquad\qquad (1.95) \\
& + 4.82x_2x_3x_4 - 25.55x_1x_2x_3x_4\}c^2 \\
& \quad (1.95) \qquad\qquad (8.73)
\end{aligned}
$$

                                                                        (3.69)

The numbers in parentheses below the coefficient estimates are the estimated standard errors of the coefficient estimates based on MSE $= 0.159$.

The specific terms in Eq. (3.69) with coefficient estimates that are significantly ($P < 0.05$) different from zero (i.e., estimates producing a $t > 2.132$) are as follows:

1. $-0.65x_1x_3$, $-9.56x_1x_4$, $6.13x_2x_3$, $-4.25x_2x_4$
2. $-4.43x_1x_2x_3$, $-7.72x_1x_2x_4$
3. $6.47x_1x_3x_4$, $11.13x_2x_3x_4$
4. $2.78x_1x_4$, $2.49x_2x_3$, $2.67x_2x_4$ (3.70)
5. $4.82x_2x_3x_4$
6. $-25.55x_1x_2x_3x_4$

When explained in terms of the DRs of the blends of slow and fast dissolving minor resins (x_1 and x_2 of type R_1 and x_3 and x_4 of type R_2) as well as the proportions c_1

and c_2 of the R_1 and R_2 resins, respectively, the significance of the individual terms listed in cases (1) to (6) above are as follows:

1. The DRs of the binary blends x_1x_3, x_1x_4, and x_2x_4 were significantly higher (i.e., $g_{13} = 15.545$, $g_{14} = 38.585$, and $g_{24} = 50.357$) when $c_1 = 0.25 < c_2 = 0.75$ than when $c_1 = 0.75 > c_2 = 0.25$ (where $g_{13} = 14.256$, $g_{14} = 19.465$, and $g_{24} = 41.868$), whereas the DR of the x_2x_3 blend was significantly higher when $c_1 = 0.75 > c_2 = 0.25$ (i.e., $g_{23} = 32.922$) than when $c_1 < c_2$ ($g_{23} = 20.657$).

2. The NLB of the R_1 minor resins x_1 and x_2, in the presence of either R_2 minor resin, was significantly less than zero when $c_1 = 0.75 > c_2 = 0.25$ (i.e., $g_{123} = -6.804$ and $g_{124} = -7.297$), but was positive when $c_1 = 0.25 < c_2 = 0.75$ ($g_{123} = 2.063$ and $g_{124} = 8.136$).

3. The NLB of the R_2 minor resins x_3 and x_4, in the presence of either R_1 minor resin, was significantly greater than zero when $c_1 = 0.75 > c_2 = 0.25$, but was significantly lower than zero when $c_1 = 0.25 < c_2 = 0.75$.

4. The DRs of the binary blends x_1x_4, x_2x_3, and x_2x_4 were significantly lower at $c_1 = c_2 = 0.50$ (where $g_{14} = 26.251$, $g_{23} = 24.300$, and $g_{24} = 43.441$) than would be expected from linear blending of the R_1 and R_2 resins. In other words, at $c_1 = c_2 = 0.50$, the binary blends consisting of one or both fast dissolving minor resins had lower DR values than the average of their respective DRs at $c_1 > c_2$ and $c_1 < c_2$ ($g_{14} = 29.025$, $g_{23} = 26.789$, and $g_{24} = 46.441$). Here one witnesses the effect of interresin NLB on the DR of the binary blends consisting of at least one fast dissolving minor resin.

5. There was a significant effect of interresin NLB on the intraresin NLB of x_3 and x_4 in the presence of x_2.

6. The joint negative NLB of x_1 with x_2 and x_3 with x_4 is significantly less at $c_1 = c_2 = 0.50$ (where $g_{1234} = -4.968$) than is expected under the assumption of interresin linear blending of R_1 and R_2 ($g_{1234} = -30.516$). Here again one witnesses the presence of interresin NLB, only this time, as it impacts the joint intraresin NLB of x_1 with x_2 and x_3 with x_4.

Additional discussion on the nonlinear blending of the minor resins along with plots depicting the NLB of the minor resins over each of the three planes is presented by Cornell and Ramsey (1998). Comparisons are made between the fits of the standard Scheffé quadratic and special cubic models and the fits of the double–Scheffé models.

SUMMARY

In this chapter additional constraints on the component proportions in the form of lower bounds, upper bounds, or both lower and upper bounds are considered. L-pseudocomponents were introduced for those cases where lower bounds only are placed on the proportions of some or all of the components. It was shown that when the subregion of the simplex is itself a simplex, the use of L-pseudocomponents simplifies the construction of designs by allowing the $\{q, m\}$ simplex-lattice or simplex-centroid designs to be used in the L-pseudocomponent system.

When both upper and lower bounds are placed on some or all of the component proportions, the constrained region (or region of interest) takes the form of a convex polyhedron. In most cases the shape of the polyhedron is considerably more complicated than that of the simplex because the polyhedron has more than q vertices and more than q edges. To determine the coordinates of the extreme vertices of the constrained region, some or all of which are used as design points for fitting first- and second-degree models, initially the lower- and upper-bound constraints are checked to see whether they are consistent or not. If they are not consistent, meaning one or more of the bounds cannot be attained, the constraints are adjusted to be consistent, after which a formula is used to enumerate the number of extreme vertices, edges, faces, and so on, of the constrained region. Calculating the coordinates of the extreme vertices is the next order of business. One algorithm is presented and that is McLean and Anderson's extreme-vertices algorithm. It can be used to calculate the coordinates of the extreme vertices of the constrained region and is illustrated using real experimental situations. Upper-bound pseudocomponents, or U-pseudocomponents, are also introduced and can be used to calculate the coordinates of the extreme vertices.

Multiple component constraints of the form $C_j \le A_{1j}x_1 + A_{2j}x_2 + \cdots + A_{qj}x_q \le D_j$, $j = 1, 2, \ldots, h$ are possible in many blending situations. Two algorithms, CONVRT and CONAEV, can be used to calculate the coordinates of the extreme vertices as well as the coordinates of the centroids of the edges, faces, and so on, of the constrained region defined by the presence of multiple component constraints. The generation of the candidate list of design points by CONVRT and CONAEV for four different constrained region examples is presented.

Categorized-components mixtures are blends of components from different categories. Each category, called a major component, must be represented in every mixture by one or more of its minor components. Modeling the component blending means modeling the inter- or between-category blending, especially how these blending properties impact or influence the intra- or within-category blending properties. Experimental designs and model forms for studying the intracategory blending of the minor components are generalizations of the simplex designs and the standard Scheffé-type models used in the single-category mixture problem. The analysis of data from a photoresist-coating experiment, in which two types of resins make up two major components and each resin type consists of two minor resins, serves to illustrate the impact that intercategory blending can have on the intracategory blending of the minor resins.

REFERENCES AND RECOMMENDED READING

Anderson, M. J., and P. J. Whitcomb (1996). Optimization of paint formulations made easy with computer-aided design of experiments for mixtures. *J. Coat. Tech.*, **68**, 71–75.

Cornell, J. A. (2002). *Experiments with Mixtures: Designs, Models, and the Analysis of Mixture Data*, 3rd ed. Wiley, New York.

Cornell, J. A., and I. J. Good (1970). The mixture problem for categorized components. *J. Am. Stat. Assoc.*, **65**, 339–355.

Cornell, J. A., and J. M. Harrison (1997). Models and designs for experiments with mixtures. Part II: Exploring a subregion of the simplex and the inclusion of other factors in mixture experiments. *Bulletin 899, Agricultural Experiment Station*, Institute of Food and Agricultural Sciences, University of Florida, Gainesville, pp. 1–78.

Cornell, J. A., and P. J. Ramsey (1998). A generalized mixture model for categorized-component problems with an application to a photoresist-coating experiment. *Technometrics*, **40**, 48–61.

Cornell, J. A., and D. C. Montgomery (1996). Interaction models as alternatives to low-order polynomials. *J. Qual. Technol.*, **28**, 163–176.

Crosier, R. B. (1984). Mixture experiments: geometry and pseudocomponents. *Technometrics*, **26**, 209–216.

Crosier, R. B. (1986). The geometry of constrained mixtures experiments. *Technometrics*, **28**, 95–102.

Gorman, J. W. (1970). Fitting equations to mixture data with restraints on compositions. *J. Qual. Technol.*, **2**, 186–194.

Kennard, R. W., and L. A. Stone (1969). Computer aided design of experiments. *Technometrics*, **11**, 137–148.

Khuri, A. I., J. M. Harrison, and J. A. Cornell (1999). Using quantile plots of the prediction variance for comparing designs for a constrained mixture region: an application involving a fertilizer experiment. *Appl. Stat. C*, **49**, 521–532.

Koons, G. F. (1989). Effect of sinter composition on emissions: a multi-component highly constrained mixture experiment. *J. Qual. Technol.*, **21**, 261–267.

Kurotori, I. S. (1966). Experiments with mixtures of components having lower bounds. *Ind. Qual. Control*, **22**, 592–596.

Lambrakis, D. P. (1968). Experiments with mixtures: a generalization of the simplex-lattice design. *J. R. Stat. Soc. B*, **30**, 123–136.

Lambrakis, D. P. (1969). Experiments with mixtures: estimated regression function of the multiple-lattice design. *J. R. Stat. Soc. B*, **31**, 276–284.

Marquardt, D. W., and R. D. Snee (1974). Test statistics for mixture models. *Technometrics*, **16**, 533–537.

McLean, R. A., and V. L. Anderson (1966). Extreme vertices design of mixture experiments. *Technometrics*, **8**, 447–454.

Mitchell, T. J. (1974). An algorithm for the construction of D-optimal experimental designs. *Technometrics*, **16**, 203–210.

Nigam, A. K., S. C. Gupta, and S. Gupta (1983). A new algorithm for extreme vertices designs for linear mixture models. *Technometrics*, **25**, 367–371.

Piepel, G. F. (1983a). Defining consistent constraint regions in mixture experiments. *Technometrics*, **25**, 97–101.

Piepel, G. F. (1983b). Calculating centroids in constrained mixture experiments. *Technometrics*, **25**, 279–283.

Piepel, G. F. (1988). Programs for generating extreme vertices and centroids of linearly constrained experimental regions. *J. Qual. Technol.*, **20**, 125–139.

Piepel, G. F. (1999). Modeling methods for mixture-of-mixtures experiments applied to a tablet formulation problem. *Pharm. Dev. Tech.*, **4**, 593–606.

Saxena, S. K., and A. K. Nigam (1977). Restricted exploration of mixtures by symmetric-simplex designs. *Technometrics*, **19**, 47–52.

Scheffé, H. (1958). Experiments with mixtures. *J. R. Stat. Soc. B*, **20**, 344–360.

Smith, W. F. (2005) *Experimental Design for Formulation*. ASA-SIAM Series on Statistics and Applied Probability. ASA, Alexandria, VA.

Smith, W. F., and J. A. Cornell (1993). Biplot displays for looking at multiple response data in mixture experiments, *Technometrics*, **35**, 337–350.

Snee, R. D. (1975). Experimental designs for quadratic models in constrained mixture spaces. *Technometrics*, **17**, 149–159.

Snee, R. D. (1979). Experimental designs for mixture systems with multicomponent constraints. *Commun. Stat.*, **A8**, No. 4, 303–326.

Snee, R. D., and D. W. Marquardt (1974). Extreme vertices designs for linear mixture models. *Technometrics*, **16**, 399–408.

Welch, W. J. (1982). Branch-and-bound search for experimental designs based on D-optimality and other criteria. *Technometrics*, **24**, 41–48.

Welch, W. J. (1984). Computer-aided design of experiments for response estimation. *Technometrics*, **26**, 217–224.

Welch, W. J. (1985). ACED: algorithms for the construction of experimental designs. *Am. Stat.*, **39**, 146.

QUESTIONS

3.1. In a three-component system, lower bounds for components 1, 2, and 3 are $L_1 = 0.10$, $L_2 = 0.15$, and $L_3 = 0.0$, respectively. Set up the L-pseudocomponents x_i', $i = 1$, 2, and 3 and construct a simplex-centroid design in the L-pseudocomponents. List the blending proportions in the original components corresponding to the simplex-centroid combinations in the L-pseudocomponents.

3.2. In the tropical beverage data of Table 3.2, the six blends numbered 1, 2, 3, 11, 12, and 13 each contain 80% watermelon. If we are asked to make final recommendations of a tropical beverage that contains exactly 80% watermelon, what suggestions do you have for the remaining 20% in terms of orange, pineapple, and grapefruit in order to produce a high flavor score?

3.3. In the formulation of powder used in marine flares, upper bounds on chemicals A, B, and C are 92%, 6%, and 4%, respectively. Let x_1, x_2, and x_3 represent the proportions associated with the chemicals A, B, and C.
   (a) Define the U-pseudocomponents u_1, u_2, and u_3.
   (b) Suggest a four-point design for fitting a first-degree model in u_1, u_2, and u_3. List the settings for the chemicals A, B, and C corresponding to the four design points.

(c) The upper bound for chemical B is raised to 10%. What effect does raising the bound for B have on the shape of the region? Draw the new region.

(d) Ten experiments can be run. It is desired to fit a second-degree model in x_1, x_2, and x_3 over the region in (c). Suggest a design for fitting the model and list the settings for x_1, x_2, and x_3.

3.4. In a three-component system, bounds on the x_i were specified as

$$0 \le x_1 \le 0.5, \quad 0.03 \le x_2 \le 0.6, \quad 0.45 \le x_3 \le 0.9$$

(a) Draw the constrained region.

(b) Check the set of constraints on the x_i for consistency. If they are not consistent, which of the bounds are unattainable?

(c) Suggest a nine-point design for fitting a second-degree model in x_1, x_2, and x_3.

3.5. A four-component tobacco experiment was planned where the constraints on the components were

$$FC = 34\% \pm 10\%, \quad B = 15\% \pm 10\%,$$
$$TB = 21\% \pm 10\%, \quad PT = 30\% \pm 10\%$$

(a) Are the constraints consistent? (*Hint:* Use x_i defined in (c).)

(b) Compute the number of extreme vertices, edges, and two-dimensional faces of the constrained region.

(c) Suggest a 15-point design for fitting a second-degree model in x_1, x_2, x_3, and x_4, where $x_1 = FC/100\%, x_2 = B/100\%, x_3 = TB/100\%$, and $x_4 = PT/100\%$.

3.6. List the coordinate settings of the seven extreme vertices and the midpoints of the eleven edges of the constrained region in Figure 3.10.

3.7. The minor components from three M-components are to be combined. In M-component 1 are the two m-components denoted by X_{11} and X_{12} where $c_1 = \frac{1}{2}$? In M-component 2 are the three m-components X_{21}, X_{22}, and X_{23} and $c_2 = \frac{1}{4}$, and in M-component 3 are the two m-components X_{31} and X_{32} where $c_3 = \frac{1}{4}$? List all of the blends that comprise the $\{2, 3, 2; m_1, m_2, m_3\}$ triple-lattice where $m_1 = 1, m_2 = 2$, and $m_3 = 2$.

3.8. Shown are data values at the seven blends of a three-component simplex-centroid design. High values of the response are more desirable than low values. Indicate the type of blending, linear or nonlinear, synergistic or antagonistic, that is present between the components,

(a) A and B: _____ and is _____ .

(b) A and C: _____ and is _____ .

(c) B and C: _____ and is _____ .

(d) A and B and C: _____ and is _____ .

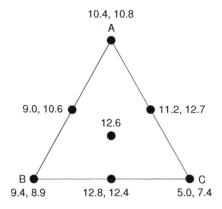

Hint: $\text{var}(b_i) = \sigma^2/2$, $\text{var}(b_{ij}) = 24\sigma^2/2$, and $\text{var}(b_{ijk}) = 1188\sigma^2$. Also SST = 63.79, SSR = 58.22, and SSE = 5.57.

CHAPTER 4

The Analysis of Mixture Data

In Section 2.6, the results of a blending experiment were presented in terms of how the polymers polyethylene, polystyrene, and polypropylene behaved singly and in combination as measured by the thread elongation of spun yarn. Briefly, a second-degree model was fitted to observed elongation values that were collected from blends of the polymers specified by the points of a $\{3, 2\}$ simplex-lattice design. Synergistic blending was assumed to be present between polyethylene and each of the other constituents because high elongation values were observed on blends containing polyethylene and a high elongation value was considered to be desirable. Polystyrene and polypropylene were suspected of being antagonistic when they were blended together without polyethylene.

In Section 2.13, the results of an experiment were discussed in which four chemical pesticides had been sprayed on strawberry plants in an attempt to control the number of mites on the plants. The data were collected at the points of a simplex-centroid design. The 15-term model seemed to fit the response values adequately, although neither an analysis of variance table nor an F-test for the fitted model was provided. The objective of the fitted model exercise was to see whether any of the multiple-component blends were as effective as the pure chemicals. Several two-component blends appeared to be as effective as the pure chemicals.

In Section 3.5, the constrained experimental region was defined from the placing of an upper bound on one of the component proportions and formulas were presented for estimating the parameters in a second-degree model. The estimation was illustrated using data from an experiment involving the formulation of a tropical beverage. Juices from watermelon (x_1), orange (x_2), pineapple (x_3), and grapefruit (x_4) were combined to make the tropical beverage. The 10-term prediction equation, Eq. (3.19), for sensory flavor score was thought to be satisfactory because the residual mean square from the analysis of the 16 flavor scores was not significantly different from the pure-error mean square that was calculated from the 40 scores of each blend. In the remainder of

Table 4.1   Three Sets of Texture Measurements Taken on Duplicate Fish Patties Where Each Set of Patties Was Processed (Prepared and Cooked) Differently from the Patties in the Other Sets

Fish Component Proportions			Texture Readings (grams $\times$ 10^{-3} of force required to puncture the patty surface)					
x_1	x_2	x_3	Set I		Set II		Set III	
1	0	0	1.98	1.70	3.12	2.89	2.34	2.30
0	1	0	0.68	0.67	1.18	1.24	0.97	0.97
0	0	1	1.53	1.48	2.36	2.27	2.11	2.13
$\frac{1}{2}$	$\frac{1}{2}$	0	1.18	1.40	1.96	1.90	1.48	1.43
$\frac{1}{2}$	0	$\frac{1}{2}$	1.45	1.39	2.66	2.48	1.80	2.06
0	$\frac{1}{2}$	$\frac{1}{2}$	1.19	1.12	1.80	1.86	1.21	1.34
$\frac{1}{3}$	$\frac{1}{3}$	$\frac{1}{3}$	1.65	1.54	2.09	1.79	1.53	1.56

Chapter 3, methods were presented for designing and analyzing data when additional restrictions were placed on the component proportions.

In this chapter, several additional techniques used in the analysis of mixture data are discussed. The model employed will be the Scheffé-type canonical polynomials. In Chapter 5, alternative models are presented.

4.1   TECHNIQUES USED IN THE ANALYSIS OF MIXTURE DATA

Let us begin by imagining that we are faced with the task of analyzing a set of data from a mixture experiment. Initially we ask ourselves whether the objective is the fitting of some proposed model for the purpose of describing the shape of the response surface over the simplex factor space or whether we are more interested in determining the roles played by (i.e., measuring the effects of) the individual components. Most of the time both objectives can be attained from the same analysis. We will illustrate the partial attainment of both objectives many times in this chapter in our discussions of the results of the analysis of many data sets.

To help us present the methodology of this section, we make use of the three sets of experimental data listed in Table 4.1. These data sets are subsets of a larger, more complete set of data that was generated during a large-scale experiment. The larger experiment is described in greater detail in Section 6.2.

The data values in Table 4.1 represent texture measurements taken on fish patties that had been formulated by blending three fish species. The species were mullet (x_1), sheepshead (x_2), and croaker (x_3). The unit of measure of the texture data is scaled (scaled value equals actual value times 10^{-3}), to facilitate the handling of the data numbers. The texture values in sets I, II, and III are listed as pairs of numbers that were collected from replicate patties of each blend. The particular blends (component proportions) correspond to the blends that are defined by the seven points of a simplex-centroid design. Each set (I, II, and III) of patties was prepared according to a specified cooking temperature and cooked for a specified length of time. The cooking temperatures and times differed for the three sets of patties.

The proposed model to which the observations in each of data sets I, II, and III will be fitted to initially is the Scheffé special cubic model

$$\eta = \sum_{i=1}^{3} \beta_i x_i + \sum \sum_{i<j}^{3} \beta_{ij} x_i x_j + \beta_{123} x_1 x_2 x_3 \tag{4.1}$$

The special cubic model in Eq. (4.1) is the simplex-centroid model of Eq. (2.39) when $q = 3$. The purpose behind fitting the special cubic equation to each data set is to illustrate an adequate fit to the data of set I as well as an overfit of the data in sets II and III. By an overfit is meant that some of the terms in Eq. (4.1) are not needed when describing the texture surface and therefore the terms can be deleted. The overfit will be discovered through the testing of hypotheses that specify zero values for some of the parameters in the model. In other words, the dropping of terms from the complete model is analogous to accepting an hypothesis that states that certain terms in Eq. (4.1) are equal to zero and thus are unimportant.

Before we discuss the testing of hypotheses concerning specific parameters in the model, we will review very briefly the strategy employed in asking questions about which of the components' effects are likely to be present in the data. The anticipated answers to the questions prompted us to try the model form presented by Eq. (4.1).

In trying to decide on the particular form of the model to be fitted to data collected at the points of a $q = 3$ simplex-centroid design, we first recognize that the special cubic model or simplex-centroid model of Eq. (4.1) is chosen over the lower degree models because the terms in the special cubic model not only provide a measure of each pure blend, but provide measures of the binary blends and a measure of the three-component blend as well. Nevertheless, we mention the following important and relevant questions that probably were in the minds of the experimenters prior to performing the experimental runs and collecting the data at the composition points.

Question: Is the response (fish patty texture) surface likely to be planar over the triangle, or are combinations of two and three fish likely to blend nonlinearly, that is, cause departures from linearity in the surface shape? If the blending of multiple-component mixtures is not additive, which pairs of components are likely to have synergistic effects, antagonistic effects? Are complete blends (three-fish blends) likely to be firmer (have higher texture values) than the binary or pure blends? Also, if we assume the texture surface is planar, should the texture values be collected at the pure blends (vertices of the triangle) only, or should we use complete mixtures consisting of all three fish types that are located very near the vertices of the triangle, such as $x_i = 0.95$ and $x_j = x_k = 0.05/2$, $i, j, k = 1, 2$, and $3, i \neq j \neq k$? Furthermore, even if we assume the surface is planar, shouldn't additional observations be collected at other locations inside (or on the boundary of) the triangle to enable us to check our assumption of a planar surface (i.e., to check the adequacy of the fitted model)?

Answer: If the objective is to model the response surface above the triangle, it is imperative that observations be taken at a sufficient number of blends that will not only support the fitted model but will also allow the fitted model to be tested for adequacy of fit (see Section 2.11). If the planar first-degree model is fitted to vertex points, for example, one should collect additional observations at several interior points or, better yet, at the midpoints of the edges of the triangle (enabling the continuation to the fitting of the second-degree model if desired). This latter strategy of collecting data at the midpoints of the edges is advantageous in the sense that if a second-degree model is required, the estimates of the coefficients β_{ij}, $i < j$, in the second-degree model will have smaller variances than if data at interior points are used to estimate the β_{ij}.

Question: If the $\{3, 2\}$ simplex-lattice arrangement is chosen because of the distinct possibility that the surface is not planar, should additional observations be collected at interior points of the triangle for the purpose of checking the model fit inside the triangle? If so, where inside the triangle should the extra points (blends) be located?

Answer: Additional points inside the simplex should always be considered whenever a $\{q, 2\}$ simplex-lattice design is to be used because the $\{q, 2\}$ lattices are strictly boundary designs. Now, if we can afford only one interior point in addition to the $\{3, 2\}$ simplex-lattice, this ternary blend should be the centroid of the triangle. This blend has the highest power of any single interior blend for testing the usefulness of the term $\beta_{123}x_1x_2x_3$ that is present in the special cubic model but is absent in the quadratic model.

Quite naturally, then, once a model of the form in Eq. (4.1) has been chosen, even though it was felt initially that the assumption of a planar surface was very realistic, the next step is to collect a set of data, similar in form to each of sets I, II, or III, obtain the fitted model, and proceed to scrutinize the fitted model as described in the next section. Note that in collecting observations, an attempt should be made to collect replicate observations at some points of the design. The replicate observations will enable an estimate of the observation variance, σ^2, to be realized from which estimates of the variances of the estimated model parameters can be obtained. Tests on the magnitudes of the individual parameter estimates as well as on the adequacy of the overall fitted model can then be performed.

Before we proceed with model fitting and hypothesis testing, it seems only natural to mention that fitting and evaluating mixture models is no different than fitting and evaluating nonmixture models. Only the forms of the models are slightly different, but the strategy is the same. With each form of the fitted model, the model summary statistics should be calculated such as (1) R^2 and R_A^2, (2) the value of the PRESS statistic and the corresponding R_{PRESS}^2, and (3) the plotted residuals, especially the studentized residuals versus the predicted values $\hat{y}$.

Most software packages such as DESIGN-EXPERT by Stat-Ease (1998) and Minitab (1999) provide the statistics mentioned above to the modeler. Piepel (1997) provides a summary of commercial software packages with mixture experiment capabilities.

We now discuss the construction of test statistics for the purpose of testing hypotheses concerning the usefulness of terms in the Scheffé models. Rather than build the model sequentially by starting with the first-degree polynomial and working toward the special cubic model, we will begin with the complete special cubic fitted to the data in each of sets I, II, and III and work backward. We will work backward by first testing the usefulness of the cubic term, then the cross-product or binary terms, and finally the similarity of the linear blending terms in the model. This approach is taken not only because the data sets have been collected and are known to support the special cubic model (4.1) but because many people prefer to start with the most complete model form and work backward in an attempt to arrive at a simpler model form by dropping terms from the model at each stage of the model-fitting exercise. The test statistics presented are discussed in greater detail in Marquardt and Snee (1974).

4.2 TEST STATISTICS FOR TESTING THE USEFULNESS OF THE TERMS IN THE SCHEFFÉ POLYNOMIALS

When the Scheffé polynomials are used to model the response surface as well as to provide measures of the blending characteristics of the components, most often the model will include *all* of the terms up to a given degree. Usually the final form is a polynomial that contains the q terms $\beta_i x_i$, $i = 1, 2, \ldots, q$, representing the very fundamental linear blending surface or additive blending among the components, and any additional terms of higher degree such as

$$\eta = \sum_{i=1}^{q} \beta_i x_i + \sum_{i<j}^{q} \beta_{ij} x_i x_j + \cdots$$

With second-degree and higher-degree models, all pairs of components and/or all triplets of the components, and so on, up to degree d, are considered. (One exception to this rule is the omission of terms like $\beta_{ijk} x_i x_j x_k$ in the quartic model, presented in Appendix 2B.)

In choosing the degree of the final polynomial model so that predictions of the response surface can be made, tests of hypotheses are performed on groups of parameters in the polynomial model. A group may consist of a single parameter only, but more likely the group will involve two or more parameters. For example, for the data of set I in Table 4.1, let us consider the fitting of the special cubic model to the data. The model is

$$\eta = \beta_1 x_1 + \beta_2 x_2 + \beta_3 x_3 + \beta_{12} x_1 x_2 + \beta_{13} x_1 x_3 + \beta_{23} x_2 x_3 + \beta_{123} x_1 x_2 x_3 \qquad (4.2)$$

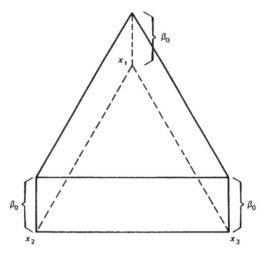

Figure 4.1 Response surface specified by the hypothesis H_0: $\beta_1 = \beta_2 = \beta_3 = \beta_0$, $\beta_{12} = \beta_{13} = \beta_{23} = \beta_{123} = 0$.

Initially we choose to test the null hypothesis (H_0), which states

> H_0: The response *does not* depend on the mixture components   (4.3)

against the alternative hypothesis (H_A), which states

> H_A: The response *does* depend on the mixture components

When the null hypothesis is true, all three linear coefficients β_1, β_2, and β_3 are equal in value (e.g., β_0), and the remaining terms in the model of Eq. (4.2) are equal to zero. For Eq. (4.2), the null hypothesis (4.3) implies

$$H_0: \beta_1 = \beta_2 = \beta_3 = \beta_0 \quad \text{and} \quad \beta_{12} = \beta_{13} = \beta_{23} = \beta_{123} = 0 \qquad (4.4)$$

so that Eq. (4.2) is more appropriately written as

$$\eta = \beta_0 x_1 + \beta_0 x_2 + \beta_0 x_3 = \beta_0 \qquad (4.5)$$

According to Eq. (4.5), the surface above the triangle is of constant height, which is denoted by β_0; see Figure 4.1. The least-squares estimate of β_0 in Eq. (4.5) is $b_0 = \sum_{u=1}^{N} y_u / N = \bar{y}$, where $\bar{y}$ is the average of all N observations collected over the simplex; that is, the estimated height of the surface for all blends is $\bar{y}$.

To test the null hypothesis stated in (4.3), the model in Eq. (4.2) is fitted to the data of set I and an F-ratio is set up:

$$F = \frac{\text{SSR}/(p-1)}{\text{SSE}/(N-p)} \qquad (4.6)$$

where $N = 14$, $p = 7$,

$$SSR = \sum_{u=1}^{N=14} (\hat{y}_u - \bar{y})^2$$

$$SSE = \sum_{u=1}^{N=14} (y_u - \hat{y}_u)^2 \qquad (4.7)$$

$$SST = \sum_{u=1}^{N=14} (y_u - \bar{y})^2$$

In the sums of squares formulas, y_u is the value of the uth observation, $\hat{y}_u$ is the predicted value of the response corresponding to the uth observation, where $\hat{y}_u$ uses the model in Eq. (4.2) with the parameter estimates substituted, and $\bar{y}$ is the average of the $N = 14$ observations. The degrees of freedom for the sources Regression and Error are $p - 1 = 7 - 1 = 6$ and $N - p = 14 - 7 = 7$, respectively, where p is the number of parameters that are estimated in the fitted model and N is the total number of observations. The value of the F-ratio in (4.6) is compared to the tabled value of $F_{(p-1, N-p, \alpha)}$ and the null hypothesis in (4.3) is rejected at the α-level of significance if the value of the F-ratio in (4.6) exceeds the tabled value. SST has $N - 1 = 13$ d.f.

Let us illustrate numerically the testing of the hypothesis in (4.3) by using the data in set I of Table 4.1. The least-squares solutions to the normal equations, the fitted model with the estimated coefficient standard errors, the predicted values of the response at the design points, and the sums-of-squares quantities in Eq. (4.7), respectively, are given by

Fitted model: *Set I:*

$$
\mathbf{b} = (\mathbf{X'X})^{-1} \quad \mathbf{X'y}
$$

$$
\begin{bmatrix} 1.84 \\ 0.68 \\ 1.51 \\ 0.13 \\ -1.01 \\ 0.26 \\ 8.75 \end{bmatrix} = \begin{bmatrix} 0.5 & 0 & 0 & -1.0 & -1.0 & 0 & 1.5 \\ & 0.5 & 0 & -1.0 & 0 & -1.0 & 1.5 \\ & & 0.5 & 0 & -1.0 & -1.0 & 1.5 \\ & & & 12.0 & 2.0 & 2.0 & -30.0 \\ & \text{(symmetrical)} & & & 12.0 & 2.0 & -30.0 \\ & & & & & 12.0 & -30.0 \\ & & & & & & 594.0 \end{bmatrix} \begin{bmatrix} 7.45 \\ 4.86 \\ 6.65 \\ 1.00 \\ 1.06 \\ 0.93 \\ 0.12 \end{bmatrix}
$$

$$\hat{y}(\mathbf{x}) = (x_1, x_2, x_3, x_1x_2, x_1x_3, x_2x_3, x_1x_2x_3)'\mathbf{b}$$

$$= 1.84x_1 + 0.68x_2 + 1.51x_3 + 0.13x_1x_2 - 1.01x_1x_3 + 0.26x_2x_3 + 8.75x_1x_2x_3$$

$\quad(0.07) \quad\quad (0.07) \quad\quad (0.07) \quad\quad (0.36) \quad\quad\quad (0.36) \quad\quad\quad (0.36) \quad\quad (2.52)$

$$(4.8)$$

Estimated response values:

$$\hat{y}(1,\, 0,\, 0) = 1.84(1) + 0.68(0) + 1.51(0) + 0.13(1)(0) - 1.01(1)(0)$$
$$+ 0.26(0)(0) + 8.75(1)(0)(0)$$
$$= 1.84$$

$$\hat{y}(0,\, 1,\, 0) = 0.68$$

$$\hat{y}(0,\, 0,\, 1) = 1.51$$

$$\hat{y}\left(\tfrac{1}{2},\, \tfrac{1}{2},\, 0\right) = 1.84\left(\tfrac{1}{2}\right) + 0.68\left(\tfrac{1}{2}\right) + 0.13\left(\tfrac{1}{2}\right)\left(\tfrac{1}{2}\right)$$
$$= 1.29$$

$$\hat{y}\left(\tfrac{1}{2},\, 0,\, \tfrac{1}{2}\right) = 1.42$$

$$\hat{y}\left(0,\, \tfrac{1}{2},\, \tfrac{1}{2}\right) = 1.16$$

$$\hat{y}\left(\tfrac{1}{3},\, \tfrac{1}{3},\, \tfrac{1}{3}\right) = 1.60$$

$$\text{SSR} = \sum_{u=1}^{14} (\hat{y}_u - \bar{y})^2 = (1.84 - 1.35)^2 + (1.84 - 1.35)^2 + (0.68 - 1.35)^2$$
$$+ \cdots + (1.60 - 1.35)^2$$
$$= 1.65$$

$$\text{SSE} = \sum_{u=1}^{14} (y_u - \hat{y}_u)^2 = (1.98 - 1.84)^2 + (1.70 - 1.84)^2 + (0.68 - 0.68)^2$$
$$+ \cdots + (1.54 - 1.60)^2$$
$$= 0.077$$

$$\text{SST} = \sum_{u=1}^{14} (y_u - \bar{y})^2 = (1.98 - 1.35)^2 + (0.68 - 1.35)^2$$
$$+ \cdots + (1.54 - 1.35)^2$$
$$= 1.73$$

where $\bar{y} = 1.35$ is the overall mean or average of the 14 values in set I.

For purposes of illustration, we are assuming the multiple observations per blend in the *sums-of-squares calculations* are replicates and not duplicates. This is because duplicate patties in all likelihood will not reflect a true measure of the error variance and we need an estimate of the error variance for our test. The value of the F-ratio (4.6) is

$$F = \frac{1.65/(7-1)}{0.077/7} = 25.0$$

and since $F = 25.0$ is greater than the tabled value $F_{(6,7, \alpha = 0.01)} = 7.19$, we reject H_0 in Eq. (4.4) and conclude that the response *does* depend on the mixture components (i.e., the magnitude of the texture or firmness measurements varies with the different fish combinations). The adjusted coefficient of determination, introduced in Section 7.4, aids us in deciding whether the model explains a sufficient amount of the variation in the response values. A sufficient amount of the variation is measured by comparing the estimate of the error variance obtained from the analysis of the fitted model against the estimate of σ^2 using the model, $y = \beta_0 + \epsilon$. Recalling the formula

$$R_A^2 = 1 - \frac{\text{SSE}/(N - p)}{\text{SST}/(N - 1)}$$

we find for the data in set I that

$$R_A^2 = 1 - \frac{0.077/7}{1.73/13} = 0.917$$

where 1.73 in the denominator is the Total SS $= \sum_{u=1}^{14}(y_u - \bar{y})^2$ with 13 d.f. Since the value of R_A^2 exceeds 0.90, this means that the error variance estimate obtained from the analysis of the fitted model is less than 10% of the error variance estimate obtained with the model $y = \beta_0 + \epsilon$. Thus we feel confident in using the fitted special cubic model of Eq. (4.8) for the purpose of predicting response values.

Typical output with software packages such as DESIGN-EXPERT of Stat-Ease (1998) also includes the following:

Actual Value	Predicted Value	Residual	Studentized Residual	Cook's Distance	Outlier t
1.70	1.840	−0.140	−1.913	0.523	−2.563
0.67	0.675	−0.005	−0.068	0.001	−0.063
1.48	1.505	−0.025	−0.342	0.017	−0.319
1.18	1.290	−0.110	−1.503	0.323	−1.691
1.39	1.420	−0.030	−0.410	0.024	−0.384
1.65	1.595	0.055	0.751	0.081	0.726
1.98	1.840	0.140	1.913	0.523	2.563
1.53	1.505	0.025	0.342	0.017	0.319
0.68	0.675	0.005	0.068	0.001	0.063
1.19	1.155	0.035	0.478	0.033	0.450
1.40	1.290	0.110	1.503	0.323	1.691
1.45	1.420	0.030	0.410	0.024	0.384
1.12	1.155	−0.035	−0.478	0.033	−0.450
1.54	1.595	−0.055	−0.751	0.081	−0.725

Rules of thumb for acceptable values of the studentized residual (see Section 7.5), Cook's distance (see Section 4.8) and outlier t are less than 3.0, less than 1.0, and less than 3.5, respectively. The outlier t-value is calculated as the quotient of the studentized residual to its estimated standard deviation, which is also calculated without the actual observation value. Montgomery, Peck, and Vining (2001) refer to this quantity as R-student.

Having fitted the Scheffé linear and quadratic models to the data of set I, an overall summary of the models' statistics is as follows:

Model	R^2	R^2_A	R^2_{PRESS}
Linear	0.8171	0.7839	0.7338
Quadratic	0.8820	0.8083	0.6867
Special cubic	0.9566	0.9170	0.8263

For the special cubic model, the value of $R^2_{PRESS} = 0.8263$ (see Section 7.5) is in reasonable agreement with $R^2_A = 0.9170$ and thus is the model of choice with set I.

A similar exercise in model fitting and in testing the hypothesis in Eq. (4.4) with the data of sets II and III produced the following results:

Fitted model: Set II:

$$\hat{y}(\mathbf{x}) = 3.01x_1 + 1.21x_2 + 2.32x_3 - 0.71x_1x_2 - 0.36x_1x_3 + 0.27x_2x_3$$

$$\phantom{\hat{y}(\mathbf{x}) = }(0.08)\quad(0.08)\quad(0.08)\quad(0.41)\quad(0.41)\quad(0.41)$$

$$\phantom{\hat{y}(\mathbf{x}) = }- 3.99x_1x_2x_3 \tag{4.9}$$

$$\phantom{\hat{y}(\mathbf{x}) = }(2.87)$$

$$F = \frac{4.01/6}{0.097/7} = 48.2 \quad \text{Reject } H_0 \text{ in Eq. (4.4)}$$

$$R^2_A = 1 - \frac{0.097/7}{4.11/13} = 0.956$$

Fitted model: Set III:

$$\hat{y}(\mathbf{x}) = 2.32x_1 + 0.97x_2 + 2.12x_3 - 0.76x_1x_2 - 1.16x_1x_3 - 1.08x_2x_3$$

$$\phantom{\hat{y}(\mathbf{x}) = }(0.06)\quad(0.06)\quad(0.06)\quad(0.28)\quad(0.28)\quad(0.28)$$

$$\phantom{\hat{y}(\mathbf{x}) = }-2.03x_1x_2x_3 \tag{4.10}$$

$$\phantom{\hat{y}(\mathbf{x}) = }(1.95)$$

$$F = \frac{2.80/6}{0.045/7} = 72.6 \quad \text{Reject } H_0 \text{ in Eq. (4.4)}$$

$$R^2_A = 1 - \frac{0.045/7}{2.84/13} = 0.971$$

With each of the three sets of data, the null hypothesis, which states that the texture value of the patties does not depend on the fish (i.e., the patty texture is constant for all combinations of the three fish), is rejected. Following the rejection of H_0, we next try to see whether all three of the fish types influence the patty texture, and if so, how. To this end, we look for the form of the polynomial model that fits the data values best. Before we proceed, however, let us rework the data in set II and test the null hypothesis of (4.3) for the following cases. In case 1, the fitted model is quadratic and in case 2, the fitted model is linear or of the first degree.

Case 1. H_0: $\beta_1 = \beta_2 = \beta_3 = \beta_0$ and $\beta_{12} = \beta_{13} = \beta_{23} = 0$

$$\hat{y}(\mathbf{x}) = 3.02x_1 + 1.22x_2 + 2.33x_3 - 0.91x_1x_2 - 0.56x_1x_3 - 0.07x_2x_3 \quad (4.11)$$

$$(0.09) \qquad (0.09) \qquad (0.09) \qquad (0.40) \qquad\qquad (0.40) \qquad\qquad (0.40)$$

$$F = \frac{3.98/5}{0.124/8} = 51.4$$

$$R_A^2 = 1 - \frac{0.124/8}{4.11/13} = 0.951$$

In case 1, since $F = 51.4$ exceeds $F_{(5, 8, 0.01)} = 6.63$, we reject H_0 and conclude that not only are the textures of the pure fish patties not the same but the textures of the two-fish patties appear to be different from the simple average of the respective textures of the single-fish patties. The presence of both pure and binary blending effects is investigated in the next section on model reduction.

Case 2. H_0: $\beta_1 = \beta_2 = \beta_3 = \beta_0$

$$\hat{y}(\mathbf{x}) = 2.89x_1 + 1.16x_2 + 2.30x_3 \quad (4.12)$$

$$(0.08) \qquad (0.08) \qquad (0.08)$$

$$F = \frac{3.87/2}{0.23/11} = 92.5$$

$$R_A^2 = 1 - \frac{0.23/11}{4.11/13} = 0.934$$

As in case 1, for case 2 we reject H_0, since $F = 92.5 > F_{(2, 11, 0.01)} = 7.21$, and conclude that the texture of the patties made from each of the three individual fish types is not constant (i.e., the heights b_i, of the plane at the vertices $x_i = 1$ are not the same). We can only infer at this point in the analysis that it appears as though the pure mullet (x_1) patties are firmer than the pure sheepshead (x_2) patties. The texture of the pure croaker (x_3) patties is somewhere between the textures of the others.

We now question ourselves concerning the fitting of the special cubic models and the subsequent F-tests of hypothesis of the parameters in the models for data sets I, II, and III, respectively:

It is granted that the fit of the special cubic model to each of the three sets of data is significant in the sense that in each case we rejected the hypothesis that the mixture components do not influence the response. But the question we ask is: Is the special cubic model the only tool that can be used to describe the response or is it possible that a lower degree model such as a quadratic model or even a linear first degree model fitted to the data will describe the shape of the response surface over the factor space of the components as well as the special cubic model? If a lower degree model does as well as the special cubic model in fitting the data, we would prefer to use the lower degree model, not only because it would be easier to handle when predicting the response but because with a less complicated model frequently it is easier to understand just how the components blend together.

Also the variance of the predicted values is lower with a lower degree model consisting of a fewer number of terms (Rao, 1971).

We now consider some techniques that can be used for reducing the form of the fitted model and again refer to the data in sets I, II, and III of Table 4.1 for illustrative purposes.

4.3 MODEL REDUCTION

In the previous section on testing hypotheses, the data in set II of Table 4.1 were fitted using a special cubic model, then a quadratic model was fitted, and finally a first-degree linear model was fitted. In each case the hypothesis that stated that the texture of the patties did not depend on the fish components comprising the patties was rejected.

Now we want to determine which of the three models does the best job of describing the texture response surface for set II. In other words, which model provides the clearest description of the shape of the texture surface while at the same time requiring the fewest number of terms in order to do so? Our feeling is that the simpler the fitted model form, the easier it is to understand the type of blending that is present among the components and to use the model for prediction purposes, particularly at blends other than the blends that were used for fitting the model.

In choosing the simplest form of the fitted model, we must remember, however, that the simpler the form of the model the more limited is the range of usefulness of the model. If it is feared that the surface will be complicated in shape, normally we will not want to sacrifice the higher precision of a complete model just for simplicity of usage of the reduced model. However, for now let us assume that the surface is generated by a well-behaved system and as such the surface can be characterized by a third-degree polynomial at most. In other words, we are going to assume that a reduced model of low degree will do as good a job of describing the surface as a more complete higher degree model.

Model reduction can be accomplished in several ways. The form of the model can be reduced by simply summing terms (i.e., summing the proportions x_i and x_j) or by setting up contrasts among the terms and using the contrasts as the new terms in the model. One might also delete terms from the model rather than combine the terms. If we use any of these approaches to come up with a reduced model form, then presumably we have simplified the task of trying the interpret the mixture system.

The most obvious approach to model reduction is to remove the nonsignificant higher-degree terms. This can be illustrated using the data of set II in Table 4.1 by comparing the fit of the special cubic model in Eq. (4.9) to the fit of the second-degree model in Eq. (4.11) Briefly, the value of R_A^2 with the special cubic model is

$R_A^2 = 0.956$, whereas with the quadratic model, the value is $R_A^2 = 0.951$. Since these values are very close, the usefulness of the extra cubic-term $b_{123}x_1x_2x_3$ in Eq. (4.9) is questionable. In fact, if we check this result by comparing the value of the estimate $b_{123} = -3.99$ to its estimated standard error by setting up the t-statistic, the value of the statistic, $t = -3.99/2.87 = -1.39$, is not significantly different from zero and this supports our decision to remove or drop the cubic term from the fitted model.

A comparison can be made now between the fitted quadratic model, Eq. (4.11) and the fitted first-degree model, Eq. (4.12) The comparison involves testing the hypothesis $H_0: \beta_{12} = \beta_{13} = \beta_{23} = 0$, and if the hypothesis is not rejected, the second-degree model is discarded in favor of the first-degree model. This test is discussed in Section 7.6, where the sum of squares for regression of the complete model, Eq. (4.11) is compared against the sum of squares for regression of the reduced model, Eq. (4.12), using the F-statistic Eq. (7.22),

$$F = \frac{[\text{SSR of Eq. (4.11)} - \text{SSR of Eq. (4.12)}]/3}{\text{SSE of Eq. (4.11)}/(14-6)}$$

$$= \frac{[3.98 - 3.87]/3}{0.124/8} = 2.36$$

Since the calculated value, $F = 2.36$, is less than $F_{(3,8,0.05)} = 4.07$, we do not reject the null hypothesis, $H_0: \beta_{12} = \beta_{13} = \beta_{23} = 0$, at the 0.05 level of significance. We conclude that the fitted first-degree model, Eq. (4.12), is sufficient for fitting the texture values in set II.

Most software packages offer tests of hypotheses on the individual terms in the model of the form $H_0: \beta_{ij} = 0$ versus $H_A: \beta_{ij} \neq 0$, for all $i < j$. Such tests for the nonlinear blending terms in the quadratic model of Eq. (4.11) are:

Estimate	Standard Error	$|t\text{-Value}|$	$P > t$
-0.91	0.40	2.27	0.053
-0.56	0.40	1.40	0.201
0.07	0.40	0.17	0.869

Since none of the values of $P > t$ is less than 0.05 (although 0.053 nearly is), we infer that none of the nonlinear blending terms are important at the 0.05 level of significance, which supports the F-test above. Additional support for the linear blending model of Eq. (4.12) for fitting data set II comes from noticing very little change in the values of R^2, R_A^2 and R_{PRESS}^2 with the fitted quadratic and special cubic models. For the set II data, the summary statistics for the three fitted models are:

Model	R^2	R_A^2	R_{PRESS}^2
Linear	0.9438	0.9336	0.9142
Quadratic	0.9698	0.9510	0.9225
Special cubic	0.9763	0.9561	0.9054

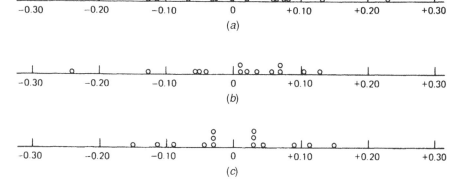

Figure 4.2 Plots of the sizes of the residuals with the (a) first-degree model, (b) second-degree model, and (c) special cubic model fitted to the data of set II. Sizes of the residuals $= y_u - \hat{y}_u$.

The closeness of the values of R^2, R_A^2, and R_{PRESS}^2 for the linear, quadratic, and special cubic models suggests the linear model is our choice.

The planar model of Eq. (4.12) with only three terms is probably the model that would be chosen by most data analysts when fitting a polynomial to the data of set II. Model simplicity and lack of evidence of nonlinearity of the surface are probably the prevailing moods with this data set. Also the residual plots of Figure 4.2 reflect very little difference in the three models. The same conclusion is reached even when the residuals are standardized by dividing them by their standard deviations. Nevertheless, before we proceed to discuss additional techniques for reducing the number of terms in the model, we recall that the t-statistic was used previously for testing the significance of the parameter estimate b_{123} in the special cubic model. Such a test suffered from the same lack of independence as the test for the size of b_{12} in that, in Eq. (4.9) the estimate b_{123} is correlated with b_i and b_{ij} in the amounts

$$\text{corr}(b_{123}, b_i) = \frac{3\sigma^2}{\sqrt{1188\sigma^2(\sigma^2)}} = \frac{1}{11.5}$$

$$\text{corr}(b_{123}, b_{ij}) = \frac{-60\sigma^2}{\sqrt{1188\sigma^2(24\sigma^2)}} = -\frac{1}{2.8}$$

In Figure 4.2, the plots of the individual residuals with each of the models in Eqs. (4.9), (4.11), and (4.12) are presented. Since the special cubic model of Eq. (4.9) is forced through the average response at each design point, the residuals at each point are highly correlated (with duplicated responses the residuals are reflections of each other with opposite signs). The plot of these residuals from the special cubic model is symmetric about zero. The plots of the residuals from the first- and second-degree models are not symmetric, and these latter groups of residuals have a slightly larger spread. In none of the groups, however, are the magnitudes of the residuals determined to be significantly correlated with the values of y_u or $\hat{y}_u$, nor are the magnitudes of the residuals related to the values of the x_i. Similar results appear with standardized residuals.

Another way of reducing the size of the model is to combine some of the terms in the model. Combining terms is achieved by summing the proportions of those components whose coefficients reflect a similar type of blending. For example, if with the fitted first-degree model

$$\hat{y}(\mathbf{x}) = b_1 x_1 + b_2 x_2 + b_3 x_3$$

the estimates b_1 and b_2 are approximately equal in magnitude and their standard errors are approximately equal as well, then the first two terms can be combined to produce the reduced model form

$$\eta = \beta_1(x_1 + x_2) + \beta_3 x_3$$

which is now a two-component model where the components are $x_1 + x_2$ and x_3.

4.4   AN EXAMPLE OF REDUCING THE SYSTEM FROM THREE TO TWO COMPONENTS

To illustrate the reduction of the number of terms in a three-component quadratic model, let us introduce the following data set, where the values represent average quality ratings (rating scale $1 - 8$ but these averages are listed in units of 0.01 for illustrative purposes because we want to use continuous data) assigned to replicate yarn samples produced from the blends specified by a $\{3, 2\}$ simplex-lattice design. The data are

$$(x_1, x_2, x_3) = (1, 0, 0) \ (\tfrac{1}{2}, \tfrac{1}{2}, 0) \ (0, 1, 0) \ (\tfrac{1}{2}, 0, \tfrac{1}{2}) \ (0, 0, 1) \ (0, \tfrac{1}{2}, \tfrac{1}{2})$$

Replicate quality averages					
= 5.77	5.49	5.97	6.17	4.36	5.86
5.61	5.90	5.67	5.84	4.56	6.14

The fitted quadratic model is

$$\hat{y}(\mathbf{x}) = b_1 x_1 + b_2 x_2 + b_3 x_3 + b_{12} x_1 x_2 + b_{13} x_1 x_3 + b_{23} x_2 x_3$$

$$= 5.69 x_1 + 5.82 x_2 + 4.46 x_3 - 0.24 x_1 x_2 + 3.72 x_1 x_3 + 3.44 x_2 x_3 \quad (4.13)$$

$$(0.15) \quad (0.15) \quad (0.15) \quad (0.71) \quad (0.71) \quad (0.71)$$

where below each coefficient estimate in parentheses is the estimated standard error of the coefficient estimate. Note that the estimated standard error, 0.15, is the same for all three linear coefficient estimates, and the same is true of the estimated standard error, 0.71, for the three bindary coefficient estimates. This is a result of using the

$\{3, 2\}$ lattice design with an equal or same number of replicates at each design point. An estimate of the observation variance obtained from the replicate average scores is $s^2 = 0.043$ with six degrees of freedom. The fitted model (4.13) produced an R^2 value of 0.9297.

The magnitudes of the coefficient estimates b_1 and b_2 are approximately the same. Also, because of the near zero value of b_{12}, we are inclined to believe that $\beta_{12} = 0$, which means that the blending of components 1 and 2 with one another is additive (i.e., the blending is neither synergistic nor antagonistic), and thus the $\beta_{12} x_1 x_2$ term is dropped from the model of Eq. (4.13). Also, since the magnitudes of the estimates $b_{13} = 3.72$ and $b_{23} = 3.44$ are approximately equal and both estimates are positive, the binary blending of components 1 and 3 is approximately identical to the binary blending of components 2 and 3. This leads us to believe that components 1 and 2 are similar in their blending properties and thus x_1 and x_2 values are summed in the model. The resulting simplified model is

$$\eta = \beta_1(x_1 + x_2) + \beta_2 x_3 + \beta_{13}(x_1 + x_2)x_3 \qquad (4.14)$$

When Eq. (4.14) is fitted to the average quality data, the resulting equation is

$$\hat{y}(\mathbf{x}) = 5.74(x_1 + x_2) + 4.46x_3 + 3.62(x_1 + x_2)x_3 \qquad (4.15)$$
$$(0.07) \qquad\qquad (0.12) \qquad (0.45)$$

The estimate of the error variance with the fitted model, Eq. (4.15) is $s^2 = 0.031$ with nine degrees of freedom and the R^2 value is $R^2 = 0.9237$ with an $R_A^2 = 0.9067$. Equation (4.15) represents an estimated surface for a two-component system, where the components are the sum of ingredients 1 and 2, and component 3. If future experimentation with these same three ingredients is performed, one must be careful not to assume that components 1 and 2 can be combined automatically when blended with component 3. Only the data for this example support this action and data from future experiments will need to be retested before combining the proportions.

Exercise 4.1 Refit the quadratic model with the average ratings at $(0, \frac{1}{2}, \frac{1}{2})$ now being 4.52 and 4.06 instead of 5.86 and 6.14. Suggest a reduced model form containing at most four terms.

Answer: $\hat{y}(\mathbf{x}) = 5.68x_1 + 5.81x_2 + 4.49x_3 + 3.56(x_1 - x_2)x_3, \ s^2 = 0.04.$

While reaching a decision on the final model form, our acceptance of the model was based on the value of R_A^2 as well as our willingness to accept the magnitude of the estimate of σ^2. With the fitted equation (4.15) the value of $s^2 = 0.03$ seemed to be acceptable to us because the magnitude of the estimate of the error variance obtained from the replicate observations was $s^2 = 0.043$. The choice of using the magnitudes of R_A^2 and of s^2 that we could live with is strictly arbitrary. For this reason we mention another criterion that can be used for deciding on the final form of the fitted model.

Another criterion called the integrated mean square error criterion is discussed in greater detail in Park (1978) and in Cornell (2002, ch.5).

4.5 SCREENING COMPONENTS

In some areas of mixture experiments, specifically in certain chemical industries, many times there is present a large number $q \geq 6$ of potentially important components that can be considered candidates in an experiment. If at all possible, a reduction in the number of necessary components to be studied is sought not only from the standpoint that it is easier to understand a system that contains only a small number of components but also for reasons of economics. Initially the strategy is to identify all the assumed potentially important components and then to single out the ones that are most important. If, to identify the components, it becomes necessary to actually perform experimental runs and decide on the most important components from the sizes of their effects, then the reduction of the number of components so that only the most important components are considered further is known as *screening the components.*

The construction of screening designs and the setting up of screening models quite often begin with the Scheffé first-degree model

$$\eta = \beta_1 x_1 + \beta_2 x_2 + \beta_3 x_3 + \cdots + \beta_q x_q \qquad (4.16)$$

When the ranges of the values of the component proportions x_i can be set as close to each other as possible, the relative effects of the components can be assessed by ranking the ratios of the parameter estimates b_i, $i = 1, 2, \ldots, q$, relative to their standard errors, that is, $b_i/\text{s.e.}(b_i)$. In a similar setting, if the ranges of the values of the x_i are equal, and the equal intervals of interest are centered at $(1/q, 1/q, \ldots, 1/q)$, then one can infer that the larger the value of b_i, relative to b_j, $i \neq j$, the more important component i is relative to component j.

Snee and Marquardt (1976) discuss several strategies that can be used to determine the most important components in a q-component blending system. Designs, called simplex screening designs, are recommended for those cases where it is possible to experiment over the total simplex region (or for the special case when the experimental region is a smaller simplex inside the original simplex). Extreme-vertices designs are suggested when the proportions of some or all of the components are restricted by upper and lower bounds. McLean and Anderson's (1966) procedure for calculating the coordinates of the extreme vertices of a subregion of the simplex is discussed in Section 3.9.

To screen out the unimportant components or to single out the important components, it is necessary to know how to measure the effects of the individual components. We present it as a definition.

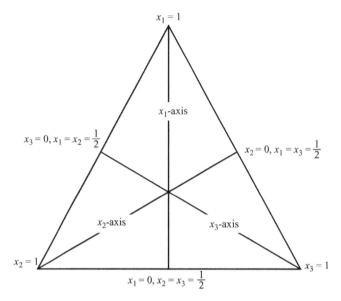

Figure 4.3 The x_i axes, $i = 1$, 2, and 3.

Definition. The effect of component i on the response is the change in the value of the response resulting from a change in the proportion of component i while holding constant the relative proportions of the other components.

The size of the change that is made in the proportion x_i is not specified at this time. The largest change that can be made in the proportion x_i is one unit when going from $x_i = 0$ to $x_i = 1$. If at the time of making this change in x_i we elect to keep the proportions x_j, $j \neq i$, of the other components equal to one another, then the change must be made along the axis of component i. (The axis of component i was defined in Section 2.14 where we introduced axial designs.) In Figure 4.3 are shown the three x_i axes inside the three-component triangle. The values of the proportions of each of the other $q - 1$ components along the x_i axis are $x_j = (1 - x_i)/(q - 1)$, for $j \neq i$.

Because along the x_i axis, the relative proportions of the other $q - 1$ components to one another are unchanged for different values of x_i, observations are collected at several points on the axis of component i, to measure the effect of component i. These data are used to calculate an estimate of the effect of component i. If we wish to measure the effects of all q components simultaneously, we could fit Cox's model of Section 5.6, to data collected at points located on all of the axes to permit the estimation of all of the parameters in Eq. (4.16).

A formula for estimating the effect of component i is derived as follows. Let us assume that an axial design of the type introduced in Section 2.14 is to be used with observations collected at the points on the axes to produce the fitted model $\hat{y}(\mathbf{x}) = b_1 x_1 + b_2 x_2 + \cdots + b_q x_q$. At the base of the x_i axis where $x_i = 0$ and

$x_j = 1/(q-1)$, $j \neq i$, the estimate of the response is

$$\hat{y}(\mathbf{x}) = \sum_{j \neq i}^{q} \frac{b_j}{q-1} = (q-1)^{-1} \sum_{j \neq i}^{q} b_j \qquad (4.17)$$

If data are collected only at the vertices of the simplex, then $\hat{y}(\mathbf{x})$ in (4.17) is the average height of the response surface associated with the other $q-1$ pure components. Similarly, at the vertex $x_i = 1$, $x_j = 0$, $j \neq i$, the estimate of the response is

$$\hat{y}(\mathbf{x}) = b_i \qquad (4.18)$$

This is the estimated height of the surface at $x_i = 1$. Recall that the effect of component i is defined as the change in the response resulting from a change in the proportion of component i while holding constant the relative proportions of other components. An intuitive comparison that could be used to estimate the effect of component i is to calculate the difference between the estimates $\hat{y}(\mathbf{x})$ in Eqs. (4.18) and (4.17). Such a comparison is particularly meaningful if observations are collected at $x_i = 0$, $x_j = 1/(q-1)$, $j \neq i$, and at $x_i = 1$. However, if observations are not collected at these points but the first-degree model (4.16) is considered to be adequate, then we can define

$$\text{Effect of component } i = b_i - (q-1)^{-1} \sum_{j \neq i}^{q} b_j, \quad i = 1, 2, \ldots, q \qquad (4.19)$$

The usefulness of formula (4.19) increases as the variances of the estimates b_i, $i = 1$, $2, \ldots, q$, are closer in magnitude to one another. Thus formula (4.19) is most useful when the ranges of the x_i are equal and the intervals of values for the x_i are centered at the centroid $(1/q, 1/q, \ldots, 1/q)$ of the simplex.

When the effects of two or more components are equal, which is when $b_i = b_j = b_k$ and their estimated standard errors are equal, the proportions of the equivalent components are summed and the sum alone is then considered to be a component. An example of two components having approximately equal effects is the three-component yarn-rating experiment of Section 4.4 where the proportions x_1 and x_2 corresponding to components 1 and 2 were summed to form the single component $(x_1 + x_2)$. As a result the six-term quadratic model in the original three components was rewritten as the simplified three-term model of Eq. (4.14) in only two components.

Formula (4.19) for measuring the effect of component i can be used for purposes of model reduction. For example, if the true surface is planar (i.e., of the first degree) and we find that $b_i = (q-1)^{-1} \sum_{j \neq i}^{q} b_j$, then component i is said to have a "zero effect" or "no effect." When this happens, the term $\beta_i x_i$ is deleted from the model and component i is removed from further consideration.

When the ranges of the component proportions x_i are not equal, which is often the case when the x_i are constrained above and/or below by upper and/or lower bounds such as $L_i \leq x_i \leq U_i$, $i = 1, 2, \ldots, q$, the formula for calculating the effect of component i contains an adjustment factor for the differences in the spreads or ranges in the values of the x_i. The adjustment consists of using the range of x_i, say, $R_i = U_i - L_i$, to weight the effect of component i in Eq. (4.19). The adjusted effect is

$$\text{Adjusted effect of component } i = R_i \left[b_i - (q-1)^{-1} \sum_{j \neq i}^{q} b_j \right] \tag{4.20}$$

With Eq. (4.20), we see that components with the largest ranges receive the greatest weight when defining their effects, where the greatest weight is of size 1.0 for those components having ranges from 0 to 1.0.

Tests of significance on the magnitudes of the effects of components can be performed in much the same way as the linear hypotheses were tested in Section 4.2. This is because the formulas for estimating the component effects can be written in matrix notation as $C\beta = 0$, where C is a $1 \times q$ matrix of scalar coefficients. For example, if we denote by E_1 the estimate of the effect of component 1, the contrast for calculating E_1 using Eq. (4.19) is

$$E_1 = \mathbf{Cb} = \left[1, \frac{-1}{q-1}, \frac{-1}{q-1}, \dots, \frac{-1}{q-1} \right] \mathbf{b}$$

and the estimate of the adjusted effect of component 1, E_{A1}, say, according to Eq. (4.20) is

$$E_{A1} = R_1 \mathbf{Cb} = R_1 \left[1, \frac{-1}{q-1}, \frac{-1}{q-1}, \dots, \frac{-1}{q-1} \right] \mathbf{b}$$

Furthermore, if the effects of all q components are to be estimated simultaneously, then C is written as a $q \times q$ matrix and the *vector* of estimated component effects is

$$\mathbf{E} = \begin{bmatrix} E_1 \\ E_2 \\ \vdots \\ E_q \end{bmatrix} = \frac{1}{q-1}[q\mathbf{I}_q - \mathbf{J}_q]\mathbf{b} = \mathbf{Cb} \tag{4.21}$$

where $\mathbf{I}_q$ is the identity matrix of order q and $\mathbf{J}_q$ is the $q \times q$ matrix of ones. The vector of adjusted component effects is

$$\mathbf{E}_A = \mathbf{RE} = [\text{diag}(R_1, R_2, \dots, R_q)]\mathbf{E} \tag{4.22}$$

where $\mathbf{R}$ is a diagonal matrix whose nonzero diagonal elements are the ranges of the x_i.

The variance–covariance matrix of the vector $\mathbf{E}$ of estimated effect contrasts is $\text{var}(\mathbf{E}) = \mathbf{C}(\mathbf{X'X})^{-1}\mathbf{C'} \, \sigma^2$, since the variance–covariance matrix of the parameter estimates $\mathbf{b}$ is $(\mathbf{X'X})^{-1} \, \sigma^2$. If the observations y_u, $u = 1, 2, \dots, N$, are assumed to be sampled from a normal population, the hypothesis $H_0 : \mathbf{E} = \mathbf{0}$ or all component effects are zero, is tested using the F-test

$$F = \frac{\mathbf{E}'[\mathbf{C}(\mathbf{X}'\mathbf{X})^{-1}\mathbf{C}']^{-}\mathbf{E}}{(q-1)s^2} \qquad (4.23)$$

The value of F in Eq. (4.23) is compared against the tabled value of $F_{(q-1, f, \alpha)}$, where f is the degrees of freedom used in estimating σ^2. Since the $q \times q$ matrix $\mathbf{C}(\mathbf{X}'\mathbf{X})^{-1}\mathbf{C}'$ is at most of rank $q-1$, $[\mathbf{C}(\mathbf{X}'\mathbf{X})^{-1}\mathbf{C}']^{-}$ for use with all of the effects is a generalized inverse matrix. To avoid the reliance on a generalized inverse matrix, one of the effects is dropped from $\mathbf{E}$ in (4.21) so that the test involves only $q-1$ of the E_i. Then the $(q-1) \times (q-1)$ matrix $\mathbf{C}(\mathbf{X}'\mathbf{X})^{-1}\mathbf{C}'$ is of full rank and the standard inverse matrix $[\mathbf{C}(\mathbf{X}'\mathbf{X})^{-1}\mathbf{C}']^{-1}$ is used.

When a test on an individual effect E_i is performed, the test suffers from the property that the E_i are not independent and thus tests on the E_i are not independent. However, if the test of the effect $E_i = \mathbf{C}_i\boldsymbol{\beta} = 0$, where $\mathbf{C}_i$ is the ith row of the matrix $\mathbf{C}$ in Eq. (4.21) is preplanned, such a test can be made using Student's t-test

$$t = \frac{\mathbf{C}_i\mathbf{b}}{\{s^2 c_{ii}\}^{1/2}} \qquad (4.24)$$

where c_{ii} is the iith element on the diagonal of the matrix $\mathbf{C}(\mathbf{X}'\mathbf{X})^{-1}\mathbf{C}'$. The number of degrees of freedom for the t-test in Eq. (4.24) is the number of degrees of freedom associated with the estimate s^2. The test in Eq. (4.24) is equivalent to the F-test in Eq. (4.23) for a single effect, but for more than one effect the F-test is the more general test statistic.

4.6   OTHER TECHNIQUES USED TO MEASURE COMPONENT EFFECTS

When the mixture component proportions are restricted by lower and upper bounds of the form $0 \le L_i \le x_i \le U_i \le 1$, the ranges $R_i = U_i - L_i$, $i = 1, 2, \ldots, q$, are seldom equal or even close to being equal to each other. Unequal ranges produce unequal values for the standard errors of the coefficient estimates and, even by using the ranges as weights in the formula (4.20) for the adjusted effects, we cannot be assured of reliable estimates of the effects of the individual components. Moreover, more often than not, the centroid of the constrained region is different from the centroid of the simplex, so the directions dictated by the component axes may not be the best directions for measuring the component effects.

When measuring the effect of component i and a reference mixture other than the centroid of the simplex is to be used, an alternative direction to that defined by the ith component axis is generally more appropriate. This alternative direction is an imaginary line projected from the reference mixture (which is usually the centroid of the constrained region) to the vertex $x_i = 1$. Such a direction to be used for measuring

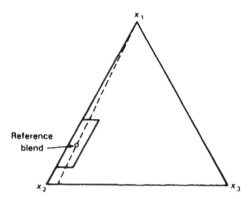

Figure 4.4 Direction indicated by the dashed line for measuring the effect of increasing the proportion x_1 of component 1.

the effect of component 1 is illustrated by the dashed line in Figure 4.4. The reference mixture in Figure 4.4 is the centroid $(x_1, x_2, x_3) = (0.25, 0.70, 0.05)$ of the constrained region defined by the constraints $0.1 \le x_1 \le 0.4$, $0.5 \le x_2 \le 0.9$ and $0 \le x_3 \le 0.1$.

Along the dashed line in Figure 4.4, as the value of x_1 increases (or decreases), the values of x_2 and x_3 decrease (or increase), but the ratio of x_2 to x_3 ($0.70/0.05 = 14:1$) remains the same. As pointed out by Hare (1985), "It is easy to visualize an experimenter pouring ingredient i into a beaker and getting a continuous reading of the response. This is the analogous situation." This direction therefore allows one to measure the effect of a component as defined in Section 4.5. This direction was introduced by Cox (1971) when he suggested that an alternative model form to Scheffé's models be used for measuring the effects of the components. We will refer to this new direction as "Cox's direction," and the dashed line as the x_i ray.

A measure of the amount of change in each of the remaining components when a change of size Δ_i is made in the proportion of component i is determined as follows. Denote the proportions of the q components at the reference mixture by $\mathbf{s} = (s_1, s_2, \ldots, s_q)$, where $s_1 + s_2 + \cdots + s_q = 1$. Suppose that the proportion of component i at s_i is now changed by an amount Δ_i (where Δ_i, could be > 0 or < 0) in Cox's direction, so that the new proportion becomes

$$x_i = s_i + \Delta_i \tag{4.25}$$

Then the proportions of the remaining $q-1$ components resulting from the change (4.25) from s_i in the ith component is

$$\begin{aligned} x_j &= s_j - \frac{\Delta_i s_j}{(1 - s_i)} \\ &= s_j \frac{(1 - s_i - \Delta_i)}{1 - s_i}, \quad j = 1, 2, \ldots, q, \ j \ne i \end{aligned} \tag{4.26}$$

The quantity, $1 - s_i$, in the denominator of (4.26) is the maximum positive value that Δ_i can take. The change Δ_i could be negative, as in the case of moving from the reference mixture away from the vertex $x_i = 1$; when this occurs, the maximum value of Δ_i is $-s_i$. Note that the ratio of the proportions for components j and k, where x_j and x_k are defined by (4.26), is

$$\frac{x_j}{x_k} = \frac{s_j(1 - s_i - \Delta_i)}{s_k(1 - s_i - \Delta_i)}$$

$$= \frac{s_j(1 - x_i)}{s_k(1 - x_i)} = \frac{s_j}{s_k}$$

which is the same value as the ratio of components j and k at the reference mixture s. In other words, when we change the proportion of component i from s_i to $x_i = s_i + \Delta_i$, the quotient of the proportions of components j and k, $j, k \neq i$, does not change from what it was at s.

4.6.1 Plotting the Response Trace

The response trace is a plot of the estimated response values, using the fitted model, along the directions defined by Eqs. (4.25) and (4.26). The steps taken in constructing the response trace are as follows:

Step 1: Select the fitted model form that you feel comfortable with in making predictions of the response over the experimental region.

Step 2: Select a reference mixture **s**. Generally this will be the centroid of the experimental region.

Step 3: On the x_i ray, increment component i by an amount Δ_i using Cox's direction by moving away from the reference mixture toward ($\Delta_i > 0$) the vertex $x_i = 1$ as well as away from ($\Delta_i < 0$) the vertex. Keep the other component proportions in the same ratios as at the reference mixture **s** [see Eq. (4.26)]. For each value of Δ_i, a set of coordinates defines a blend on the x_i ray. Choose some number of blends on the x_i ray at which to make predictions of the response, remembering that blends only within the experimental region are valid.

Step 4: Substitute the coordinates of each blend generated in step 3 into the fitted model to obtain the predicted response values along the x_i ray. Repeat steps 3 and 4 on the other x_j rays, $j = 1, 2, \ldots, q$, $j \neq i$.

Step 5: Plot the predicted response values against changes made in x_i for each i, $i = 1, 2, \ldots, q$. There will be q plots.

Example

In the formulation of vinyl used for automobile seat covers, three plasticizers ($P1$, $P2$, and P_3) are known to comprise 79.5% of the formulation. Of the 79.5%, the

Table 4.2 Vinyl Thickness Measurements at the Nine Blends of the Three-Plasticizer Experiment

Blend	x_1	x_2	x_3	Scaled Thickness Value (y)
1	0.849	0	0.151	8,7
2	0.726	0	0.274	4,6
3	0.474	0.252	0.274	12,10
4	0.597	0.252	0.151	13,10
5	0.6615	0.126	0.2125	18,21
6	0.7875	0	0.2125	12
7	0.600	0.126	0.274	13
8	0.5355	0.252	0.2125	16
9	0.723	0.126	0.151	14

individual plasticizers are restricted by the following constraints:

$$32.5\% \leq P1 \leq 67.5\%$$
$$0\% \leq P2 \leq 20.0\% \tag{4.27}$$
$$12.0\% \leq P3 \leq 21.8\%$$

When expressed in terms of the proportion of total plasticizer present, the component (plasticizer) proportions are defined as $x_i = P_i/79.5\%$ so that the restrictions (4.27) expressed in terms of the component proportions are

$$0.409 \leq x_1 \leq 0.849$$
$$0 \leq x_2 \leq 0.252 \tag{4.28}$$
$$0.151 \leq x_3 \leq 0.274$$

The experimental region in the component proportions is shown in Figure 4.5. Shown also are nine blends defined by the four extreme vertices of the region, the midpoints of the four edges of the region, and the overall centroid of the constrained region. In Table 4.2 are listed the coordinates of the nine blends along with scaled values of the response, vinyl thickness in millimeters. Two replicate measurements were taken from blends 1 to 5 representing the extreme vertices and the overall centroid of the constrained region while only a single measurement was collected from the midedge blends. The nine blends were chosen in order to fit a quadratic model in x_1, x_2, and x_3.

The fitted quadratic model in the plasticizer proportions is

$$\hat{y}(x) = -50.1x_1 - 282.2x_2 - 911.6x_3 + 317.4x_1x_2 + 1464.3x_1x_3$$
$$+ 1846.2x_2x_3 \tag{4.29}$$

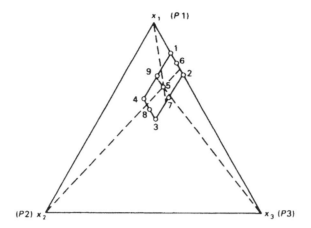

Figure 4.5 The plasticizer experimental region defined by the constraints in Eq. (4.28). Also shown are the effect directions (dashed lines) specified by Eqs. (4.25) and (4.26)

The value of R^2 with the fitted model (4.29) is $R^2 = 0.9500$, and an estimate of the error variance is MSE $= 1.79$ with eight degrees of freedom. The lack-of-fit F-test [see Eq. (2.37)] with three and five degrees of freedom in the numerator and denominator, respectively, produced a value of 0.30, which is clearly nonsignificant. Thus the fitted model (4.29) is considered to be adequate for prediction purposes.

To obtain plots of the predicted response using the model (4.29), first the overall centroid (or blend 5) with the coordinates (0.6615, 0.1260, 0.2125) is chosen as the reference blend. Second, the plotting directions are defined by rays drawn from the centroid to each vertex of the triangle as indicated by the dashed lines in Figure 4.5. Then the values of the incremental changes Δ_1, Δ_2, Δ_3 that are taken along the respective rays are chosen. For this example, we let $\Delta_1 = \Delta_2 = \Delta_3 = 0.02$.

Consider a blend along the x_1 ray in which the proportion of component 1 is increased by an amount equal to $\Delta_1 = 0.02$ over that of the reference blend. Then when x_1 is increased from 0.6615 to 0.6815, the proportion $x_2 + x_3$ is decreased by 0.02. However, along the x_1 ray, the ratio of the proportions, x_2/x_3, remains constant and equal to the value of the ratio, s_2/s_3, at the reference blend, that is, $s_2/s_3 = 0.5929$. The coordinates of the first blend defined on the x_1 ray are

$$x_1 = 0.6615 + \Delta_1(= 0.02), \qquad x_2 = 0.1260 - \frac{0.02(0.1260)}{0.3385}$$

$$= 0.6815 \qquad\qquad = 0.1186$$

$$x_3 = 0.2125 - \frac{0.02(0.2125)}{0.3385}$$

$$= 0.1999$$

and $x_2/x_3 = 0.1186/0.1999 \approx 0.5929$.

When these values of x_1, x_2, and x_3 are substituted into (4.29), the predicted thickness value is $\hat{y}(\mathbf{x}) = 19.0$. The second blend defined on the x_1 ray has the coordinates

$$x_1 = 0.6615 + 2\Delta_1 (= 0.04), \qquad x_2 = 0.1260 - \frac{0.04(0.1260)}{0.3385}$$

$$= 0.7015 \qquad\qquad\qquad = 0.1111$$

$$x_3 = 0.2125 - \frac{0.04(0.2125)}{0.3385}$$

$$= 0.1874$$

and the predicted thickness value for this blend is $\hat{y}(\mathbf{x}) = 18.3$.

In Table 4.3 are listed the coordinate values and predicted response values for nine blends taken along the x_1 ray, 13 blends taken along the x_2 ray, and 7 blends taken along the x_3 ray. Only points on the three rays falling inside the constrained region of Figure 4.5 are of interest, so that the first blend that is listed on the x_1 ray in Table 4.3 and that is farthest away from the $x_1 = 1$ vertex, has the coordinate settings ($x_1 = 0.5815$, $x_2 = 0.1558$, $x_3 = 0.2627$), while the last blend that is listed, and that is the closest blend to the $x_1 = 1$ vertex, has the coordinates ($x_1 = 0.7415$, $x_2 = 0.0962$, $x_3 = 0.1623$). In Figure 4.6 predicted vinyl thickness values using the model (4.29) are plotted (or traced) for the blends defined on each of the three rays. The values along the abscissa in Figure 4.6 represent the deviations from the reference blend as measured along the x_i ray, $i = 1, 2, 3$. The plot was generated by DESIGN-EXPERT software of Stat-Ease (1998).

The response trace plot of Figure 4.6 illustrates the effects of incrementing the proportions of the three plasticizers in the manner discussed previously on the estimated vinyl thickness as one moves away from the reference blend $(0.6615, 0.126, 0.2125)$ in the directions shown in Figure 4.5. Owing to the high thickness values, 18 and 21, at reference blend 5 in Table 4.3, the plot shows a general dropoff in thickness value as one moves away from the reference blend, with the exception of moving slightly in the direction of the $P2$ vertex, where the estimated thickness increases. The highest estimated vinyl thickness ($\hat{y} = 19.55$) appears along the x_2 ray at approximately the coordinate setting $(x_1, x_2, x_3) = (0.6388, 0.156, 0.2051)$.

The parabolic nature of the three curves in Figure 4.6 illustrates that the estimated vinyl thickness value is quite sensitive to changes that are made in the separate plasticizers. If one of the curves had been a horizontal line, say, where the vinyl thickness value did not change with changes in the specific x_i, this would imply that the vinyl thickness value was insensitive (or robust) to changes in the proportion of the ith plasticizer. When this happens we say that we have detected an *inactive ingredient.* Detecting inactive ingredients is often as profound as discovering active ingredients.

To summarize how the response trace can be useful in understanding the roles played by the different components in terms of affecting the response, we mention that (1) components with shorter ranges will have shorter response traces; (2) when fitting the linear blending model where the traces will be straight lines, components

Table 4.3 Coordinates of Blends and the Predicted Vinyl Thickness Values of the Blends Taken along the Three Rays in Figure 4.5

Component 1 Direction[a]				Component 2 Direction[b]				Component 3 Direction[c]			
x_1	x_2	x_3	$\hat{y}$	x_1	x_2	x_3	$\hat{y}$	x_1	x_2	x_3	$\hat{y}$
0.5815	0.1558	0.2627	15.4	0.7523	0.006	0.2417	10.6	0.7119	0.1356	0.1525	14.8
0.6015	0.1483	0.2502	17.1	0.7372	0.026	0.2368	12.9	0.6951	0.1324	0.1725	17.5
0.6215	0.1409	0.2376	18.3	0.7220	0.046	0.2320	14.8	0.6783	0.1292	0.1925	19.0
0.6415	0.1334	0.2251	19.0	0.7069	0.066	0.2271	16.4	*0.6615	0.1260	0.2125	19.3
*0.6615	0.1260	0.2125	19.3	0.6918	0.086	0.2222	17.7	0.6447	0.1228	0.2325	18.4
0.6815	0.1186	0.1999	19.0	0.6766	0.106	0.2174	18.7	0.6279	0.1196	0.2525	16.3
0.7015	0.1111	0.1874	18.3	*0.6615	0.126	0.2125	19.3	0.6111	0.1164	0.2725	13.1
0.7215	0.1037	0.1748	17.1	0.6464	0.146	0.2076	19.5				
0.7415	0.0962	0.1623	15.4	0.6312	0.166	0.2028	19.5				
				0.6161	0.186	0.1979	19.1				
				0.6010	0.206	0.1930	18.3				
				0.5858	0.226	0.1882	17.3				
				0.5707	0.246	0.1833	15.8				

Note: These are coordinates of the reference blend.
[a]$x_2/x_3 = 0.5929 = s_2/s_3$.
[b]$x_1/x_3 = 3.1129 = s_1/s_3$.
[c]$x_1/x_2 = 5.25 = s_1/s_2$.

having the steepest response traces are thought to have the greatest impact on the response; and (3) two or more components with nearly identical response traces, and taken over approximately the same range of values, are thought to have approximately identical effects on the response.

4.6.2 Total and Partial Effects

To measure the component effects when the mixture region is constrained, Piepel (1982) defines two types of effects: the *total effect* of component i, and the *partial effect* due to a change Δ_i in the proportion of the ith component. Both types of effects use the fitted model in the following way. Let $\hat{y}(\mathbf{x})$ denote the predicted value of the response at the point $\mathbf{x} = (x_1, x_2, \ldots, x_q)$. Then the *total effect of component i*, denoted by TE_i, is defined as the difference

$$TE_i = \hat{y}(\mathbf{x}_H) - \hat{y}(\mathbf{x}_L) \tag{4.30}$$

The points $\mathbf{x}_L$ and $\mathbf{x}_H$ represent the lowest and highest possible settings, respectively, for component i, conditioned on both points $\mathbf{x}_L$ and $\mathbf{x}_H$ being in the constrained region. The *partial effect* of the ith component, owing to a proportional change of Δ_i while moving from the point $\mathbf{x}_M$ to the point $\mathbf{x}_N$, is defined as the difference

$$PE(\Delta_i) = \hat{y}(\mathbf{x}_N) - \hat{y}(\mathbf{x}_M) = \hat{y}(\mathbf{x}_N - \mathbf{x}_M) \tag{4.31}$$

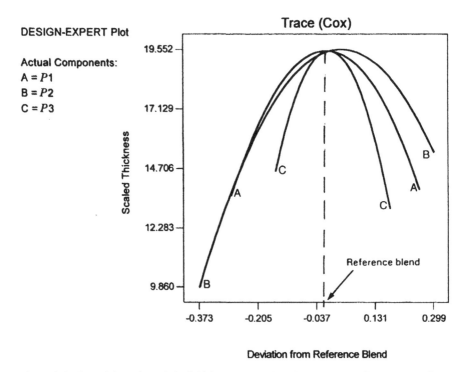

Figure 4.6 Plots of the estimated vinyl thickness values along the x_i, $i = 1, 2, 3$, rays, called Cox's directions, using model (4.29). The plots were generated by DESIGN-EXPERT of Stat-Ease (1998).

We write the coordinates of the points as $\mathbf{x}_l = (\mathbf{x}_{l1}, \mathbf{x}_{l2}, \dots, \mathbf{x}_{lq})$ for $l = M$ and N, the coordinates of the difference, $\mathbf{x}_N - \mathbf{x}_M$, when taken along Cox's direction for component i, as

$$\mathbf{x}_N - \mathbf{x}_M = (-\Delta_i x_{M1}, \ -\Delta_i x_{M2}, \dots, (1 - x_{Mi})\Delta_i,$$
$$-\Delta_i x_{Mi+1}, \dots, -\Delta_i x_{Mq})/(1 - x_{Mi}) \tag{4.32}$$

Substituting these coordinates into the fitted model $\hat{y}(\mathbf{x}_N - \mathbf{x}_M)$ produces the value in (4.31) for PE(Δ_i).

To find the coordinates of the blends used for calculating the estimates of the total effects and of the partial effects of each of the components, Piepel defines a new direction that is closely related to Cox's direction. The new direction is defined in the L-pseudocomponent system. Using only the lower-bound constraints, L_i, $i = 1, 2, \dots, q$, the L-pseudocomponent x_i' is defined as in (3.3),

$$x_i' = \frac{x_i - L_i}{1 - \sum_{i=1}^{q} L_i}, \qquad i = 1, 2, \dots, q \tag{4.33}$$

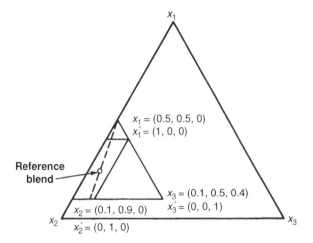

Figure 4.7 Direction from the reference blend to the x_1' vertex for measuring the partial effect of component 1. The coordinate values are for (x_1, x_2, x_3) in the original components.

The L-pseudocomponent region is a simplex with vertices defined by $x_i' = 1, x_j' = 0, j \neq i$.

The partial effect (4.31) for component i is calculated by taking $\mathbf{x}_M$ to be the centroid of the constrained region and by selecting the point $\mathbf{x}_N$ to lie on the imaginary line projected from $\mathbf{x}_M$ to the L-pseudocomponent vertex $x_i' = 1$ at a distance of Δ_i. This new direction for component 1 is illustrated by the dashed line in Figure 4.7. The total effect (4.30) of component i is obtained by substituting the range R_i, or using the difference between the highest and lowest values for x_i that fall in the constrained region, for Δ_i in (4.31) and then calculating $\hat{y}(\mathbf{x}_N - \mathbf{x}_M)$.

Let us work through Piepel's procedure for calculating the partial and total effects for the ith component, using the three-component constrained-region example of Figure 4.7. In Figure 4.7 the constrained region is defined by the constraints $0.1 \leq x_1 \leq 0.4, 0.5 \leq x_2$, and $x_3 \leq 0.1$, where the coordinates of the centroid of the region are $(0.25, 0.70, 0.05)$. Let us define the L-pseudocomponents,

$$x_1' = \frac{x_1 - 0.1}{0.4}, \quad x_2' = \frac{x_2 - 0.5}{0.4}, \quad x_3' = \frac{x_3}{0.4}$$

The L-simplex is the small triangle in Figure 4.7 whose vertices in the L-pseudocomponents correspond to the following coordinates in the original components:

$$(x_1', x_2', x_3') = (1, 0, 0) \Rightarrow (x_1, x_2, x_3) = (0.5, 0.5, 0)$$
$$(0, 1, 0) \Rightarrow \qquad\qquad = (0.1, 0.9, 0)$$
$$(0, 0, 1) \Rightarrow \qquad\qquad = (0.1, 0.5, 0.4)$$

and the coordinates of the centroid of the constrained region are

$$(s_1', s_2', s_3') = (0.375, \ 0.50, \ 0.125) \Rightarrow (s_1, \ s_2, \ s_3) = (0.25, \ 0.70, \ 0.05)$$

To get the partial effect of component i owing to a change of Δ_i in the proportion of component i, the points $\mathbf{x}_M$ and $\mathbf{x}_N$ are obtained as follow:

Define

$$\mathbf{x}_M : \quad x_{Mi} = s_i + \delta_i$$

where $\delta_i (< 0)$ is a constant that allows us to measure the effect beginning with the lowest possible value for x_i. If $\mathbf{x}_M$ is to be the reference blend (centroid of the constrained region), then $\delta_i = 0$,

$$x_{Mj} = s_j - \frac{\delta_i s_j'}{1 - s_i'}, \quad j \neq i \tag{4.34}$$

Define

$$\mathbf{x}_N : \quad x_{Ni} = x_{Mi} + \Delta_i$$

$$x_{Nj} = x_{Mj} - \frac{\Delta_i s_i'}{1 - s_i'}, \quad j \neq i \tag{4.35}$$

For example, let $i = 1$ and $\Delta_1 = 0.05$. Suppose that we choose the point $\mathbf{x}_M$ so that $x_1 = 0.1$; that is, let $\delta_1 = -0.15$. Then the coordinates of $\mathbf{x}_M$ are

$$x_{M1} = s_1 + \delta_1 = 0.25 + (-0.15) = 0.10$$

$$x_{M2} = s_2 - \frac{\delta_1 s_2'}{1 - s_1'} = 0.70 - \frac{(-0.15)(0.50)}{1 - 0.375} = 0.82$$

$$x_{M3} = s_3 - \frac{\delta_1 s_3'}{1 - s_1'} = 0.05 - \frac{(-0.15)(0.125)}{1 - 0.375} = 0.08$$

so we have $\mathbf{x}_M = (0.1, 0.82, 0.08)$. Now, corresponding to a positive change of $\Delta_1 = 0.05$ in x_1 from $\mathbf{x}_M$,

$$x_{N1} = x_{M1} + \Delta_1 = 0.1 + 0.05 = 0.15$$

$$x_{N2} = x_{M2} - \frac{\Delta_1 s_2'}{1 - s_1'} = 0.82 - \frac{(0.05)(0.50)}{1 - 0.375} = 0.78$$

$$x_{N3} = x_{M3} - \frac{\Delta_1 s_3'}{1 - s_1'} = 0.08 - \frac{(0.05)(0.125)}{1 - 0.375} = 0.07$$

so we have

$$x_N = (0.15, \ 0.78, \ 0.07).$$

The partial effect of component 1, upon moving from $\mathbf{x}_M$ to $\mathbf{x}_N$, is computed by calculating the difference

$$\text{PE}(\Delta_1 = 0.05) = \hat{y}(0.15, \ 0.78, \ 0.07) - \hat{y}(0.10, \ 0.82, \ 0.08) \tag{4.36}$$

Note that the fitted model can be any of the Scheffé-type model forms (or any of the forms discussed in Chapter 5 and is not limited to being of the first degree.

The total effect of component i is calculated as in (4.36) and by first replacing $\mathbf{x}_M$ in (4.34) with $\mathbf{x}_L$, and $\mathbf{x}_N$ in (5.48) with $\mathbf{x}_H$, respectively, where for component i,

$$x_{Li} = L_i \quad \text{(i.e., } \delta_i = L_i - s_i\text{)}$$

and

$$\Delta_i = U_i - L_i$$

If U_i is unattainable, then use the implied upper bound U_i^* to get Δ_i. For component 1, $L_1 = 0.1$, $\delta_1 = 0.1 - 0.25 = -0.15$, and $\Delta_1 = 0.4 - 0.1 = 0.3$. Then $\mathbf{x}_L = (0.1, 0.82, 0.08)$ and $\mathbf{x}_H = (0.4, 0.58, 0.02)$, so that

$$\text{TE}_1 = \hat{y}(0.4, \ 0.58, \ 0.02) - \hat{y}(0.1, \ 0.82, \ 0.08) \tag{4.37}$$

Performing tests of significance of the magnitudes of the partial and total effects of component i as well as setting up a $(1 - \alpha) \times 100\%$ confidence interval on the expected effects is possible once standard errors of these effects are calculated. Write the standard error (s.e.) of the predicted response value at $\mathbf{x}$, obtained with the fitted model, as

$$\text{s.e.}[\hat{y}(\mathbf{x})] = \{\mathbf{x}'(\mathbf{X}'\mathbf{X})^{-1}\mathbf{x}\sigma^2\}^{1/2}$$

Here $\mathbf{X}$ is the $N \times p$ matrix of component settings and cross-products between the component settings if $p > q$. Then

$$\text{s.e.}[\mathbf{PE}(\Delta_i)] = \{(\mathbf{x}_N - \mathbf{x}_M)'(\mathbf{X}'\mathbf{X})^{-1}(\mathbf{x}_N - \mathbf{x}_M)\sigma^2\}^{1/2} \tag{4.38}$$

Furthermore $\text{s.e.}[\text{TE}_i] = \{(\mathbf{x}_H - \mathbf{x}_L)'(\mathbf{X}'\mathbf{X})^{-1}(\mathbf{x}_H - \mathbf{x}_L)\sigma^2\}^{1/2}$. If σ^2 in (4.38) is unknown, then the error mean square in the analysis of variance table for the fitted model is used as an estimate of σ^2. The tests of the hypothesis H_0: $\text{PE}(\Delta_i) = 0$ versus

H_A: $\mathrm{PE}(\Delta_i) \neq 0$, or H_0: $\mathrm{TE}_i = 0$ versus H_A: $\mathrm{TE}_i \neq 0$, are performed at the α level of significance using the t-statistic:

$$t = \frac{\mathrm{PE}(\Delta_i)}{\widehat{\mathrm{s.e}}[\mathrm{PE}(\Delta_i)]} \quad \text{or} \quad t = \frac{\mathrm{TE}_i}{\widehat{\mathrm{s.e}}[\mathrm{TE}_i]} \tag{4.39}$$

The value of the calculated t-statistic in (4.39) is compared to the table value, $t_{v,\alpha/2}$, where v is the number of degrees of freedom associated with the estimate of σ^2. The null hypothesis is rejected when the computed value of t in (4.39) exceeds the table value of $t_{v,\alpha/2}$.

Smith (2005) devotes an entire chapter to component effects. This allows him to compare Cox's and Piepel's directions in greater detail than is covered in this book.

4.7 LEVERAGE AND THE HAT MATRIX

In some mixture designs one finds that a single design point or a subset of points exert a disproportionate amount of influence on the least-squares fit of the model. As a result the parameter estimates and/or predictions with the model may depend more on the influential subset than on the majority of the data. According to Montgomery and Voth (1994), a point or a set of points may be influential because of its/their location in the simplex region, or the value(s) of the response observed at the point(s), or both. The influence of a particular point or subset of points is directly related to the amount of leverage the point or points have on the least-squares fit; in other words, the higher the leverage the greater the influence. Furthermore the variance of prediction by the fitted model is directly proportional to the leverage at the point so that if the distribution of leverage is very nonuniform across the experimental region, so will the variance of prediction be nonuniform.

In Section 7.1 a review of least squares using matrix notation is presented with the properties of the parameter estimates outlined. Briefly, in matrix notation, let us write the Scheffé general linear mixture model to be fitted as

$$\mathbf{y} = \mathbf{X}\boldsymbol{\beta} + \boldsymbol{\varepsilon} \tag{4.40}$$

where $\mathbf{y}$ is an $N \times 1$ vector of observations, $\mathbf{X}$ is an $N \times p$ matrix with the uth row containing the values of the p predictor variables (with only linear blending terms $p = q$, while with linear plus binary cross-product terms $p = q(q+1)/2$), $\boldsymbol{\beta}$ is a $p \times 1$ vector of unknown parameters, and $\boldsymbol{\varepsilon}$ is an $N \times 1$ vector of random errors with $E(\boldsymbol{\varepsilon}) = \mathbf{0}$ and $\mathrm{var}(\boldsymbol{\varepsilon}) = \sigma^2 \mathbf{I}_N$. The least-squares estimator for $\boldsymbol{\beta}$ is $\mathbf{b} = (\mathbf{X'X})^{-1} \mathbf{X'y}$ and the variance–covariance matrix of $\mathbf{b}$ is $\mathrm{var}(\mathbf{b}) = (\mathbf{X'X})^{-1} \sigma^2$. The $N \times 1$ vector of predicted values at the design points is

$$\hat{\mathbf{y}} = \mathbf{Xb} = \mathbf{X}(\mathbf{X'X})^{-1}\mathbf{X'y} = \mathbf{Hy} \tag{4.41}$$

and the $N \times 1$ vector of residuals at the design points is

$$\mathbf{e} = \mathbf{y} - \hat{\mathbf{y}} = (\mathbf{I} - \mathbf{H})\mathbf{y} \tag{4.42}$$

In Eqs. (4.41) and (4.42) the idempotent matrix $\mathbf{H} = \mathbf{X}(\mathbf{X}'\mathbf{X})^{-1}\mathbf{X}'$ is known as the "hat" matrix.

A measure of the leverage of the uth design point is the diagonal element h_{uu} of the hat matrix $\mathbf{H}$. The element h_{uu} is a measure of the location of the uth design point in the simplex and while the elements $h_{uu'}$, $u' \neq u$, of $\mathbf{H}$ may be interpreted as the amount of leverage exerted by $\hat{y}_{u'}$ on y_u, generally speaking, most of the attention has been focused on the diagonal elements h_{uu} (see Belsey, Kuh, and Welsch, 1980). Furthermore, since $\sum_{u=1}^{N} h_{uu} = \text{trace } (\mathbf{H}) = \text{rank } (\mathbf{H}) = \text{rank } (\mathbf{X}) = p$, the average size of the diagonal elements h_{uu}, $u = 1, 2, \ldots, N$, of the $\mathbf{H}$ matrix is p/N. A rule of thumb taken from the nonmixture regression literature for the value of h_{uu} in order for the uth design point to be considered a high-leverage point is $h_{uu} > 2p/N$ as given in Montgomery and Voth (1994). However, Montgomery and Voth go on to say that such a rule should not be applied to mixture designs because p is often large relative to N, resulting in $2p/N > 1$. They suggest a point has high leverage if the value of $h_{uu} \geq 0.75$.

Montgomery and Voth (1994) discuss the nonuniformity of leverage of design points even for configurations that are considered good designs such as the $\{q,m\}$ simplex-lattice and simplex-centroid designs. They suggest that one should try to make the distribution of leverage more uniform, if at all possible, by replicating observations at the high leverage points, particularly when the objective of the experiment is to use the prediction equation for describing the shape of the mixture response surface. Also, since it is possible to evaluate the leverages of points in any of the standard simplex-shaped designs as well as with computer-generated designs before actually selecting a design, they recommend doing so as a standard practice.

There are several other statistics that are useful in diagnosing the influence of points. One that appears in most of the more popular software packages is Cook's distance measure

$$D_u = \frac{(\mathbf{b} - \mathbf{b}_{(u)})'(\mathbf{X}'\mathbf{X})(\mathbf{b} - \mathbf{b}_{(u)})}{p\,\text{MSE}}$$

for $u = 1, 2, \ldots, N$. The vector $\mathbf{b}_{(u)}$ is the least-squares estimator of $\boldsymbol{\beta}$ with the uth observation removed. Computationally

$$D_u = \frac{r_u^2 h_{uu}}{p(1 - h_{uu})}$$

where $r_u = e_u /\{\text{MSE } (1 - h_{uu})\}^{1/2}$ is the uth studentized residual (see Montgomery, Peck, and Vining, 2001). Large values of Cook's distance ($D_u > 1$) can occur when a point has high leverage (large h_{uu}) or when there is a large studentized residual value indicating an unusual y value.

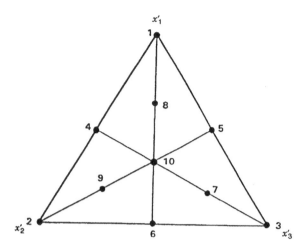

Figure 4.8 Ten points of the augmented simplex-centroid design.

4.8   A THREE-COMPONENT PROPELLANT EXAMPLE

The following three-component example is taken from Montgomery and Voth (1994). The three components are fuel (x_1), oxidizer (x_2), and binder (x_3). Together they form a propellant used in aircrew escape systems. As with most mixture experiments, several responses are of interest, and in this example three are listed. They are the average burning rate, y_1, measured by testing several samples of propellant from the same blend and recording the average (cm/s); the standard deviation of the recorded burning rates for each blend, y_2; and a manufacturability index, y_3, that reflects the cost and difficulty associated with producing a particular mixture or blend. Upper- and lower-bound constraints are placed on the component proportions of the form

$$0.30 \leq x_1 \leq 0.50, \quad 0.20 \leq x_2 \leq 0.40, \quad 0.20 \leq x_3 \leq 0.40$$

where $x_1 + x_2 + x_3 = 0.9$. In other words, the three components make up exactly 90% of the propellant. Consequently, if we wish to study how changing the proportion x_1, x_2, and x_3 affects any and all of the three responses, the components comprising the remaining 10% must be held fixed for all blends studied.

In order to illustrate the nonuniformity of leverage of points associated with a good design, Montgomery and Voth (1994) selected an augmented simplex-centroid design of the type shown earlier in Figure 2.17*b* and in Figure 4.8 as a basic 10-point design. The six-term Scheffé quadratic model is the model of choice and in order to be able to test the fitted model for adequacy of fit, or what amounts to the same, to test for lack of fit, the next question is, "At which of the design points should replicate observations be collected in order to obtain an estimate of the *pure experimental error variance* for testing lack of fit of the model?" Listed in Table 4.4 are the leverage values at the design points for three separate designs. Design A is the

Table 4.4   Leverage Values Associated with Three Versions of an Augmented Simplex-Centroid Design in Three Components for Fitting a Second-Order Model

Design Point (Type[a])	Design A: One Replicate h_{uu}	Design B: Two Replicates h_{uu}	Design C: h_{uu}
1 (V)	0.930	0.465	0.479
2 (V)	0.930	0.465	0.479
3 (V)	0.930	0.465	0.479
4 (E)	0.736	0.368	0.694
5 (E)	0.736	0.368	0.694
6 (E)	0.736	0.368	0.694
7 (A)	0.251	0.125	0.186
8 (A)	0.251	0.125	0.186
9 (A)	0.251	0.125	0.186
10 (C)	0.250	0.125	0.162

[a]V is vertex, E is midedge, A is axial, C is centroid

Source: © 1994 American Society for Quality, Reprinted with permission.

single replicate ten-point design ($N = 10$). Design B is design A with two replicate observations collected at each point ($N = 20$). Design C is design A with each of the vertices replicated a second time and the centroid point replicated three times ($N = 15$). The coordinates of the design points are listed in Table 4.5 in the metric of L-pseudocomponents, where

$$x_1' = \frac{x_1 - 0.30}{0.9 - (0.3 + 0.2 + 0.2)}, \quad x_j' \frac{x_j - 0.20}{0.9 - (0.3 + 0.2 + 0.2)}, \quad j = 2, 3 \quad (4.43)$$

so that $x_1' + x_2' + x_3' = 1$. In the denominator $0.9 = x_1 + x_2 + x_3$.

From Table 4.4 we notice with the equally replicated designs A and B that the highest values of the diagonal elements h_{uu} of the **H** matrix occur at the vertices followed by the midedge points and then the interior points. In other words, observations collected at or near the boundary of the experimental region exert greater influence on the coefficient estimates as well as the variance of prediction with the fitted model than do observations collected well inside the region. Furthermore, since the maximum value that h_{uu} can take is 1.0, the value of $h_{uu} = 0.93$ at the vertices of the single replicate design A is especially troublesome.

Replicating observations at the design points will always reduce the leverage of the points. This is noticed by comparing the h_{uu}'s of designs A and B, where with design B, $h_{uu} = \frac{1}{2} (h_{uu}$ of A). But it is not necessary to collect replicate observations at all of the design points in B to reduce the influence of the high leverage points in design A and that is where design C comes into play. With design C, the high leverage vertices of A are replicated two times while the centroid point is replicated three times. The end result is a more uniform leverage design at a lower cost ($N = 15$)

Table 4.5 Data for the Aircraft Propellant Experiment

Design Point	Pseudocomponents Fuel x_1'	Oxidizer x_2'	Binder x_3'	Average Burning Rate (y_1)	Standard Deviation of Burn Rate (y_2)	Manufacturability Index (y^3)
1	1	0	0	32.5	4.1	32
	1	0	0	37.9	3.7	25
2	0	1	0	54.5	8.9	18
	0	1	0	32.5	9.2	21
3	0	0	1	64.0	14.0	14
	0	0	1	78.5	13.0	16
4	$\frac{1}{2}$	$\frac{1}{2}$	0	44.0	6.8	20
5	$\frac{1}{2}$	0	$\frac{1}{2}$	63.2	4.7	18
6	0	$\frac{1}{2}$	$\frac{1}{2}$	94.0	4.5	17
7	$\frac{1}{3}$	$\frac{1}{3}$	$\frac{1}{3}$	112.5	4.6	19
	$\frac{1}{3}$	$\frac{1}{3}$	$\frac{1}{3}$	98.5	3.5	20
	$\frac{1}{3}$	$\frac{1}{3}$	$\frac{1}{3}$	103.6	3.0	18
8	$\frac{2}{3}$	$\frac{1}{6}$	$\frac{1}{6}$	67.1	3.5	20
9	$\frac{1}{6}$	$\frac{2}{3}$	$\frac{1}{6}$	73.0	5.2	22
10	$\frac{1}{6}$	$\frac{1}{6}$	$\frac{2}{3}$	87.5	7.0	17

than would be required with design B. Thus design C was the choice of the engineers running the experiments.

Listed in Table 4.5 are the coordinates of the 15-point design C in the L-pseudocomponents of Eq. (4.43) along with the values of the average burning rate (y_1), standard deviation of the burning rate (y_2), and manufacturability index (y_3). When fit to the values of the average burning rate data, the fitted Scheffé quadratic model is

$$\hat{y}_1(\mathbf{x}') = 33.7x_1' + 41.0x_2' + 68.6x_3' + 83.9x_1'x_2' + 104.2x_1'x_3' + 204.7x_2'x_3'$$
$$\quad (9.5) \quad\quad (9.5) \quad\quad (9.5) \quad\quad (52.6) \quad\quad (52.6) \quad\quad (52.6)$$
$$(4.44)$$

with an $R^2 = 0.8247$ and MSE $= 186.11$. Other relevant model summary statistics are (from Section 4.2) PRESS $= 8284.9$, $R_A^2 = 0.7274$, and $R_{PRESS}^2 = 0.1332$. The value of $R_{PRESS}^2 = 0.1332$ is considerably lower than $R_A^2 = 0.7274$ and $R^2 = 0.8247$, which suggests the model (4.44) is not adequate. If we believe that is true and we fit the special cubic model, we obtain

$$\hat{y}_1(\mathbf{x}') = 35.5x_1' + 42.8x_2' + 70.4x_3' + 16.0x_1'x_2' + 36.3x_1'x_3' + 136.8x_2'x_3' + 855.0x_1'x_2'x_3'$$
$$\quad (6.1) \quad\quad (6.1) \quad\quad (6.1) \quad\quad (38.3) \quad\quad (38.3) \quad\quad (38.3) \quad\quad (229.2)$$
$$(4.45)$$

with $R^2 = 0.9360$ and MSE $= 76.43$. Other relevant model summary statistics are PRESS $= 3297.0$, which is only 40% of the PRESS with Eq. (4.44), $R_A^2 = 0.8881$,

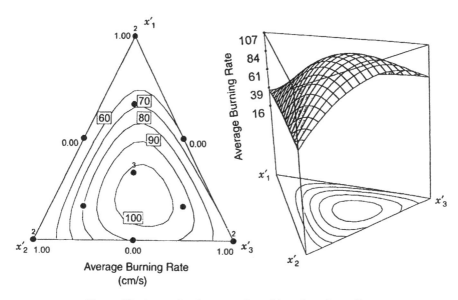

Figure 4.9 Average burning rate surface of the estimated propellant.

and $R^2_{\text{PRESS}} = 0.6550$. These improvements in the model summary statistics suggest that the special cubic model is a better model than the quadratic model of Eq. (4.44) and a plot of the estimated burning rate surface contours and a three-dimensional surface plot are shown in Figure 4.9. Montgomery and Voth (1994) likewise claim the special cubic model to be the better model for the burning rate response.

4.9 SUMMARY

In this chapter several techniques that are used in the analysis of mixture data were presented. In Section 4.1 three data sets were presented that served to help us initiate our model-fitting scheme. The modeling procedure that was suggested consisted of fitting the complete Scheffé special cubic model and the subsequent testing of the usefulness of the terms in the model. The strategy of starting with the special cubic model and working to simplify its form by testing the usefulness of the higher degree terms is a reversal of the approach that many experimenters might propose. This is because generally one starts with the simplest form of the model and adds on terms until the surface is fitted closely (or represented adequately), as described by Draper and St. John (1977a) and in Section 5.1. Our reverse strategy was outlined because the data sets had already been collected, enabling us to proceed by fitting the special cubic model initially.

Screening out the unimportant components from a large group of components by measuring and detecting the effects of the most important components is another topic that is presented. The purpose behind screening out the unimportant components is to reduce the number of essential components in the mixture system in the hope of reducing the complexity of the system. Formulas are provided for estimating the component effects so as to discover the most important components. It was pointed out that the formulas are generally most useful when simplex screening designs are used. Model reduction is possible by eliminating the nonsignificant terms in the model, either by deleting components or by combining component proportions. This is discussed in greater detail in Chapter 5 (Cornell 2002).

When the data are collected from a highly constrained region, plotting the estimated response is recommended for studying the behavior in the response while changing the proportion of a specific component and allowing the remaining components to take up the slack. A three-plasticizer example is presented where the response—vinyl thickness—is seen to be quite sensitive to changes made in each plasticizer proportion. The use of the response trace as well as the computing of partial and total effects of the components can be accomplished quite easily even with systems involving a large ($q > 5$) number of components.

The final two sections of this chapter contain a discussion on the impact that one or more design points, based on their position in the experimental region, might have on the least-squares fit of the model. The impact or influence of a particular point is directly related to and is thus measured by the amount of leverage the point exerts where the higher the leverage the greater the influence. Furthermore the variance of prediction of the fitted model is affected by the leverage of a particular point or set of points. It is shown that the leverage of a point or a set of points can be reduced by collecting replicate observations at the particular point or set of points. This information is available to the modeler through the study of the elements of the "hat" matrix before any experiments are performed.

REFERENCES AND RECOMMENDED READING

Becker, N. G. (1978). Models and designs for experiments with mixtures. *Austral. J. Stat.*, **20**, 195–208.

Belsey, D. A., E. Kuh, and R. E. Welsch (1980). *Regression Diagnostics: Identifying Influential Data and Sources of Collinearity*. Wiley, New York.

Cornell, J. A. (1975). Some comments on designs for Cox's mixture polynomial. *Technometrics*, **17**, 25–35.

Cornell, J. A. (2002). *Experiments with Mixtures: Design, Models, and the Analysis of Mixture Data*, 3rd ed. Wiley, New York.

Cornell, J. A., and L. Ott (1975). The use of gradients to aid in the interpretation of mixture response surfaces. *Technometrics*, **17**, 409–424.

Cox, D. R. (1971). A note on polynomial response functions for mixtures. *Biometrika*, **58**, 155–159.

Hare, L. B. (1985). Graphical display of the results of mixture experiments. In: *Experiments in Industry*, edited by R. D. Snee, L. B. Hare, and J. R. Trout. American Society for Quality Control, Milwaukee, WI, pp. 99–109.

Marquardt, D. W., and R. D. Snee (1974). Test statistics for mixture models. *Technometrics*, **16**, 533–537.

Minitab (1999). *User's Guide 2: Data Analysis and Quality Tools, Release 13*. Minitab Inc., State College, PA.

Montgomery, D. C., and S. R. Voth (1994). Multicollinearity and leverage in mixture experiments. *J. Qual. Technol.*, **26**, 96–108.

Montgomery, D. C., E. A. Peck, and G. G. Vining (2001). *Introduction to Linear Regression Analysis*, 3rd ed. Wiley, New York.

Park, S. H. (1978). Selecting contrasts among parameters in Scheffé's mixture models: screening components and model reduction. *Technometrics*, **20**, 273–279.

Park, S. H., and J. I. Kim (1988). Slope-rotatable designs for estimating the slope of response surfaces in experiments with mixtures. *J. Korean Stat. Soc.*, **17**, 121–133.

Piepel, G. F. (1982). Measuring component effects in constrained mixture experiments. *Technometrics*, **24**, 29–39.

Piepel, G. F. (1990). Screening designs for constrained mixture experiments derived from classical screening designs. *J. Qual. Technol.*, **22**, 23–33.

Piepel, G. F. (1991). An addendum. *J. Qual. Tech.*, **23**, 96–101.

Piepel, G. F. (1997). Survey of software with mixture experiment capabilities. *J. Qual. Tech.*, **29**, 76–85.

Scheffé, H. (1963). The simplex-centroid design for experiments with mixtures. *J. R. Stat. Soc. B*, **25**, 235–263.

Smith, W. F. (2005). Experimental Design for Formulation. *ASA-SIAM Series on Statistics and Applied Probability*, ASA, Alexandria, VA.

Snee, R. D. (1973). Techniques for the analysis of mixture data. *Technometrics*, **15**, 517–528.

Snee, R. D., and D. W. Marquardt (1976). Screening concepts and designs for experiments with mixtures. *Technometrics*, **18**, 19–29.

Stat-Ease (1998). *DESIGN-EXPERT Software for Response Surface Methodology and Mixture Experiments, Version 6.0*. Stat-Ease, Inc., Minneapolis, MN.

QUESTIONS

4.1. In a three-component system where data are collected at the seven points of a simplex-centroid design, a decision is to be made to fit the special cubic model to the data values collected at the points. List some advantages to fitting the complete model as well as some advantages to fitting a reduced model form.

4.2. To the data of set III in Table 4.1 fit the first-degree model $y = \beta_1 x_1 + \beta_2 x_2 + \beta_3 x_3 + \varepsilon$ and test the hypothesis $H_0 : \beta_1 = \beta_2 = \beta_3 = \beta_0$. Are you satisfied with a planar model fitted to this data set? If not, explain.

4.3. The following data set was collected in an experiment involving the blending of a saltwater fish (x_1) with two brands (A and B) of soy protein supplement filler. Fish patties were formulated and the scores represent the sum of flavor plus texture.

$x_1 = 100\%$ F	$x_2 = 70\%$ F:30% S_A	$x_3 = 70\%$ F:30% S_B	Replicated Scores
1.0	0	0	6.8, 6.2
0	1.0	0	3.7, 5.7
0.5	0.5	0	9.5, 10.6, 10.2
0	0	1.0	2.9, 3.5
0	0.5	0.5	8.2, 8.0, 6.2
0.5	0	0.5	11.6, 11.9, 14.3

(a) Fit a first-degree model to the scores and compute the analysis of variance quantities, including R_A^2. Test $H_0: \beta_1 = \beta_2 = \beta_3 = $ constant.

(b) Fit a second-degree model to the scores and compute the increase in the regression sum of squares with this model over the model in (a). Is the increase significant? What is the value of R_A^2 with the second-degree model? Calculate the C_p values associated with the first- and second-degree models using $s^2 = 1.09$.

4.4. Shown are data values at the seven blends of a 3-component simplex-centroid design. High values of the response are more desirable than low values. Indicate the type of blending, linear or nonlinear, synergistic or antagonistic, that is present between the components:

(a) A and B: _____ and is _____ .

(b) A and C: _____ and is _____ .

(c) B and C: _____ and is _____ .

(d) A and B and C: _____ and is _____ .

Hint: var $(b_A) = $ MSE/2, var$(b_{AB}) = \frac{24}{2}$(MSE), var$(b_{ABC}) = $ 958.5(MSE). Also SST $= 63.79$, SSR $= 58.22$, and SSE $= 5.57$.

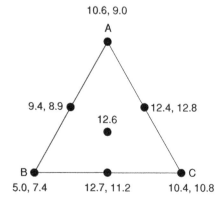

10.6, 9.0
A

9.4, 8.9

12.4, 12.8

12.6

B
5.0, 7.4

12.7, 11.2

10.4, 10.8
C

4.5. When calculating the total effect of component 1 by increasing the proportion of component 1 from $x_{L1} = 0.1$ to $x_{H1} = 0.4$ and using Eq. (4.37), isn't it possible the magnitude of TE_1 depends not only on $\Delta_1 = U_1 - L_1 = 0.3$ but also on changes that occur among the other components when taken along Piepel's direction?

Show and describe what is happening with the ratio of x_2/x_3 upon increasing x_1 by 0.3. Could this affect the value of the response? Explain.

CHAPTER 5

Other Mixture Model Forms

In Chapter 4 nearly all of the techniques that were suggested for analyzing data from mixture experiments were developed around the fitting of the Scheffé polynomials. The use of polynomial models was possible because the data evolved from (or at least were assumed to evolve from) systems that were well behaved. Typically well-behaved systems have surfaces that are expressible with a functional form that is continuous in the mixture component proportions, and within the ranges of the experimental data, the degree of the functional form is usually less than or equal to three. For modeling well-behaved systems, generally the Scheffé polynomials are adequate.

In this chapter we investigate other types of systems for which functional forms, other than the Scheffé polynomials, are more appropriate than the Scheffé models. We begin in the next section with a slight modification made to the Scheffé polynomials, namely the inclusion of terms that are reciprocals (or inverse terms) x_i^{-1} of the component proportions. Such terms were introduced in mixture models by Draper and St. John (1977a).

5.1   THE INCLUSION OF INVERSE TERMS IN THE SCHEFFÉ POLYNOMIALS

To model an extreme change in the response behavior as the value of one or more components tends to a boundary of the simplex region (i.e., where one or more $x_i \rightarrow 0$), the following equations containing inverse terms have been proposed:

$$\eta = \sum_{i=1}^{q} \beta_i x_i + \sum_{i=1}^{q} \beta_{-i} x_i^{-1} \tag{5.1}$$

$$\eta = \sum_{i=1}^{q} \beta_i x_i + \sum_{i<j}^{q} \sum \beta_{ij} x_i x_j + \sum_{i=1}^{q} \beta_{-1} x_i^{-1} \tag{5.2}$$

A Primer on Experiments with Mixtures, By John A. Cornell
Copyright © 2011 John Wiley & Sons, Inc. Published by John Wiley & Sons, Inc.

$$\eta = \sum_{i=1}^{q} \beta_i x_i + \sum_{i<j}^{q} \beta_{ij} x_i x_j + \sum_{i<j<k}^{q} \beta_{ijk} x_i x_j x_k + \sum_{i=1}^{q} \beta_{-i} x_i^{-1} \quad (5.3)$$

$$\eta = \sum_{i=1}^{q} \beta_i x_i + \sum_{i<j}^{q} \beta_{ij} x_i x_j + \sum_{i<j<k}^{q} \beta_{ijk} x_i x_j x_k$$

$$+ \sum_{i<j}^{q} \gamma_{ij} x_i x_j (x_i - x_j) + \sum_{i=1}^{q} \beta_{-i} x_i^{-1} \quad (5.4)$$

The models presented as Eqs. (5.1) to (5.4) are augmentations of the Scheffé polynomials with the additional terms of the form $\beta_{-1} x_i^{-1}$ included to account for the possible extreme change in the response as x_i approaches zero. It is assumed that the value of x_i never reaches zero but, that the value could be very close to zero; that is, $x_i \to \varepsilon_i > 0$, where ε_i is some extremely small quantity that is defined for each application of these models. Also, although the models presented as Eqs. (5.1) to (5.4) contain x_i^{-1} terms for all $i = 1, 2, \ldots, q$, if only a couple of the components are likely to produce extreme changes in the response as $x_i \to \varepsilon_i$, then only these terms are included in the model.

When lower-bound constraints of the form shown below are considered (as in Section 3.1),

$$0 \le L_i \le x_i, \qquad \text{for all } i, \ \sum_{i=1}^{q} L_i < 1$$

L-pseudocomponents $x_i' = (x_i - L_i)/(1 - \sum_{i=1}^{q} L_i)$ are suggested for use in place of the x_i. If an extreme change in the response occurs as x_i approaches the bound L_i, to model this change, we define $L_i' = L_i - \varepsilon_i$, where $L_i > \varepsilon_i > 0$. So, instead of the previous L-pseudocomponent definition, the L-pseudocomponent is now

$$x_i' = \frac{x_i - L_i'}{1 - \sum_{i=1}^{q} L_i'} \quad (5.5)$$

When $x_i = L_i$ we have $x_i' = [\varepsilon_i/(1 - \sum_{i=1}^{q} L_i')] > 0$; that is, x_i never quite reaches L_i but rather x_i is implied to reach $L_i + \varepsilon_i$ so that the lower bound for x_i' is not zero. Terms such as $(x_i')^{-1}$, when included in the L-pseudocomponent model, help pick up extreme changes that occur near (but not on) the zero boundary.

When, according to the data collection scheme, some of the blends have $x_i = 0$, then in order to include a term such as x_i^{-1} in the model, we must add a small positive amount, say, c_i, to each value of x_i. This is equivalent to working again with the pseudocomponents x_i', where this time they are defined as

$$x_i' = \frac{(x_i + c_i) - L_i}{1 - \sum_{i=1}^{q} (L_i - c_i)} \quad (5.6)$$

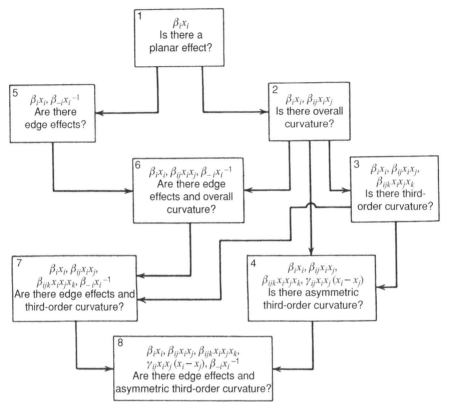

Figure 5.1 Sequential model fitting. Reproduced from Draper and St. John (1977a), p. 40, Fig. 1, with permission of the authors.

that is, $c_i/[1 - \sum_{i=1}^{q} (L_i - c_i)]$ is the lower bound for x_i'. Draper and St. John (1977a) suggest that, when working with the pseudocomponent model, a rule of thumb for values of the c_i would be somewhere between 0.02 and 0.05, and ideally to let $c_1 = c_2 = \cdots = c_q = c$. These c_i values are suitable generally for most problems.

The need to include terms like x_i^{-1} in the model is most likely discovered during a program of sequential model fitting. Such a program might resemble the "family tree of mixture models" presented in Figure 5.1. Each box in the figure contains the types of terms contained in the various models and the arrows indicate the possible paths that are taken when going from one model to the next. The sequential program of building up a model beginning with the Scheffé first-degree polynomial and advancing all the way if necessary to a cubic polynomial can be staged as stated by Draper and St. John:

> We assume that an experimenter would often follow a sequential approach when fitting Scheffé's polynomial. That is, an experimenter would first fit the linear canonical polynomial and make a judgement about the adequacy of

this model to fit his data. If he deemed the model inadequate, he would then fit the second order polynomial. If he also considered this model inadequate he would then fit either the cubic canonical polynomial or the special cubic canonical polynomial. (© 1977, American Statistical Association, p. 39, used by permission of the authors)

We now illustrate the sequential model-fitting approach along with the use of inverse terms in the fitted model using data from a gasoline-blending experiment.

5.2 FITTING GASOLINE OCTANE NUMBERS USING INVERSE TERMS IN THE MODEL

Research octane numbers were recorded for nine blends of an olefin (x_1), an aromatic (x_2), and a saturate (x_3) at 1.5 milliliters of lead per gallon. The octane numbers and the blending compositions, along with the pseudocomponent settings, are presented in Table 5.1.

Particularly noticeable are the three octane numbers (111.5, 101.3, and 107.0) in Table 5.1 that exceed 100.0 and the respective component proportions of the blends that yielded the high octane numbers. In particular, $x_1 = 0.010$ with 111.5, $x_2 = 0.000$ with 101.3, and $x_1 = 0.007$ and $x_3 = 0.036$ with 107.0. Since we cannot use the term x_2^{-1} to model these extreme responses, nor can we use an L-pseudocomponent model with inverse terms $(x_i')^{-1}$, where the value of $x_i' = 0$ when $x_i = L_i$, we choose the lower bound for the x_i' to be approximately 0.02. Letting $c_1 = c_2 = c_3 = 0.02$ in Eq. (5.6) so that the denominator for x_i' in (5.6) is $1.017 = 1 - (0.007 - 0.02) - (0 - 0.02) - (0.036 - 0.02)$, we have for the ith L-pseudocomponent,

$$x_i' = \frac{(x_i + 0.02) - L_i}{1.017} \qquad (5.7)$$

Table 5.1 Research Octane Numbers

Blending Components			Pseudocomponents			Research Octane Number at 1.5 mL Pb/gal
x_1	x_2	x_3	x_1'	x_2'	x_3'	
0.010	0.870	0.120	0.023	0.875	0.102	111.5
0.541	$L_2 = 0.000$	0.459	0.545	0.020	0.435	101.3
0.427	0.061	0.512	0.433	0.080	0.487	80.6
0.022	0.464	0.514	0.034	0.476	0.490	91.0
$L_1 = 0.007$	0.957	$L_3 = 0.036$	0.020	0.961	0.019	107.0
0.414	0.278	0.308	0.420	0.293	0.287	97.0
0.648	0.030	0.322	0.650	0.049	0.301	98.6
0.162	0.514	0.324	0.172	0.525	0.303	92.2
0.008	0.068	0.924	0.021	0.087	0.892	77.8

For example, the L-pseudocomponent proportions corresponding to the blend $(x_1, x_2, x_3) = (0.01, 0.87, 0.12)$ in Table 5.1 that produced $y_1 = 111.5$ are

$$x_1' = \frac{(0.010 + 0.02) - 0.007}{1.017} = 0.023$$

$$x_2' = \frac{(0.870 + 0.02) - 0.000}{1.017} = 0.875$$

$$x_3' = \frac{(0.120 + 0.02) - 0.036}{1.017} = 0.102$$

The results of the sequential buildup of the fitted Scheffé L-pseudocomponent polynomial with and without inverse terms is displayed in Table 5.2. In terms of producing a high value of R_A^2 (or a low value for the estimate of the error variance MSError), the linear-plus-complete-inverse-terms model, model 6, appears to perform best. Models 4 and 5 are next in the hierarchy of the best prediction equations with the remaining models following.

The model with the single inverse term $(x_2')^{-1}$, model 4, appears to be an improvement over the use of the simple linear model, model 1. This improvement is present in the increase of 0.1369 in the value of R_A^2 in going from 0.6598 to 0.7967 (or similarly a decrease in the value of MSError of $42.5 - 25.4 = 17.1$ or approximately 40%). The single-inverse-term model, model 4, also appears to be a better fitted model than the Scheffé second-degree polynomial, model 2. Thus with this set of data, we have an example where a second-degree equation does not improve on the fit of the surface obtained with the first-degree model but the addition of inverse terms to the first-degree model does improve the fit.

Exercise 5.1 Calculate the value of the C_p statistic (see Question 2.10 for each of the models 1 to 6 in Table 5.2 using MSError $= 0.20$. Are the results the same with C_p as we found with R_A^2? Are these statistics related?

The modeling of the data using the equation corresponding to model 6 in Table 5.2 is an example of the use of an equation other than a polynomial. Another phenomenon that cannot be modeled very well by polynomial equations is a system where one or more components exhibit linear blending (i.e., their effects are additive) in an otherwise nonlinear (or nonplanar surface) system. Such nonpolynomial forms can be discovered in a typical modeling exercise, as we show next.

5.3 AN ALTERNATIVE MODEL FORM FOR MODELING THE ADDITIVE BLENDING EFFECT OF ONE COMPONENT IN A MULTICOMPONENT SYSTEM

The occurrence of a single component exhibiting linear blending might be expected in systems where the component i serves as a diluent. As a diluent, component i

Table 5.2 Values of R_A^2 and MSError Corresponding to the Different Models Fitted to the Octane Numbers

Model	Fitted Model[a]	R_A^2	MSError
1	$\hat{y}(\mathbf{x}') = 108.5x_1' + 109.5x_2' + 71.5x_3'$	0.6598	42.5
2	$\hat{y}(\mathbf{x}') = 140.6x_1' + 112.3x_2' + 74.1x_3' - 87.8x_1'x_2' - 71.5x_1'x_3' - 1.0x_2'x_3'$	0.5262	59.2
3	$\hat{y}(\mathbf{x}') = 117.3x_1' + 99.1x_2' + 61.6x_3' + 0.3(x_1')^{-1}$	0.6926	38.4
4	$\hat{y}(\mathbf{x}') = 95.9x_1' + 111.3x_2' + 67.4x_3' + 0.3(x_2')^{-1}$	0.7967	25.4
5	$\hat{y}(\mathbf{x}') = 95.4x_1' + 115.6x_2' + 65.7x_3' + 0.4(x_2')^{-1} - 0.1(x_3')^{-1}$	0.7831	27.1
6	$\hat{y}(\mathbf{x}') = 109.9x_1' + 107.0x_2' + 52.2x_3' + 0.3(x_1')^{-1} + 0.3(x_2')^{-1} - 0.3(x_3')^{-1}$	0.8815	14.8

[a]Fitted model obtained using the values of x_i', $i = 1, 2, 3$, from Table 5.1.

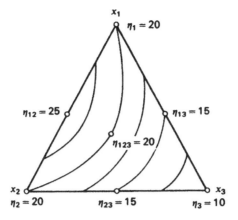

Figure 5.2 Surface exhibiting the complete linear blending of component 3.

has the effect of diluting the mixture in the sense that as the proportion x_i increases, the effect of the remaining components on the response diminishes in proportion. Thus the mean or average response changes linearly as x_i is varied between 0 and 1.0 along every line in the simplex region, while fixing or holding constant the relative proportions of the remaining components.

An example of a three-component system where component 3 blends linearly, but synergism is present between components 1 and 2, is shown in Figure 5.2 where the response drops off linearly along the edges connecting components 1 and 3, and components 2 and 3. On first sight one might expect that such a system could be represented by

$$\eta = \beta_1 x_1 + \beta_2 x_2 + \beta_2 x_3 + \beta_{12} x_1 x_2$$

which, according to the numbers in Figure 5.2 would be

$$\eta = 20x_1 + 20x_2 + 10x_3 + 20x_1 x_2 \tag{5.8}$$

Although Eq. (5.8) does suffice for the true binaries involving components 1 and 3 as well as components 2 and 3, as seen by substituting x_1 or $x_2 = x_3 = \frac{1}{2}$ into Eq. (5.8) to get $\eta = 15 = \eta_{13} = \eta_{23}$, the ternary blends involving component 3 with mixtures of components 1 and 2 are not modeled correctly. A more correct model for the system is a nonpolynomial model of the form

$$\eta = \beta_1 x_1 + \beta_2 x_2 + \beta_3 x_3 + \delta_{12} \frac{x_1 x_2}{x_1 + x_2} \tag{5.9}$$

where the term $x_1 x_2/(x_1 + x_2)$ takes the value of zero whenever $x_1 + x_2 = 0$ (i.e., at the vertex $x_3 = 1$).

To understand why Eq. (5.9) is better than the special cubic model in an additive environment, we show first how the special cubic model fails to depict the linear blending of one of the components in an otherwise nonlinear or curvilinear system.

Let us assume that a three-component system has complete linear blending with respect to component 3, accompanied by the curvilinear blending of components 1 and 2. Initially the system is modeled with the special cubic model

$$\eta = \beta_1 x_1 + \beta_2 x_2 + \beta_3 x_3 + \beta_{12} x_1 x_2 + \beta_{13} x_1 x_3 + \beta_{23} x_2 x_3 + \beta_{123} x_1 x_2 x_3$$

Since component 3 only blends linearly with components 1 and 2 (this means that component 3 has an additive effect when in combination with either component 1 or 2), then $\beta_{13} = 0$ and $\beta_{23} = 0$. However, components 1 and 2 blend curvilinearly, that is, $\beta_{12} \neq 0$, and so the model becomes

$$\eta = \beta_1 x_1 + \beta_2 x_2 + \beta_3 x_3 + \beta_{12} x_1 x_2 + \beta_{123} x_1 x_2 x_3 \tag{5.10}$$

Next let us impose the condition that component 3 blends linearly with any mixture containing *both* components 1 and 2. Then along any arbitrary ray such as defined by *ab* in Figure 5.3 the response is expressible as a linear equation in x_3,

$$\eta = (1 - x_3)\eta' + \beta_3 x_3 \tag{5.11}$$

where η' is the response to the blend $(x_1', x_2', x_3 = 0)$. However, curvilinear blending is present with components 1 and 2, so that

$$\eta' = \beta_1 x_1' + \beta_2 x_2' + \beta_{12} x_1' x_2' \tag{5.12}$$

and η' is constant along ray *ab*. A decreasing value of η along ray *ab* as x_3 goes from 0 to 1 is shown in Figure 5.4.

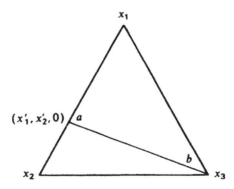

Figure 5.3 Ray *ab* from $x_3 = 0$ to $x_3 = 1$. Reproduced from Cornell and Gorman (1978), with permission of the Biometric Society.

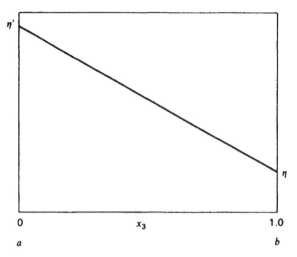

Figure 5.4 Decreasing linear response along ray ab.

To express Eq. (5.10) in the form of Eq. (5.11), we note that along ray ab the ratio x_1/x_2 is a constant; that is, the relative proportions of the components other than 3 remain fixed. If along the ray, $(x_1/x_2) = c$, then in terms of x_3,

$$x_1 = \frac{c(1 - x_3)}{1 + c}, \quad x_2 = \frac{(1 - x_3)}{1 + c} \tag{5.13}$$

Upon substituting these expressions for x_1 and x_2 into Eq. (5.10), a modified form of the special cubic model, which is applicable along ray ab, is

$$\eta = (1 - x_3)\left\{ \frac{\beta_1 c}{1 + c} + \frac{\beta_2}{1 + c} + \frac{\beta_{12}(1 - x_3)c}{(1 + c)^2} + \frac{\beta_{123}x_3(1 - x_3)c}{(1 + c)^2} \right\} + \beta_3 x_3 \tag{5.14}$$

In Eq. (5.14) the multiplier of $(1 - x_3)$ in curly brackets corresponds to η' in Eq. (5.11), and η' is required to be constant along ray ab. However, the quantity in curly brackets is constant only when both β_{12} and β_{123} are zero, since both are multipliers of x_3 and x_3 varies from 0 to 1 along the ray. Furthermore, since curvilinear blending is assumed to exist between components 1 and 2, that is, $\beta_{12} \neq 0$, this means the reduced special cubic polynomial equation (5.10) cannot satisfy the complete linear blending or additive blending of component 3 and at the same time account for curvilinear blending of components 1 and 2.

To derive the particular model form presented in Eq. (5.9), we notice from Eq. (5.13) that at point a in Figure 5.3, $x_1' = c/(c + 1)$. Thus η' in Eq. (5.11) can be expressed as $\eta' = [c\beta_1 + \beta_2 + c\beta_{12}/(c + 1)]/(c + 1)$ and at any point along ray ab,

$$x_1' = \frac{c}{c + 1} = \frac{x_1/x_2}{(x_1/x_2) + 1} = \frac{x_1}{x_1 + x_2},$$

$$x_2' = \frac{1}{c + 1} = \frac{1}{(x_1/x_2) + 1} = \frac{x_2}{x_1 + x_2}$$

Upon substituting these expressions for x_1' and x_1' into Eq. (5.12), we have for η'

$$\eta' = \frac{\beta_1 x_1}{x_1 + x_2} + \frac{\beta_2 x_1}{x_1 + x_2} + \frac{\beta_{12} x_1 x_2}{(x_1 + x_2)^2}$$

so the linear blending of component 3 along the ray is expressed as

$$\eta = (1 - x_3)\eta' + \beta_3 x_3$$

$$= (1 - x_3)\left\{\frac{\beta_1 x_1}{x_1 + x_2} + \frac{\beta_2 x_2}{x_1 + x_2} + \frac{\beta_{12} x_1 x_2}{(x_1 + x_2)^2}\right\} + \beta_3 x_3$$

Since $x_1 + x_2 + x_3 = 1$ everywhere on the simplex, then $(1 - x_3) = x_1 + x_2$, and therefore the nonpolynomial form in Eq. (5.9) is

$$\eta = \beta_1 x_1 + \beta_2 x_2 + \delta_{12}\frac{x_1 x_2}{x_1 + x_2} + \beta_3 x_3 \tag{5.15}$$

Equation (5.15) is used to express the complete linear blending (or additive effect) of component 3 with all blends of components 1 and 2.

We ask ourselves, Are there clues in terms of the coefficients of the special cubic model, that tell us the simpler model of Eq. (5.15) is the more likely alternative model? The answer is yes, particularly when the special cubic model for $q = 3$ is fitted to the responses at the points of the lattice design. In this case, the additive blending of component 3 at the points $x_1 = x_3 = \frac{1}{2}$, $x_2 = x_3 = \frac{1}{2}$, for example, forces $\beta_{13} = \beta_{23} = 0$. Furthermore, if the expected responses at the design points are denoted by η_i, η_{ij}, and η_{ijk}, as presented in Figure 5.2, then it can be shown (e.g., see Cornell and Gorman, 1978, app. C) that the expected response at the centroid η_{123} should be related to η_3 and η_{12} by the relation

$$\eta_{123} = \tfrac{2}{3}\eta_{12} + \tfrac{1}{3}\eta_3 \tag{5.16}$$

Also the condition presented in Eq. (5.16) would require that the coefficients in the special cubic model be related by

$$\beta_{123} = \tfrac{3}{2}\beta_{12} \tag{5.17}$$

Thus, in the analysis of a special cubic design when $q = 3$, if two of the three b_{ij} are zero and b_{123} is $\frac{3}{2}$ times the nonzero b_{ij}, an equation of the form presented in

Eq. (5.15) is a possible model. Of course, when the observations are subject to error, the relations presented in Eqs. (5.16) and (5.17) are approximate, holding only in expectation.

For illustrative purposes, let us refer to Figure 5.2 and fit the special cubic model to the data at the lattice points. We have

$$\beta_1 = \eta_1 = 20, \quad \beta_2 = \eta_2 = 20, \quad \beta_3 = \eta_3 = 10$$
$$\beta_{12} = 4\eta_{12} - 2(\eta_1 + \eta_2) = 4(25) - 2(20 + 20) = 20$$
$$\beta_{13} = 4(15) - 2(20 + 10) = 0, \quad \beta_{23} = 0$$
$$\beta_{123} = 27\eta_{123} - 12(\eta_{12} + \eta_{13} + \eta_{23}) + 3(\eta_1 + \eta_2 + \eta_3)$$
$$= 27(20) - 12(25 + 15 + 15) + 3(20 + 20 + 10) = 30$$

and the special cubic model, which fits the seven data points exactly, is

$$\eta = 20x_1 + 20x_2 + 10x_3 + 20x_1x_2 + 30x_1x_2x_3 \tag{5.18}$$

Unfortunately, Eq. (5.18) produces an artificial buckling in all of the ternary blends involving component 3 as shown by the curved line in Figure 5.7. This buckling effect is a maximum along the ray known as the x_3 axis, which goes from $x_3 = 1$, through the centroid design point, to $x_3 = 0$ at the midpoint of the opposite side. Hence we are prompted to try the model form

$$\eta = 20x_1 + 20x_2 + 10x_3 + 20 \left(\frac{x_1x_2}{x_1 + x_2} \right) \tag{5.19}$$

since this model does not produce the buckling effect shown in Figure 5.7. Also, if a single observation only is collected at each of the seven points in the special cubic design, an estimate of the error variance with three degrees of freedom is available if needed since the model presented in Eq. (5.19) contains only four terms.

Although we have chosen to discuss the $q = 3$ case in detail, the ideas presented in this section are easily extended to q components. With q components, if one component, say x_k, exhibits complete linear blending, the nonpolynomial form becomes

$$\eta = \sum_{i=1}^{q} \beta_i x_i + \frac{\sum \sum_{\substack{1 \le i < j \\ i, j \ne k}}^{q} \beta_{ij} x_i x_j}{\sum_{\substack{i=1 \\ i \ne k}}^{q} x_i}$$

If, in addition, the components that blend nonlinearly require cubic or quartic models, the divisors $D = \sum_{i=1}^{q} x_i$, $i \ne k$, of the polynomial numerators are raised to a power one less than the degree of the polynomial term in the numerator. An example is the full cubic model in $q = 4$ components with the blending of the single component x_1 being completely additive with the other components, so the model is

$$\eta = \sum_{i=1}^{4} \beta_i x_i + \sum_{2 \le i < j}^{4} \frac{\beta_{ij} x_i x_j}{1 - x_1} + \sum_{2 \le i < j}^{4} \frac{\gamma_{ij} x_i x_j (x_i - x_j)}{(1 - x_1)^2} + \frac{\beta_{234} x_2 x_3 x_4}{(1 - x_1)^2}$$

We now present a three-component example where two components represent liquid pesticides and the third component represents a wettable powder pesticide. The pesticides are used to control mite numbers on plants.

5.4 A BIOLOGICAL EXAMPLE ON THE LINEAR EFFECT OF A POWDER PESTICIDE IN COMBINATION WITH TWO LIQUID PESTICIDES USED FOR SUPPRESSING MITE POPULATION NUMBERS

An experiment was designed to measure the suppression of the population numbers of mites on strawberry plants. During the experimental program three pesticides were applied biweekly to the strawberry plants, which had been infested with the mites prior to the start of the experiment. The data in Table 5.3 represent the seasonal average mite population numbers determined on a per-leaf basis, where the averages were calculated using data recorded from five plants sampled from each plot and taken at six biweekly dates. Prior to spraying the plants at the beginning of the experiment, 10 plants were positioned in every experimental plot and 50 mites were placed on five leaves of each of the 10 plants.

Table 5.3 Strawberry Mite Experimental Data

Run	Component Proportions			Observed Numbers y_u	Predicted Numbers Using Special Cubic Equation (5.20)	Predicted Numbers Using Nonpolynomial Equation (5.21)
	x_1	x_2	x_3			
1	1.00	0	0	49.8	49.7	49.2
2	0.50	0.50	0	35.8	36.8	35.4
3	0	1.00	0	84.2	83.4	83.5
4	0	0.50	0.50	52.4	52.1	52.3
5	0	0	1.00	20.1	20.2	21.0
6	0.50	0	0.50	34.7	34.3	35.1
7	0.20	0.20	0.60	26.1	28.2	26.8
8	0	0.75	0.25	66.0	67.9	67.9
9	0	0.25	0.75	39.4	36.2	36.6
10	0.25	0	0.75	28.8	27.1	28.0
11	0.75	0	0.25	41.3	41.8	42.1
12	0.40	0.40	0.20	32.7	30.7	32.5
13	0.30	0.30	0.40	29.6	28.8	29.6
14	0.25	0.25	0.50	27.9	28.5	28.2
15	0.10	0.10	0.80	23.3	26.3	23.9

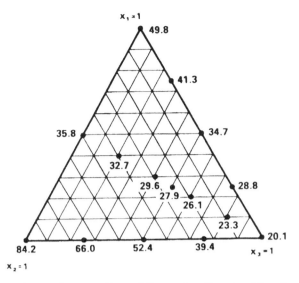

Figure 5.5   Observed seasonal average mite population numbers. Reproduced from Cornell and Gorman (1978), with permission of the Biometric Society.

Initially the plan was to observe the average mite numbers corresponding to the first seven mixtures in Table 5.3 only. However, because component 3 is a wettable powder and it was suspected that the effect of the powder would be additive when used in combination with the two liquids whose proportions are represented by x_1 and x_2, eight additional spray combinations, listed as run numbers 8 to 15 in Table 5.3, were prepared. The design and observed mite numbers are presented in Figure 5.5.

A special cubic model was fitted to all 15 values, resulting in

$$\hat{y}(\mathbf{x}) = 49.70x_1 + 83.42x_2 + 20.21x_3 - 119.14x_1x_2 - 2.55x_1x_3 + 1.24x_2x_3$$

$$\quad (2.03) \qquad (2.03) \qquad (1.80) \qquad (10.14) \qquad (9.02) \qquad (9.02)$$

$$-232.54x_1x_2x_3 \hspace{6cm} (5.20)$$

$$\quad (54.30)$$

The quantities in parentheses below the coefficient estimates are the estimated standard errors of the estimates and the analysis of variance is presented in Table 5.4.

Table 5.4   Analysis of Variance for the Mite Data with the Special Cubic Equation (5.20)

Source of Variation	Degrees of Freedom	Sum of Squares	Mean Square	R_A^2
Regression	6	4187.23	697.87	0.985
Residual	8	37.24	4.65	
Total	14	4224.47		

At first glance, Eq. (5.20) appears to be as good as any other model form, because with Eq. (5.20) $R_A^2 = 0.985$. From Table 5.4 an estimate of the error variance is $s^2 = 4.65$. However, particularly noticeable in Eq. (5.20) are the extreme differences in the magnitudes of the estimates of the binary parameters β_{12}, β_{13}, and β_{23}, where $b_{12} = -119.14$, $b_{13} = -2.55$, and $b_{23} = 1.24$. Also the magnitude of the estimate $b_{123} = -232.54$, which is significantly less than zero, is approximately twice the magnitude of b_{12}. The relative sizes of b_{123} and b_{12} are a little disturbing, since the ratio $b_{123}/b_{12} = 1.95$ suggests the existence of a ternary synergistic effect that is twice the size of the largest binary effect, and we are nearly sure there should not be much of a ternary effect at all. Hence, since b_{13} and b_{23}, are both nearly zero and $b_{123} \approx 2b_{12}$, which approximates the relation shown in Eq. (5.17), it was decided to fit the nonpolynomial model in Eq. (5.9).

Fitting the nonpolynomial equation (5.9) to the 15 observation values in Table 5.3 produced

$$\hat{y}(\mathbf{x}) = 49.18x_1 + 83.54x_2 + 20.98x_3 - 123.82\frac{x_1x_2}{x_1 + x_2} \qquad (5.21)$$

$$(0.89) \qquad (0.89) \qquad (0.65) \qquad (3.90)$$

With this fitted model, the error variance estimate is $s^2 = 1.40$, and the adjusted R_A^2 value with Eq. (5.21) is $R_A^2 = 0.995$. A contour plot of the estimated surface using predicted values from Eq. (5.21) is presented in Figure 5.6.

The linear terms for the models of Eqs. (5.20) and (5.21) are similar in size, and so are the coefficients for the x_1x_2 term of the special cubic model and the

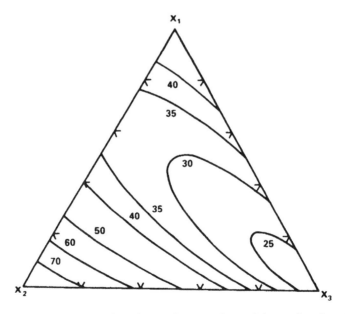

Figure 5.6 Contours of constant estimated seasonal average mite population numbers. Reproduced from Cornell and Gorman (1978), with permission of the Biometric Society.

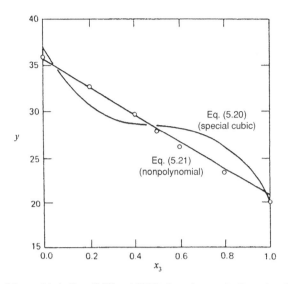

Figure 5.7 Fit of the models in Eqs. (5.20) and (5.21) along the x_3 axis. Reproduced from Cornell and Gorman (1978), with permission of the Biometric Society.

$x_1 x_2/(x_1 + x_2)$ term of the nonpolynomial form. Thus calculated values (predicted response values) from each model form agree reasonably well for binary mixtures but differ considerably for three-component blends at points along the x_3 axis as shown in Figure 5.7. On the x_3 axis the special cubic model tends to fit the points near $x_3 = 0$, $\frac{1}{2}$, and 1.0, but the model exhibits a cubic buckling between these points. Therefore to model the additive effect of a component, a model other than the special cubic equation is recommended.

Several other types of model forms exist that would fit the data just as well as the model presented in Eq. (5.21). Two equation forms that come to mind are models that are homogeneous of degree 1, where in Eq. (5.21) the term $\delta_{12}x_1 x_2/(x_1 + x_2)$ is replaced by $\delta'_{12} \min (x_1, x_2)$ or by $\delta'_{12}(x_1 x_2)^{1/2}$. Each of these models would produce a different set of surface contours for Figure 5.6 as we show in the next section, while providing approximately equivalent fits to the data, however. The discrepancies in the contours arise because data in the example were taken along the three rays emanating from the $x_3 = 1$ vertex only. If additional data had been taken on the $x_3 = 0$ side resulting in different values to be realized by the terms $x_1 x_2$, $\min(x_1, x_2)$, and $(x_1 x_2)^{1/2}$, then some differences among the models might be observed also. In fact data at these points could be used to determine which of the three models most closely resembles the true surface.

5.5 THE USE OF RATIOS OF COMPONENTS

In some areas of mixture experimentation, one is likely to be interested in one or more of the components, not so much from the standpoint of their proportions in the mixtures but from their relationship to the other components in the mixtures in the form of ratios of their proportions. In the manufacturing of a particular type of porcelain glass, for example, it is sometimes more meaningful to think of the ratio of silica to soda rather than to look at the actual proportions of each in a blend. A third ingredient, lime, might also be studied by looking at the ratio of silica to lime, or by looking at the ratio of soda to lime.

The use of ratios of components can be handled in a variety of ways. With three components whose proportions are denoted by x_1, x_2, and x_3, several possible sets of transformations from the x_i to the ratio variables r_1 and r_2 are

$$\text{Set I}: \quad r_1 = \frac{x_2}{x_1}, \qquad r_2 = \frac{x_2}{x_3}$$

$$\text{Set II}: \quad r_1 = \frac{x_1}{x_2}, \qquad r_2 = \frac{x_2}{x_3} \qquad (5.22)$$

$$\text{Set III}: \quad r_1 = \frac{x_1}{x_2 + x_3}, \quad r_2 = \frac{x_2}{x_3}$$

In each set of ratios, each ratio r_i, $i = 1$ or 2, contains at least one of the components used in the other ratio of the same set. The number of ratios r_i in each set should be less than the number of components in the system. If the number of ratios in a set equals q, the ratios form a redundant set because the sum of the component proportions is unity. Note that any type of ratio can be used in a set as long as there is a tie-in with a component in one of the other ratios in the same set.

A simple first-degree model in r_1 and r_2 is

$$\eta = \alpha_0 + \alpha_1 r_1 + \alpha_2 r_2$$

which becomes in x_1, x_2, and x_3, with the first set in Eqs. (5.22)

$$\eta = \alpha_0 + \alpha_1 \left(\frac{x_2}{x_1} \right) + \alpha_2 \left(\frac{x_2}{x_3} \right)$$

The model is fitted to data collected at design-point settings chosen in the r_i. The corresponding settings of the x_i in terms of r_1 and r_2 are then

$$x_1 = \frac{r_2}{r_1 + r_1 r_2 + r_2}$$

$$x_2 = \frac{r_1 r_2}{r_1 + r_1 r_2 + r_2}$$

$$x_3 = \frac{r_1}{r_1 + r_1 r_2 + r_2}$$

Let us illustrate with a numerical example the construction of a factorial design in the ratios r_1 and r_2 and the subsequent fitting of a second-degree model in r_1

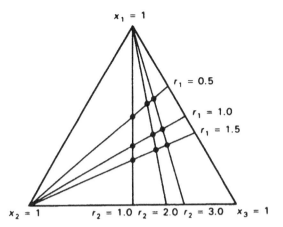

Figure 5.8 Design points in the region defined by values of the ratio variables $r_1 = x_3/x_1$ and $r_2 = x_3/x_2$.

and r_2 to data collected at the design points. Let us define the two ratio variables of interest as

$$r_1 = \frac{x_3}{x_1}, \quad r_2 = \frac{x_3}{x_2}$$

Paths of values for the ratio variables r_1 and r_2 are defined along the rays emanating from the vertex $x_2 = 1$ for r_1 and from the vertex $x_1 = 1$ for r_2. These rays are drawn in Figure 5.8. Now suppose that we choose to work with values of r_1 of 0.5, 1.0, and 1.5 in combination with values of r_2 of 1.0, 2.0, and 3.0. A second-degree model of the form

$$y = \alpha_0 + \alpha_1 r_1 + \alpha_2 r_2 + \alpha_{11} r_1^2 + \alpha_{22} r_2^2 + \alpha_{12} r_1 r_2 + \varepsilon \tag{5.23}$$

can be fitted to observations collected at the $3^2 = 9$ design points.

The settings (proportions) in the mixture components corresponding to the nine factorial settings in r_1 and r_2 are determined as follows. Corresponding to the combination $(r_1, r_2) = (0.5, 1.0)$, we have only to solve the equations

$$r_1 = \frac{x_3}{x_1} = 0.5, \quad r_2 = \frac{x_3}{x_2} = 1.0, \quad x_1 + x_2 + x_3 = 1.0$$

to get $x_1 = 0.5$, $x_2 = 0.25$, -and $x_3 = 0.25$. In the same manner as $(r_1, r_2) = (1.5, 2.0)$, we can set up the equations

$$\frac{x_3}{x_1} = 1.5, \quad \frac{x_3}{x_2} = 2.0, \quad x_1 + x_2 + x_3 = 1.0$$

to get $x_1 = 4/13$, $x_2 = 3/13$, and $x_3 = 6/13$. The seven compositions corresponding to the remaining seven (r_1, r_2) combinations are presented in Table 5.5 along

Table 5.5 Octane Numbers for Ratios of Components

Design Point	Ratio Variables		Component Proportions			Octane Number (y)
	$r_1 = x_3/x_1$	$r_2 = x_3/x_2$	x_1	x_2	x_3	
1	0.5	1.0	0.50	0.25	0.25	82.8
2	1.0	1.0	0.33	0.33	0.33	87.3
3	1.5	1.0	0.25	0.37	0.37	84.9
4	0.5	2.0	0.57	0.14	0.29	85.6
5	1.0	2.0	0.40	0.20	0.40	93.0
6	1.5	2.0	0.31	0.23	0.46	89.3
7	0.5	3.0	0.60	0.10	0.30	87.2
8	1.0	3.0	0.43	0.14	0.43	89.6
9	1.5	3.0	0.33	0.17	0.50	90.6

with response values observed from the ratios. The nine compositions are noted in Figure 5.8.

The second-degree model presented as Eq. (5.23) can be fitted to the data of Table 5.5 using the values of r_1 and r_2 as listed, or the values of r_1 and r_2 can be scaled to enable us to utilize orthogonal design settings. Since the values of r_1 are equally spaced, as are the values of r_2, we can rewrite Eq. (5.23) as

$$y = \gamma_0 + \gamma_1 z_1 + \gamma_2 z_2 + \gamma_{11} z_1^2 + \gamma_{22} z_2^2 + \gamma_{12} z_1 z_2 + \varepsilon \tag{5.24}$$

where $z_1 = (r_1 - 1.0)/0.5$ and $z_2 = (r_2 - 2.0)/1.0$. The second-degree equation (5.24) was fitted to the nine octane values, resulting in the prediction model

$$\hat{y}(\mathbf{z}) = 91.46 + 1.53 z_1 + 2.07 z_2 - 3.23 z_1^2 - 2.23 z_2^2 + 0.33 z_1 z_2$$
$$(1.23) \quad (0.67) \quad (0.67) \quad (1.17) \quad (1.17) \quad (0.83)$$

The value of $R_A^2 = 0.72$ and the estimate of the error variance is $s^2 = 2.73$.

The corresponding model in the ratios r_1 and r_2 is

$$\hat{y}(\mathbf{r}) = 63.69 + 27.63 r_1 + 10.35 r_2 - 12.93 r_1^2 - 2.23 r_2^2 + 0.65 r_1 r_2 \tag{5.25}$$

$$(6.27) \quad (10.00) \quad (5.00) \quad (4.67) \quad (1.17) \quad (1.65)$$

which, when written in the mixture components, is

$$\hat{y}(\mathbf{x}) = 63.69 + 27.63\left(\frac{x_3}{x_1}\right) + 10.35\left(\frac{x_3}{x_2}\right) - 12.93\left(\frac{x_3}{x_1}\right)^2 - 2.23\left(\frac{x_3}{x_2}\right)^2 + 0.65\left(\frac{x_3^2}{x_1 x_2}\right)$$

$$(6.27) \quad (10.00) \qquad (5.00) \qquad (4.67) \qquad (1.17) \qquad (1.65)$$

From this latter model (as well as with Eq. (5.25)) we see that increasing the value of x_3 relative to both x_1 and x_2 up to a certain level produces higher estimated octane numbers (since both $\hat{\gamma}_1$ and $\hat{\gamma}_2$ are positive). However, above certain values of r_1 and r_2, increasing the value of x_3 results in a reduction (both $\hat{\gamma}_{11}$ and $\hat{\gamma}_{22}$ are negative) in the value of the estimate of the octane number.

We must remember that these inferences apply only to blends belonging to the experimental region shown in Figure 5.8 where $0.5 \leq r_1 = x_3/x_1 \leq 1.5$ and $1.0 \leq r_2 = x_3/x_2 \leq 3.0$.

Problems where the number of components is greater than 3 can be handled by the use of additional ratios. As a reminder, the number of ratios in a set should be kept equal to $q - 1$. We also remember that each ratio in the set should contain at least one of the components used in at least one of the other ratios belonging to the set.

5.6   COX'S MIXTURE POLYNOMIALS: MEASURING COMPONENT EFFECTS

Let us express the first- and second-degree polynomial models in more general forms.

First-degree model:

$$\eta(\mathbf{x}) = \beta_0 + \sum_{i=1}^{q} \beta_i x_i \tag{5.26}$$

Second-degree model:

$$\eta(\mathbf{x}) = \beta_0 + \sum_{i=1}^{q} \beta_i x_i + \sum_{i=1}^{q} \sum_{j=1}^{q} \beta_{ij} x_i x_j, \quad \beta_{ij} = \beta_{ji} \tag{5.27}$$

where $\eta(\mathbf{x})$ is written to mean the expected response at the point $\mathbf{x}$. The individual parameters in Eqs. (5.26) and (5.27) are given different interpretations from those with the Scheffé polynomials by imposing constraints on their values. This is done in the paper by Cox (1971), where the parameters are said to represent relative changes in the measured response that is accomplished by comparing the value of the response at points in the simplex against the value of the response taken at a standard mixture.

The parameter β_i in Eq. (5.26) can be made to represent the effect on the response of changing the proportion of component i, keeping in mind that when component i is changed so also is the proportion of at least one of the remaining $q - 1$ components. To see this, let us select a standard mixture at some point in the simplex and denote it by $\mathbf{s}$. Let us select another point $\mathbf{x}$ in the simplex that is different from $\mathbf{s}$. If we choose the point $\mathbf{x}$ to lie on the ray that connects the vertex $x_i = 1$ with the point $\mathbf{s}$, then at $\mathbf{x}$ the proportion for component i is $x_i = s_i + \Delta_i$ ($\Delta_i > 0$), where s_i is the proportion for component i at $\mathbf{s}$; see Figure 5.9. Since $\mathbf{x}$ lies on the ray from $\mathbf{s}$ to $x_i = 1$, the other $q - 1$ components are in the same relative proportions to one another as at $\mathbf{s}$. These proportions are $x_j = s_j - \Delta_i s_j/(1 - s_j) = s_j(1 - x_i)/(1 - s_i)$, $j \neq i$. If $\mathbf{x}$ lies on the opposite side of $\mathbf{s}$ from the vertex $x_i = 1$, then $\Delta_i < 0$.

Let us write the expected response using the first-degree equation (5.26). Then the change in the expected response, which is measured by taking the difference $\eta(\mathbf{x}) - \eta(\mathbf{s})$, is

$$
\begin{aligned}
\Delta\eta(\mathbf{x}) &= \eta(\mathbf{x}) - \eta(\mathbf{s}) \\
&= \beta_0 + \sum_{i=1}^{q} \beta_i x_i - \beta_0 - \sum_{i=1}^{q} \beta_i s_i \\
&= \beta_i \Delta_i + \sum_{j \neq i}^{q} \beta_j(x_j - s_j)
\end{aligned}
$$

where $\Delta_i = x_i - s_i$. If we force the constraint $\sum_{i=1}^{q} \beta_i s_i = 0$ on the values of β_i, then the change in the expected response is expressed as

$$
\Delta\eta(\mathbf{x}) = \frac{\beta_i \Delta_i}{1 - s_i} \tag{5.28}
$$

The adoption of this constraint means simply that the model for the response at $\mathbf{s}$ becomes β_0, whereas the change in response at any other point $\mathbf{x}$ is $\beta_i \Delta_i/(1 - s_i)$. Rewriting the equality (5.28) so that β_i is expressed as a function of the change $\Delta\eta(\mathbf{x})$, we have $\beta_i = \Delta\eta(\mathbf{x})(1 - s_i)/\Delta_i$ and an estimate of β_i, which we call an estimate of the effect of component i, is

$$
b_i = \frac{(1 - s_i)}{\Delta_i}\{y(\mathbf{x}) - y(\mathbf{s})\}, \qquad i = 1, 2, \ldots, q \tag{5.29}
$$

where $y(\mathbf{x})$ and $y(\mathbf{s})$ are values of the response observed at $\mathbf{x}$ and at $\mathbf{s}$.

In the formula for b_i, the difference in the heights of the surface at $\mathbf{x}$ and at $\mathbf{s}$ is weighted by the incremental change Δ_i made in the proportion of component i, relative to the amount $(1 - s_i)$ that is an upper bound for the value of Δ_i. The weight $(1 - s_i)/\Delta_i$ is greater than or equal to unity. An example of a weight of approximately 2 units is shown in Figure 5.9 for the system consisting of components, i, j, and k.

When the surface is more appropriately represented by the second-degree polynomial presented in Eq. (5.27), the change $\Delta\eta(\mathbf{x})$ in the expected response contains an additional quantity $\beta_{ii}\Delta_i^2/(1 - s_i)^2$. In other words, if $\eta(\mathbf{x})$ is more correctly written as Eq. (5.27), the change expressed in Eq. (5.28) would read as $\Delta\eta(\mathbf{x}) = \eta(\mathbf{x}) - \eta(\mathbf{s}) = \beta_i\Delta_i/(1 - s_i) + \beta_{ii}\Delta_i^2/(1 - s_i)^2$, since in addition to placing the constraint $\sum_{i=1}^{q} \beta_i s_i = 0$ on the coefficients, the q constraints $\sum_{k=1}^{q} \beta_{jk}s_k = 0$, for $j = 1, 2, \ldots$ q, are imposed also.

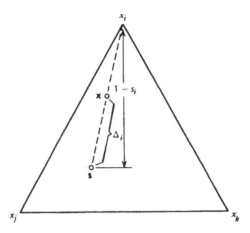

Figure 5.9 Incremental change Δ_i in the proportion of component i and the upper bound $1 - s_i$ for Δ_i.

Another way to view the difference $\eta(\mathbf{x}) - \eta(\mathbf{s}) = \beta_i \Delta_i/(1 - s_i) + \beta_{ii} \Delta_i^2/(1 - s_i)^2$ is to express the response at $\mathbf{x}$ in the form of a quadratic polynomial in Δ_i as

$$\eta(\mathbf{x}) = \eta(\mathbf{s}) + \left(\frac{\beta_i}{1 - s_i} \right) \Delta_i + \frac{\beta_{ii}}{(1 - s_i)^2} \Delta_i^2$$

$$= \eta(\mathbf{s}) + A_i \Delta_i + B_i \Delta_i^2$$

which Smith and Beverly (1997) point out makes it easier to understand several useful properties:

1. The quantity $A_i \Delta_i + B_i \Delta_i^2$ is a measure of the expected change in the response when the proportion of the ith component is changed by an amount Δ_i.
2. The gradient, or change in the response value per unit change in x_i, at $\mathbf{s}$ along the Cox-effect direction for x_i is $\beta_i/(1 - s_i)$ and is called the *effect of x_i*, provided x_i is free to range from 0 to 1.
3. The gradient at any point along the Cox-effect directions can be estimated from the equation for $\eta(\mathbf{x})$ above as

$$\frac{\partial \eta(\mathbf{x})}{\partial \Delta_i} = A_i + 2B_i \Delta_i \quad \text{for all} \quad i = 1, 2, \ldots, q$$

4. From the second-degree model $\eta(\mathbf{x})$ above, one can infer what the shape of the response surface is along the Cox-effect direction for x_i based on the signs of β_i and β_{ii} taken together and whether the reference mixture $\mathbf{s}$ is located before or after a maximum or before or after a minimum as one moves toward the vertex $x_i = 1, x_j = 0, i, j = 1, 2, \ldots, q, i = j$.

β_i	β_{ii}	
$-$	$-$	Past a maximum
$+$	$-$	Before a maximum
$-$	$+$	Before a minimum
$+$	$+$	Past a minimum

Let us now consider estimating the parameters associated with all of the q components simultaneously with either the first- or the second-degree model, subject to the constraints being placed on the parameter estimates. A general formula can be written for estimating the restricted parameters in Cox's models, since the Cox estimates can be written as functions of the parameter estimates in the corresponding Scheffé polynomial model. To show this we will use matrix notation.

Let us write the matrix formula for estimating the coefficients in Scheffé's model as $\mathbf{g}_l = (\mathbf{X}'_l\mathbf{X}_l)^{-1}\mathbf{X}'_l\mathbf{y}$, where the subscript l denotes the degree of the model, $l = 1$ or 2. When $l = 1$, the model is $\mathbf{y} = \mathbf{X}_1\boldsymbol{\gamma}_1 + \boldsymbol{\varepsilon}$, where $\mathbf{y}$ is an $N \times 1$ vector of observations, $\mathbf{X}_1$ is an $N \times q$ matrix of component proportions, $\boldsymbol{\gamma}_1$ is a $q \times 1$ vector of unknown parameters, $\mathbf{g}_1$ is the estimator of $\boldsymbol{\gamma}_1$, and $\boldsymbol{\varepsilon}$ is an $N \times 1$ vector of random errors, where $E(\boldsymbol{\varepsilon}) = \mathbf{0}$ and $E(\boldsymbol{\varepsilon}\boldsymbol{\varepsilon}') = \sigma^2\mathbf{I}$. The vector of estimates $\mathbf{b}_l$ associated with the lth-degree Cox model is

$$\mathbf{b}_l = \mathbf{B}_l\begin{bmatrix} \mathbf{g}_l \\ \mathbf{0} \end{bmatrix} = \mathbf{B}_l\begin{bmatrix} (\mathbf{X}'_l\mathbf{X}_l)^{-1}\mathbf{X}'_l\mathbf{y} \\ \mathbf{0} \end{bmatrix}, \quad l = 1, 2 \tag{5.30}$$

where if $l = 1$, the matrix $\mathbf{B}_1$ is $(q + 1) \times (q + 1)$ and $\mathbf{0}$ is 1×1, and if $l = 2$, the matrix $\mathbf{B}_2$ is $(q + 1)(q + 2)/2 \times (q + 1)(q + 2)/2$ and $\mathbf{0}$ is $(q + 1) \times 1$.

The elements of the matrices $\mathbf{B}_1$ and $\mathbf{B}_2$ in Eq. (5.30) are the coefficients in the linear relations of the β_i to the γ_i. For example, when $l = 1$, the β_i in terms of the γ_i, subject to the constraint $\sum_{i=1}^{q} \beta_i s_i = 0$, are

$$\beta_0 = \sum_{i=1}^{q} \gamma_i s_i, \quad \beta_i = \gamma_i(1 - s_i) - \sum_{j \neq i}^{q} \gamma_j s_j = \gamma_i - \beta_0, \quad i = 1, 2, \ldots, q$$

(Note that if the γ_i are replaced by the estimates $\hat{\gamma}_i$ obtained with the fitted Scheffé linear blending model, then $\hat{\beta}_0$ is an estimate of the response *at the standard mixture* s. Furthermore $\hat{\beta}_i = \hat{\gamma}_i - \hat{\beta}_0$ is the difference between the estimated responses at $x_i = 1$ and the standard mixture s.) The form of the matrix $\mathbf{B}_1$ is

$$\mathbf{B}_1 = \begin{bmatrix} s_1 & s_2 & \cdots & s_q & 0 \\ & \mathbf{I} - \mathbf{1}\mathbf{s}' & & & \mathbf{0} \end{bmatrix} \tag{5.31}$$

where $\mathbf{1}$ and $\mathbf{0}$ are $q \times 1$ vectors of 1's and 0's, respectively, and $\mathbf{s} = (s_1, s_2, \ldots, s_q)'$. In a similar way, the elements of the matrix $\mathbf{B}_2$ can be found. A special case is presented in Appendix 5A.

The variance–covariance matrix of the restricted estimates of Eq. (5.30) is

$$\text{var}(\mathbf{b}_l) = \mathbf{B}_l \begin{bmatrix} (\mathbf{X}_l'\mathbf{X}_l)^{-1} & \mathbf{0} \\ \mathbf{0} & \mathbf{0} \end{bmatrix} \mathbf{B}_l' \sigma^2 \tag{5.32}$$

The elements of the matrix $\text{var}(\mathbf{b}_l)$ are functions of the component proportions in $\mathbf{X}_l$. Of particular interest to an experimenter are the magnitudes of the diagonal elements of the matrix $\text{var}(\mathbf{b}_i)$, which are the variances of the individual b_i.

In choosing a design configuration for fitting the first-degree polynomial presented in Eq. (5.26), several suggestions are made in the paper by Cornell (1975). Specifically, axial designs of the type introduced in Section 2.14 are recommended. Two approaches were taken by Cornell in searching for a design, but we discuss briefly only the case where the standard mixture $\mathbf{s}$ is positioned at the center of the simplex and changes in the sizes of the elements of $\text{var}(\mathbf{b}_1)$ are brought about by varying the distances to the design points, that is, by varying the elements in $\mathbf{X}_1$.

Suppose we are interested in estimating the effects of each of the q components simultaneously. An axial design is proposed, and the points are positioned on the axes a distance Δ from the centroid $\mathbf{s}$ either toward each of the q vertices (max $\Delta = (q-1)/q$) or away from (max $\Delta = -1/q$) each of the q vertices. If r observations are taken at each point in the design (but none at $\mathbf{s}$), then the form of matrix $(\mathbf{X}_1'\mathbf{X}_1)$ in Eq. (5.30) is

$$\frac{1}{r}(\mathbf{X}_1'\mathbf{X}_1) = b\mathbf{I} + c\mathbf{J}, \quad \text{where } b = \frac{\Delta^2 q^2}{(q-1)^2}, \quad c = \frac{q^2 - \Delta^2 q^2 - 2q + 1}{q(q-1)^2}$$

and where the matrix $\mathbf{I}$ is an identity matrix of order q and the matrix $\mathbf{J}$ is a $q \times q$ matrix of 1's. The inverse $(\mathbf{X}_1'\mathbf{X}_1)^{-1}$ is

$$r(\mathbf{X}_1'\mathbf{X}_1)^{-1} = d\mathbf{I} + e\mathbf{J}$$

where $d = 1/b$, $e = -c/b(b + cq)$, and the form of the variance matrix of the estimates is

$$\text{var}(\mathbf{b}_1) = \frac{\sigma^2}{rq} \begin{bmatrix} 1 & \mathbf{0}' \\ \mathbf{0} & \dfrac{(q-1)^2}{\Delta^2 q}\left\{\mathbf{I} - \dfrac{1}{q}\mathbf{J}\right\} \end{bmatrix} \tag{5.33}$$

The form of the matrix $\text{var}(\mathbf{b}_1)$ in Eq. (5.33) suggests that, as the magnitude of Δ increases, the variances of the individual b_i, $i = 1, 2, \ldots, q$, decrease. In fact, if the points are placed at the vertices of the simplex resulting in the Scheffé $\{q, 1\}$ lattice, then $\Delta = (q-1)/q$ and $\text{var}(b_i) = \sigma^2(q-1)/qr$, $i = 1, 2, \ldots, q$. Also we notice that even though the Scheffé estimates are uncorrelated when the extreme-vertices design is used, because of the restriction $\sum_{i=1}^{q} b_i s_i = \sum_{i=1}^{q} b_i/q = 0$, the estimates b_i and

b_j in the Cox polynomial are dependent; that is, the covariance between b_i, and b_j is $\text{cov}(b_i, b_j) = -\sigma^2/qr$.

Exercise 5.2   Write the formula for $\text{var}(b_i)$ when the design points are placed at the bases of the q component axes so that $\Delta = -1/q$. Compare the magnitudes of the variances of these estimates to the variances of the estimates of the same coefficients obtained with the $\{q, 1\}$ simplex-lattice.

5.7   AN EXAMPLE ILLUSTRATING THE FITS OF COX'S MODEL AND SCHEFFÉ'S POLYNOMIAL

In Section 2.6 we discussed briefly the experimental results of blending the components polyethylene (x_1), polystyrene (x_2), and polypropylene (x_3) on the response, thread elongation of spun yarn. Here we again discuss the blending of the same ingredients, but this time the response is the knot strength of the yarn. We assume that the components polystyrene and polypropylene are approximately the same in terms of cost and availability, and we choose to measure the effect of the first component, polyethylene, relative to the others owing to the high availability and low cost of polyethylene. It is assumed that the importance of polyethylene can be measured by fitting a second-degree polynomial, since the knot strength values are suspected of behaving in a curvilinear fashion for increasing percentages of polyethylene. However, for purposes of illustrating the fitting of the complete Cox polynomial as well as for discussing and comparing the interpretations given to the parameter estimates in the Cox and Scheffé polynomials, initially we will fit the first-degree models and then extend the discussion to the second-degree models.

Let us assume that the cost of polyethylene is less than half the cost of either polystyrene or polypropylene, so that only mixtures are considered where the polyethylene proportion is greater than or equal to 0.50. Also, since only blends consisting of all three components simultaneously (interior points of the simplex) are practical and we would like at least 2% of the mixture to be made up of both polystyrene (x_2) and polypropylene (x_3), that is, $x_i \geq 0.02$, $i = 2$ and 3, then the original components are transformed to L-pseudocomponents (see Section 3.2) so that a lattice or simplex-centroid design in the L-pseudocomponents can be selected. The feasible blends comprise the upper interior triangular region (whose sides are the dashed lines) displayed in Figure 5.10, where $x_1 \geq 0.50$, $x_2 \geq 0.02$, and $x_3 \geq 0.02$.

Since the purpose of this example is the comparison of the fits of the Cox and Scheffé models, the position of the standard mixture will be the centroid of the triangular factor space as in our discussion in the previous section. The coordinates of the standard mixture in the L-pseudocomponent system are $(s_1', s_2', s_3') = \left(\frac{1}{3}, \frac{1}{3}, \frac{1}{3}\right)'$.

The design in the L-pseudocomponent system at whose points the data are collected in order to estimate the parameters in the first-degree models is the $\{3,1\}$ simplex-lattice. The choice of the $\{3, 1\}$ lattice is made because the points are positioned at the vertices of the interior triangle, and thus the parameter estimates possess

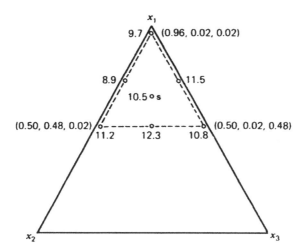

Figure 5.10 Design of the three-component yarn experiment with the observed knot strength values at the design points.

minimum variance. Furthermore, the $\{3, 1\}$ lattice can be augmented easily to the $\{3, 2\}$ simplex-lattice for use in fitting the second-degree models.

The proportions for L-pseudocomponents 1, 2, and 3 are found using

$$x_1' = \frac{x_1 - 0.50}{1 - 0.54}, \quad x_2' = \frac{x_2 - 0.02}{1 - 0.54}, \quad x_3' = \frac{x_3 - 0.02}{1 - 0.54} \qquad (5.34)$$

where $0.54 = 0.50 + 0.02 + 0.02$ is the sum of the lower bounds for x_1, x_2, and x_3. Corresponding to the L-pseudocomponent proportions $x_1', = 1, x_2' = x_3' = 0$, and $x_1' = 0, x_2' = 1, x_3' = 0$, and $x_1' = 0, x_2' = 0, x_3' = 1$, which are the vertices of the L-pseudocomponent triangle (the dashed-line triangle in Figure 5.10), the original component proportions are $x_1 = 0.96, x_2 = x_3 = 0.02$, and $x_1 = 0.50, x_2 = 0.48, x_3 = 0.02$, and $x_1 = 0.50, x_2 = 0.02, x_3 = 0.48$, respectively. Later for the $\{3, 2\}$ lattice we will additionally have the L-pseudocomponent values $x_i' = x_j' = 0.5, x_k' = 0, k \neq j \neq i$. The mixture component settings of the x_i' and the x_i at the six design points of the $\{3, 2\}$ lattice plus at the standard mixture s are listed in Table 5.6 where the first three points, 1, 2, and 3, represent the $\{3, 1\}$ lattice. Also listed in Table 5.6 are the knot strength values at the compositions where the strength values are measured in pounds of force exerted on the knots before rupture of the yarn threads. To enable us to calculate the standard errors of the parameter estimates, an independent estimate of the observation variance is given to be $s^2 = 0.30$.

Corresponding to the knot strength values at the three vertices, which are listed as points 1, 2, and 3, the fitted response equations using the Cox polynomial as well as the Scheffé polynomials in the original components x_i and in the L-pseudocomponents x_i' are

Table 5.6   Spun Yarn Example

	Pseudocomponents			Original Components			Observed Knot Strength
Point	x_1'	x_2'	x_3'	x_1	x_2	x_3	(lb force)
1	1	0	0	0.96	0.02	0.02	9.7
2	0	1	0	0.50	0.48	0.02	11.2
3	0	0	1	0.50	0.02	0.48	10.8
4	0.5	0.5	0	0.73	0.25	0.02	8.9
5	0	0.5	0.5	0.50	0.25	0.25	12.3
6	0.5	0	0.5	0.73	0.02	0.25	11.5
7	0.33	0.33	0.33	0.653	0.173	0.173	10.5

Cox polynomial:

$$\hat{y}(\mathbf{x}) = b_0 + \sum_{i=1}^{3} b_i x_i = 10.56 - 0.97x_1 + 2.29x_2 + 1.42x_3 \tag{5.35}$$

$$(0.32) \quad (0.50) \quad\quad (1.28) \quad\quad (1.28)$$

Scheffé polynomial:

$$\hat{y}(\mathbf{x}) = \sum_{i=1}^{3} g_i x_i = 9.59x_1 + 12.85x_2 + 11.98x_3 \tag{5.36}$$

$$(0.60) \quad\quad (1.31) \quad\quad (1.31)$$

$$\hat{y}(\mathbf{x}) = \sum_{i=1}^{3} g_i' x_i' = 9.7x_1' + 11.2x_2' + 10.8x_3' \tag{5.37}$$

$$(0.55) \quad (0.55) \quad (0.55)$$

In the L-pseudocomponent model of Eq. (5.37), the coefficient estimates g_i' represent the heights of the knot strength planar surface at the vertices of the L-pseudocomponent triangle inside the original triangle. The coefficient estimate g_i' is the knot strength value observed at the vertex of the L-pseudocomponent triangle; that is, $g_i' = y_u$ at $x_i' = 1$, $x_j' = 0$, $j \neq i$, $i = 1, 2$, and 3.

For the Scheffé polynomial model (5.36) in the original components, the coefficient estimate g_i represents the height of the knot strength surface extrapolated to the vertices $x_i = 1$ of the original component triangle. The coefficient g_i also represents the *difference* between the height of the surface at the ith L-pseudocomponent vertex ($x_1' = 1$) and the weighted sum of the heights at the three L-pseudocomponent vertices (where the weights are the lower bounds of the original components) divided by the linear size of the L-pseudocomponent triangle (Figure 5.11). Computationally,

$$g_i = \frac{g_i' - [0.5g_1' + 0.02g_2' + 0.02g_3']}{1 - 0.54} \tag{5.38}$$

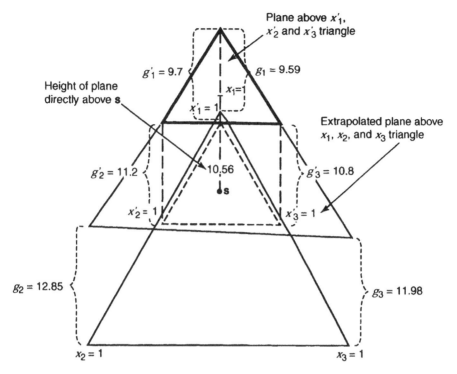

Figure 5.11 Plane defined by Eq. (5.37) directly above the L-pseudocomponent triangle. Also shown are extrapolations of the heights of the plane to the vertices of the triangle in the original components. These heights of the extrapolated plane are g_1, g_2, and g_3.

where the denominator $1 - 0.54 = 0.46$ also represents the length of the L-pseudocomponent axes relative to a length of unity in the original component proportions.

With the Cox polynomial model represented by Eq. (5.35), the coefficient estimate b_0 represents the estimate of the response at the standard mixture

$$b_0 = \sum_{i=1}^{3} s_i g_i = 0.653(9.59) + 0.173(12.85) + 0.173(11.98)$$

$$= 10.56$$

The estimate b_i in Eq. (5.35) represents a contrast among the parameter estimates in the Scheffé model of Eq. (5.36), or between the parameter estimate g_i and b_0 above, that is, the difference between the estimated responses at $x_i = 1$ and at the standard mixture **s**,

$$b_i = g_i - \sum_{i=1}^{3} s_i g_i = (1 - s_i)g_i - \sum_{j \neq i} s_j g_j, \quad \text{or} \quad b_i = g_i - b_0$$

For example, $b_2 = 12.85 - 10.56 = 2.29$. The matrix formulas for obtaining these estimates are shown by Eqs. (5.30) and (5.31). A check on the b_i values is $b_1 s_1 + b_2 s_2 + b_3 s_3 = 0$.

Fitting all seven response values in Table 5.6 with a second-degree model, the estimated response equation using Cox's restrictions as well as the Scheffé polynomial in the original components and the L-pseudocomponents, respectively, are as follows:

Cox polynomial:

$$\hat{y}(\mathbf{x}) = 10.82 - 2.43x_1 + 0.96x_2 + 8.22x_3 + 1.36x_1^2 + 16.82x_2^2 - 17.78x_3^2$$
$$(0.34) \quad (0.68) \quad (1.75) \quad (1.75) \quad (1.06) \quad (6.78) \quad (6.78)$$

$$+20.41x_2 x_3 + 4.02x_1 x_3 - 14.32x_1 x_2 \qquad\qquad (5.39)$$
$$(10.86) \qquad (4.91) \qquad (4.91)$$

Scheffé polynomial:

$$\hat{y}(\mathbf{x}) = 9.75x_1 + 28.59x_2 + 1.26x_3 + 21.37x_2 x_3 + 20.43x_1 x_3 - 32.50x_1 x_2$$
$$(0.70) \quad (6.21) \quad (6.21) \quad (11.86) \qquad (11.86) \qquad (11.86)$$
$$(5.40)$$

$$\hat{y}(\mathbf{x}') = 9.7x_1' + 11.2x_2' + 10.8x_3' + 4.5x_2' x_3' + 4.3x_1' x_3' - 6.9x_1' x_2' \qquad (5.41)$$
$$(0.54) \quad (0.54) \quad (0.54) \quad (2.51) \quad (2.51) \quad (2.51)$$

The estimated coefficients in Eq. (5.39) were obtained using a value of $a = 3.77$ in the matrix $\mathbf{B}_2$ of Cornell (2002, app. 6A). Since an observation value at $\mathbf{s}$ was used in obtaining the parameter estimates in the models shown in Eqs. (5.39) through (5.41), it is difficult to give precise interpretations to the coefficient estimates g_i', g_{ij}', $i < j$, and g_i, g_{ij}, $i < j$, in the Scheffé Eqs. (5.41) and (5.40), respectively. If the points of the $\{3, 2\}$ lattice on the boundary of the L-pseudocomponent triangle only had been used for obtaining the estimates in Eq. (5.41), the g_i' would be estimates of the heights of the surface at the three vertices $x_i' = 1$, $x_j' = 0$, $j \neq i$, and the g_{ij}' would represent departures of the surface from the plane defined by $g_1' x_1' + g_2' x_2' + g_3' x_3'$. Since an observation was taken at $\mathbf{s}$ and was used in the analysis, however, these interpretations are relaxed slightly, particularly if the observation at $\mathbf{s}$ does not fall on the plane specified by $g_1' x_1' + g_2' x_2' + g_3' x_3'$. Nevertheless, one often chooses to interpret the parameter estimates as if interior points had not been included, for reasons of simplicity in the interpretation.

The parameter estimates in the original mixture component model of Eq. (5.40) are linear functions of the parameter estimates in the L-pseudocomponent model. Since the simplex space is restricted by lower bounds on the x_i, the coefficient estimate g_i, $i = 1, 2,$ and 3, approximately represents the difference between the height of the surface at the L-pseudocomponent vertex $x_i' = 1$, $x_j' = 0$, $j \neq i$, and the weighted sum of the heights of all of the vertices as in Eq. (5.38) plus some measure of departure from the plane. The g_{ij} represent only departure from the plane scaled by the square of the length of the L-pseudocomponent axes in the original units.

Table 5.7 Estimated Standard Errors ($\hat{\sigma}^2 = 0.30$) of the Cox Model Coefficient Estimates at Different Locations of the Standard Mixture s on the x_1 Axis

	$s = (0.54, 0.23, 0.23)$	$(0.6534, 0.1733, 0.1733)$	$(0.75, 0.125, 0.125)$	$(0.90, 0.05, 0.05)$
$b_0 =$	0.359	0.341	0.330	0.321
$b_1 =$	1.650	0.694	0.443	0.383
$b_2 =$	2.079	1.747	2.263	4.566
$b_3 =$	2.079	1.747	2.263	4.566
$b_{11} =$	2.012	1.068	0.547	0.095
$b_{22} =$	5.696	6.760	8.051	10.866
$b_{33} =$	5.696	6.760	8.051	10.866
$b_{23} =$	8.186	10.826	12.719	17.239
$b_{13} =$	6.042	4.927	3.866	1.893
$b_{12} =$	6.042	4.927	3.866	1.893

With the Cox polynomial, the estimates b_0 and b_i, $i = 1, 2$, and 3, represent linear combinations of the g_i and the g_{ij} according to the top four rows of the matrix $\mathbf{B}_2$ in Cornell (2002, app. 6A). The estimates b_{ij}, $i \leq j = 1, 2$, and 3, are linear functions only of the g_{ij} and hence only of the g'_{ij} and thus represent departure from the plane. For example, the estimate b_{11} is calculated approximately as $b_{11} = [g_{23} - 2(g_{13} + g_{12})]/(a + 2)^2$ and is a comparison of the binary blending of components 2 and 3 (polystyrene and polypropylene) versus the binary blending of polyethylene with each of these components. Note that for increasing values of a (where $s_1 = as_2 = as_3$), while the contrast inside the brackets is unchanged, less weight is assigned to the contrast owing to the reduction in the value of $1/(a + 2)$. Furthermore, as the location of the standard mixture s is placed closer to the $x_1 = 1$ vertex (a is large), the smaller are the estimated standard errors (assuming $s^2 = 0.30$ remains constant) of b_1, b_{11}, and $b_j, j = 2, 3$, relative to the estimated standard errors of b_j, b_{jj}, and b_{23}, $j = 2, 3$. Table 5.7 lists the estimated standard errors of the estimated coefficients in Cox's second-degree model for different locations of s on the x_1 axis.

A final note on the comparison of Cox's model and Scheffé's model concerns the estimated response contours corresponding to both models. While few will argue that Cox's model is better suited than Scheffe's model for measuring the effects of the components on the response, particularly along the Cox-effect directions, nevertheless for a given degree, Scheffé's and Cox's models are equivalent. This means simply that estimated response surface contours generated by the two model forms are identical. Finally, Smith and Beverly (1997) provide instructions for obtaining from StatLib a FORTRAN listing that generates estimates for the Scheffé and Cox linear and quadratic models for 2 up to 10 components. Three examples are presented by Smith and Beverly to illustrate the usefulness of measuring surface gradients with the Cox model.

5.8 FITTING A SLACK-VARIABLE MODEL

Designating one of the mixture components as the slack variable and leaving it out of the fitted model is known as the *slack-variable* (SV) approach. Many practitioners and researchers alike profess to have been successful using the SV approach. Reference to the slack-variable form of the mixture model appears in Snee and Marquardt (1974), Snee and Rayner (1982), Cain and Price (1986), and Piepel and Cornell (1994), while examples of fitting a SV model can be found in articles by Fonner, Buck, and Banker (1970) and Soo, Sander, and Kess (1978).

The slack variable is usually the component whose range of proportionate values is the largest, or, the component having the largest proportion ($x_i \geq 0.80$) in the mixture, or, the component presumed to be most inert or inactive (e.g., H_2O). Then there may be a situation where one of the components has a very small range, say, $R_i = U_i - L_i = 0.005$, and rather than run the risk of creating collinearity among those terms in the model involving the component x_i, one chooses to not include x_i in the model. Whatever the reason for choosing to use the SV approach, there can be risks involved with doing so. Before mentioning such risks, let us define the SV model forms. First, however, we offer two definitions.

Definition. A *complete* model of degree d is one that contains all of the terms up to and including those of degree d.

The definition above quite naturally leads to the following definition:

Definition. A *reduced* model is one that contains only a subset of the terms of the complete model.

Generally, in a mixture setting, when we refer to a reduced model of degree d, it will contain all of the terms up to degree $d - 1$ but only a subset of the terms of degree d.

The *complete* or *full* Scheffé-type and the corresponding SV models are equivalent model forms. To show this, let the number of components be q and let x_q be the slack variable. Then the Scheffé and SV models up to the special cubic are

	Scheffé	Slack Variable	
Linear:	$y = \sum_{i=1}^{q} \beta_i x_i + \varepsilon$	$y = \alpha_0 + \sum_{i=1}^{q-1} \alpha_i x_i + \varepsilon$	(5.42)

Quadratic:
$$y = \sum_{i=1}^{q} \beta_i x_i \qquad\qquad y = \alpha_0 + \sum_{i=1}^{q-1} \alpha_i x_i$$
$$+ \sum_{i<j}^{q} \beta_{ij} x_i x_j + \varepsilon \qquad\qquad + \sum_{i\leq j}^{q-1} \alpha_{ij} x_i x_j + \varepsilon \qquad (5.43)$$

Special cubic: $\quad y = $ Quadratic $\qquad\qquad\qquad y = $ Quadratic
$$+ \sum_{i<j<k}^{q} \beta_{ijk} x_i x_j x_k + \varepsilon \qquad\qquad + \sum_{i<j<k}^{q-1} \alpha_{ijk} x_i x_j x_k$$
$$+ \sum_{i<j}^{q-1} \alpha_{ij} x_i x_j (x_i + x_j) + \varepsilon$$
$$(5.44)$$

The equivalence of the respective equations means the coefficients (and their estimates) in the Scheffé models are simple functions of the coefficients (and their estimates) in the SV models, and vice versa. More specifically, with the linear or first-degree models in Eq. (5.42)

$$\alpha_0 = \beta_q, \quad \alpha_i = \beta_i - \beta_q, \qquad i = 1, 2, \ldots, q - 1$$

while with the quadratic models in Eq. (5.43),

$$\alpha_0 = \beta_q, \quad \alpha_i = \beta_i - \beta_q + \beta_{iq},$$
$$\alpha_{ii} = -\beta_{iq}, \quad \alpha_{ij} = \beta_{ij} - (\beta_{iq} + \beta_{jq}), \qquad i, j = 1, 2, \ldots, q - 1, i \neq j$$

Letting $q = 3$, for example,

$$\alpha_0 = \beta_3, \quad \alpha_1 = \beta_1 - \beta_3 + \beta_{13}, \quad \alpha_2 = \beta_2 - \beta_3 + \beta_{23}$$
$$\alpha_{11} = -\beta_{13}, \quad \alpha_{22} = -\beta_{23}, \quad \alpha_{12} = \beta_{12} - (\beta_{13} + \beta_{23})$$

(5.45)

We see, for example, when fitting the quadratic or second-degree SV model of Eq. (5.43), that the coefficient estimate of the pure quadratic term $(\alpha_{ii}x_i^2)$ is simply a measure of the nonlinear blending between the ith component and the slack variable, while the coefficient estimate of the cross-product term $(\alpha_{ij}x_ix_j)$ is a measure of the difference between the nonlinear blending of components i and j, β_{ij}, and the sum of the nonlinear blending of each with the slack variable, $\beta_{iq} + \beta_{jq}$. Hereafter in this section and the next when referring to any slack-variable model with component x_q being the slack component, we will abbreviate the model using SLK-x_q.

The choice of which of the q components to designate as the slack variable has not been defended from either a theoretical or a practical point of view. Recently Cornell (2000) addressed the question, "Does it matter which component is designated the slack variable when it comes to fitting the slack-variable model since x_i is not included in the model?" He then showed that if one tries to build a model using such variable-selection techniques as stepwise regression, it can make a difference which component is chosen. It can make a difference, that is, in terms of the form or number of terms that are included in final model selected. Before we proceed into the next section to illustrate this further, let us work through a small numerical example.

Let us recall the three-component yarn elongation example data set in Table 2.3. where the fitted Scheffé quadratic model of Eq. (2.24) was

$$\hat{y}(\mathbf{x}) = 11.7x_1 + 9.4x_2 + 16.4x_3 + 19.0x_1x_2 + 11.4x_1x_3 - 9.6x_2x_3$$

(5.46)

$$(0.60) \quad (0.60) \quad (0.60) \quad (2.61) \quad (2.61) \quad (2.61)$$

$$[t = 7.28] \; [t = 4.37] \; [t = -3.68]$$

where the quantities in parentheses are the estimated standard errors of the coefficient estimates and the quantities in brackets are the calculated t-values obtained by dividing the coefficient estimate by its estimated standard error. The tabled t-value for the two-sided test of hypothesis at the $\alpha/2 = 0.025$ level of significance, with nine degrees of freedom, is $t_{9,0.025} = 2.262$. The corresponding SLK-x_i fitted models are

$$\text{SLK-}x_1: \quad \hat{y}(x_2, x_3) = 11.7 + 16.7x_2 + 16.1x_3 - 19.0x_2^2 - 11.4x_3^2 - 40.0x_2x_3$$
$$\quad\quad (2.74) \quad (2.74) \quad (2.61) \quad (2.61) \quad\quad (4.18)$$
$$[t = 6.09] \quad [5.87] \quad [-7.28] \quad [-4.37] \quad\quad [-9.56]$$

$$(5.47)$$

$$\text{SLK-}x_2: \quad \hat{y}(x_1, x_3) = 9.4 + 21.3x_1 - 2.6x_3 - 19.0x_1^2 + 9.6x_3^2 - 2.0x_1x_3$$
$$\quad\quad (2.74) \quad (2.74) \quad (2.61) \quad (2.61) \quad\quad (4.18)$$
$$[t = 7.76] \quad [-0.95] \quad [-7.28] \quad [3.68] \quad\quad [-0.48]$$

$$(5.48)$$

$$\text{SLK-}x_3: \quad \hat{y}(x_1, x_2) = 16.4 + 6.7x_1 - 16.6x_2 - 11.4x_1^2 + 9.6x_2^2 + 17.2x_1x_2$$
$$\quad\quad (2.74) \quad (2.74) \quad (2.61) \quad (2.61) \quad\quad (4.18)$$
$$[t = 2.44] \quad [-6.05] \quad [-4.37] \quad [3.68] \quad\quad [4.11]$$

$$(5.49)$$

Now, in the discussion of the results of fitting the Scheffé quadratic model of Eq. (2.24) previously in Section 2.6, which is Eq. (5.46) above, we inferred the blending of each of the components 2 or 3 with component 1 was synergistic. This was because the average elongation value of the yarn containing components 1 and 2 and containing components 1 and 3 was higher than the average of the elongation values of the single-component yarns. However, when components 2 and 3 were combined, the resulting yarn had a lower average elongation value than would be expected by averaging the elongation values of the single-component blends. What we would like to know is, "Do the slack-variable models of Eqs. (5.47)–(5.49) convey the same type of information that the Scheffé model does?"

To try and answer this question from a component-blending point of view, let us see what type of information is conveyed by the fitted SLK-x_1 model in Eq. (5.47) as an example. The significance of the coefficient estimate $c_2 = 16.7 (= b_2 - b_1 + b_{12} = 9.4 - 11.7 + 19.0)$ of the $16.7x_2$ term coupled with the coefficient estimate $c_{22} = -19.0 (= -b_{12})$ of the $-19.0x_2^2$ term suggests that the average elongation increases linearly and then decreases as one moves away from the $x_2 = 0$ edge of the triangular composition space toward the $x_2 = 1$ vertex. In a similar way the positive coefficient of the $16.1x_3$ term coupled with the negative coefficient of the $-11.4x_3^2$ suggests the average elongation value increases and then decreases (or behaves quadratically) as one moves from the $x_3 = 0$ edge of the triangular composition space toward the $x_3 = 1$ vertex. Furthermore one could just as easily ask, "Is the significance of the coefficient estimate 16.7 of the $16.7x_2$ term owing to the difference $b_2 - b_1$ or to the nonlinear blending (b_{12}) of components 1 and 2?" Having just fitted both the Scheffé and the SLK-x_1 quadratic models enables us to answer that question confidently. But, if we had only fit the SLK-x_1 model and had not fit the Scheffé model of Eq. (5.46), we probably could not have answered the question above, saying that it is probably due to the nonlinear blending of components 1 and 2; the absence of x_1 in the model of Eq. (5.47) would have made us focus on the roles that x_2 and x_3 play on the response and ignore the role of x_1. And yet the two models of Eqs. (5.46) and (5.47) are equivalent, which means a contour plot of the estimated yarn elongation surface over the three-component triangle, generated by Eq. (5.46), would look exactly like the contour plot of the estimated surface generated by Eq. (5.47).

In the next section two approaches are taken in an effort to compare the forms of the fitted Scheffé and slack-variable models. The first approach compares the results of fitting all possible regression equations using PROC REG/selection = RSQUARE in SAS (1998) followed by obtaining the "best" fitted equations according to the MAXR and MINR criteria. Each of the three procedures (RSQUARE, MAXR, and MINR) begins by adding nonlinear blending terms one at a time to the forced fitted linear blending model.

The second approach presented is that of obtaining reduced quadratic models consisting of the linear blending terms and all or possibly some of the nonlinear blending terms of the complete quadratic model using stepwise regression. Both DESIGN-EXPERT of Stat-Ease (1998) and SAS (1998) were used to generate the models. The main stepwise regression procedures are *forward selection* of terms, *backward elimination* of terms, and just *stepwise* regression. This latter procedure is a modification of the forward selection procedure, where at each step in the model-fitting exercise all nonlinear blending terms entered previously into the model are reassessed as to their importance. Cornell (2000) chose stepwise regression and stated he believed it to be the most popular of the three procedures. Furthermore, since the stated objective by Cornell (2000) was to see if and how final model forms may differ when different components were chosen as the slack variable, the assignment of the level of significance for a term to enter the model as well as to stay in the model was liberally assigned at 0.15.

5.9 A NUMERICAL EXAMPLE ILLUSTRATING THE FITS OF DIFFERENT REDUCED SLACK-VARIABLE MODELS: TINT STRENGTH OF A HOUSE PAINT

Tint strength is a property of a house paint known to be affected by varying the relative proportions of two known pigments (P_1 and P_2) and a specific solvent (S) in the paint. The volume percent of the two pigments and the solvent is fixed at 8% of the paint for all mixes or blends. Scaling by the total percent of the three components, it is possible to write, for all mixes studied, $A + B + C = 1$, by letting $A = P_1/8\%$, $B = P_2/8\%$, and $C = S/8\%$ represent the proportions of the two pigments and solvent, respectively.

Listed in Table 5.8 are recorded tint strength values for seven blends of paint, where the coordinate settings of A, B, and C comprise a simplex-centroid design. The tint strength values represent average color-coded scores, where a negative score signifies the paint color was less desirable than that of the standard-colored paint to

Table 5.8   Tint Strength Values (Scaled) for the Seven House Paint Blends

Blend	$A(P_1)$	$B(P_2)$	$C(S)$	Tint Strength
1	1	0	0	1.19
2	0	1	0	0.97
3	0	0	1	-5.68
4	$\frac{1}{2}$	$\frac{1}{2}$	0	0.13
5	$\frac{1}{2}$	0	$\frac{1}{2}$	-4.02
6	0	$\frac{1}{2}$	$\frac{1}{2}$	-4.18
7	$\frac{1}{3}$	$\frac{1}{3}$	$\frac{1}{3}$	$-3.36, -3.69, -3.04$

Source: Copyright 2000 American Society for Quality. Reprinted with permission.

which it was compared. Blends 1 to 3 represent single pigment/solvent blends while blends 4 to 7 represent two- and three-component blends. The centroid blend, number 7, was replicated three times, producing the three color scores.

When fitted to the $N = 9$ tint strength values in Table 5.8, the Scheffé quadratic model and the equivalent SLK-A quadratic model are

$$\hat{y}_{\text{Scheffé}} = b_A A + b_B B + b_C C + b_{AB} AB + b_{AC} AC + b_{BC} BC$$
$$= 1.21A + 0.99B - 5.66C - 4.19AB - 7.49AC - 7.69BC \qquad (5.50)$$
$$ (1.23) \qquad (1.23) \qquad (1.23)$$

and

$$\hat{y}_{\text{SLK}-A} = a_0 + a_B B + a_C C + a_{BB} B^2 + a_{CC} C^2 + a_{BC} BC$$
$$= 1.21 - 4.41B - 14.36C + 4.19B^2 + 7.49C^2 + 3.99BC \qquad (5.51)$$
$$ (1.23) \qquad (1.23) \qquad (2.17)$$

where the quantities in parentheses below the coefficient estimates of the nonlinear blending terms in Eqs. (5.50) and (5.51) are the estimated standard errors of the coefficient estimates. Each of the fitted models of Eqs. (5.50) and (5.51) has an $R^2 = 0.9952$ and mean-square error (MSE) $= 0.08$.

The coefficient estimates of each nonlinear blending term in the Scheffé model of Eq. (5.50), when divided by its estimated standard error, exceeds the tabled value of $t_{0.025,3} = 3.182$. Thus we would infer that each coefficient estimates a quantity that is significantly ($P < 0.05$) different from zero and that all three nonlinear blending terms are important. But this is not the case with the $3.99BC$ term in the slack-variable model of Eq. (5.51) because when it is divided by 2.17, the result is a quotient less than 3.182. Hence, if we were to choose to reduce the number of terms in Eq. (5.51) by dropping the $3.99BC$ term and refitting the remaining five terms, the resulting model would be different from Eq. (5.51) and would no longer be equivalent to Eq. (5.50).

Owing to the equivalence of the two models in Eqs. (5.50) and (5.51), it is possible to express the coefficient estimates in the SLK-A model of Eq. (5.51) as functions of the coefficient estimates in the Scheffé model of Eq. (5.50). This is done as

$$a_0 = 1.21 = b_A$$
$$a_B = -4.41 = b_B - b_A + b_{AB} = 0.99 - 1.21 + (-4.19)$$
$$a_C = -14.36 = b_C - b_A + b_{AC} = -5.66 - 1.21 + (-7.49)$$
$$a_{BB} = 4.19 = -b_{AB}, \quad a_{CC} = 7.49 = -a_{AC}$$

and

$$a_{BC} = 3.99 = b_{BC} - (b_{AB} + b_{AC}) = -7.69 - (-4.19 - 7.49)$$

Listed in Table 5.9 are the results of selecting the best $p = 4$ and $p = 5$ term-reduced fitted Scheffé and slack-variable models as obtained by using all possible regressions (RSQUARE) and the MAXR (maximum R^2) and MINR (minimum R^2) criteria in PROC REG in SAS (1998). Also listed are the values of R^2, Mallow's (1973) C_p statistic, PRESS, and MSE for each fitted model. For a particular model, the closer the value of R^2 is to one, the closer the value of C_p is to p, and the lower

Table 5.9 Best Subset p-Term Models as Determined by RSQUARE, MAXR, and MINR Criteria for Models Fitted to the Tint Strength Values in Table 5.8

Term	$p = 4$				$p = 5$			
	Scheffé	SLK-*A*	SLK-*B*	SLK-*C*	Scheffé	SLK-*A*	SLK-*B*	SLK-*C*
Constant		*	*	*		*	*	*
$b_1 A$	*		*	*	*		*	*
$b_2 B$	*	*		*	*	*		*
$b_3 C$	*	*	*		*	*	*	
$b_{12} AB$								
$b_{13} AC$					*			
$b_{23} BC$	*				*			
$a_{11} A^2$							*	*
$a_{22} B^2$				*		*		*
$a_{33} C^2$		*	*			*	*	
C_p	53.0	13.6	13.6	43.1	15.6	7.4	8.1	29.7
R^2	0.9133	0.9766	0.9766	0.9301	0.9767	0.9898	0.9887	0.9544
PRESS	34.94	27.21	27.21	73.25	34.84	15.83	17.54	60.55
MSE	0.87	0.23	0.23	0.70	0.29	0.13	0.14	0.57
Ranking	4	1.5	1.5	3	3	1	2	4
(1 = best,								
4 = worst)								

* Signifies term that remained in or became part of the final model form.

Table 5.10 Coefficient Estimates and Summary Statistics for the Complete Scheffé Quadratic Model and Reduced Quadratic SLK-A, SLK-B, or SLK-C Models Fitted to the Tint Strength Values in Table 5.8

Term	Scheffé	Reduced SLK-A	Full SLK-B^a	Reduced SLK-$B(0.10)^b$	Reduced SLK-C
Constant		0.96	0.99	0.59	−6.53
$b_1 A$	1.21		−3.97	0.24	7.22
$b_2 B$	0.99	−2.86			1.23
$b_3 C$	−5.66	−12.81	−14.34	−14.03	
$b_{12} AB$	−4.19				n.s.
$b_{13} AC$	−7.49		4.39	n.s.	
$b_{23} BC$	−7.69	n.s.			
$a_{11} A^2$			4.19	n.s.	n.s.
$a_{22} B^2$		2.91			6.38
$a_{33} C^2$		6.21	7.69	7.84	
P	6	5	6	4	4
C_p	6.00	7.37	6.00	13.60	43.09
R^2	0.9952	0.9898	0.9952	0.9767	0.9299
PRESS	7.63	15.83	7.63	27.21	73.25
MSE	0.08	0.13	0.08	0.23	0.70

Source: Copyright 2000 American Society for Quality. Reproduced with permission.
[a]The fitted quadratic SLK-B model did not reduce at the $\alpha = 0.15$ level of significance to enter or to exit.
[b]The fitted reduced SLK-B model at the $\alpha = 0.10$ to enter and to exit.

the values of PRESS and MSE, the better we feel about the usefulness of the model. For each value of $p = 4$ and $p = 5$, a subjective rating of the four models is provided at the bottom of the table, where $1 = $ best and $4 = $ worst. For both $p = 4$ and 5, the SLK-A or SLK-B model is preferred to the reduced Scheffé or SLK-C models.

Listed in Table 5.10 are the coefficient estimates of the fitted Scheffé quadratic model and the significant ($P < 0.15$) terms in the reduced slack-variable models. The reduced models were generated using the stepwise regression option in DESIGN-EXPERT of Stat-Ease (1998) and also were verified using PROC REG in SAS (1998). Values of R^2, C_p, PRESS, and MSE are listed for each model also.

Upon looking at the results of the stepwise model fitting exercise in Table 5.10 we note that the Scheffé quadratic model did not reduce in form when subjected to the model reduction algorithm. The reason for this is that each of the coefficient estimates for the three nonlinear blending terms is significantly different from zero at the 0.05 level. Shown in Figures 5.12a–d are contour plots of the estimated tint strength surfaces corresponding to the Scheffé quadratic model and the reduced quadratic SLK-A, SLK-B, and SLK-C models, respectively. These plots were generated by DESIGN-EXPERT.

The reduced SLK-A, SLK-B, and SLK-C models in Table 5.10 contain five, four, and four terms, respectively. From Figures 5.12a–d it is clear that the degree of similarity of the surface contours of each of the estimated surfaces generated by the

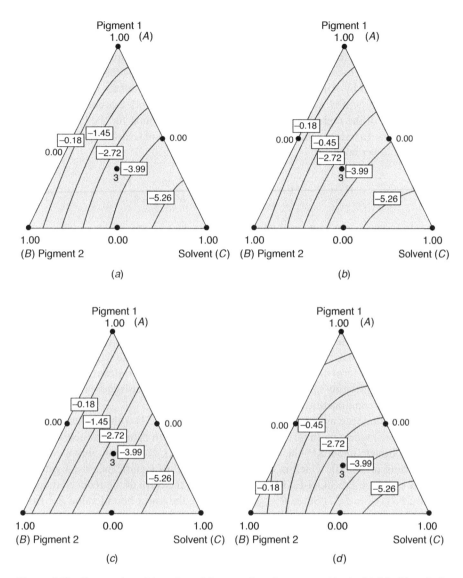

Figure 5.12 Contour plots of the estimated tint strength surface generated by the (*a*) Scheffé quadratic, (*b*) reduced SLK-*A*, (*c*) reduced SLK-*B*, and (*d*) reduced SLK-*C* models. Copyright 2000 American Society for Quality. Reprinted with permission.

slack-variable models lessens when compared to the Scheffé model surface as the number of nonlinear blending terms in the respective slack-variable model decreases. Of the three reduced slack-variable models, the fitted five-term SLK-*A* model with its estimated tint strength surface contour plot of Figure 5.12*b* most closely reflects the fit of the Scheffé quadratic model whose estimated surface contour plot is shown in Figure 5.12*a* but this is just a subjective judgment at this point.

How can we decide if any of the reduced slack-variable models are not adequate or do not provide a satisfactory fit when compared to the fit of the complete Scheffé quadratic model or to the fits of their respective six-term complete slack-variable models? One such measure of adequacy of fit of a subset model can be traced back to a test proposed by Aitkin (1974). In an effort to choose a subset of the regressor variables that provided an adequate fit relative to the fit of the complete model containing all of the regressors, Aitkin defined a lower bound for the coefficient of determination of a reduced model containing only a subset of terms from the complete model. The lower bound is

$$R_0^2 = 1 - (1 - R_{\text{complete}}^2)\left(1 + \frac{k F_{\alpha, k, N-k-1}}{N - k - 1}\right) \tag{5.52}$$

where N is the total number of observations, $k + 1$ is the number of terms in the complete model (including the constant term), R_{complete}^2 is the coefficient of determination for the full or complete model, and $F_{\alpha, k, N-k-1}$ is the upper α percentage point of the F distribution having k and $N-k-1$ degrees of freedom in the numerator and denominator, respectively. Aitkin refers to a subset of regressors producing an $R^2 \geq R_0^2$ an R^2-*adequate* (α) subset, which means that such a subset produces a model with an adequate R^2 value. In other words, if we find that any of the fitted reduced slack-variable models in Table 5.10 has an R^2 value less than R_0^2, then it is probably safe or correct to infer that such models are not adequate when compared to their complete six-term model. This suggests that one cannot arbitrarily select one of the mixture components to serve as the slack variable and thus ignore it but include all of the other components in the fitted reduced model.

Let us illustrate the use of the lower bound R_0^2 in Eq. (5.52) with the fitted reduced slack-variable models in Table 5.10. With the tint strength data set in Table 5.8 $N = 9$. Each complete quadratic slack-variable model has $k = 5$ terms plus a constant term. Letting $\alpha = 0.10$, which is lower than the level of significance used with the stepwise option in DESIGN-EXPERT in arriving at the form of the reduced models, we have $F_{0.10,5,3} = 5.31$. Since $R_{\text{complete}}^2 = 0.9952$, the lower-bound value $R_0{}^2$ from Eq. (5.52) is

$$R_0^2 = 1 - (1 - 0.9952)\left(1 + \frac{5(5.31)}{9 - 5 - 1}\right) = 0.9527$$

Hence any subset or reduced slack-variable model possessing an R^2 value lower than 0.9527 is considered not to be an R^2-*adequate* (0.10) model. From Table 5.10 both reduced quadratic SLK-A and SLK-B models have R^2 values that exceed 0.9527, while the fitted reduced SLK-C model does not. Hence, if component C is chosen to be the slack variable and we choose to reduce the fitted quadratic model in components A and B using stepwise regression, we would end up with a four-term model that does not satisfy the criterion of R^2-*adequacy* (0.10) of fit relative to the full quadratic model even though the full quadratic model contains two terms that were rejected by stepwise regression at the $\alpha = 0.15$ level.

In summary, whereas it is well known that *only upon fitting a complete model form can one feel rest assured that it does not matter which component is chosen to be the slack variable*, we now see that the final model form, if reduced, may differ with different components being selected as the slack variable. The reason for this lies not only in the shape of the mixture surface to be approximated by the fitted model but also in the location and size of the experimental region inside the simplex when constraints are placed on the component proportions.

5.10 SUMMARY

The use of model forms other than the standard polynomials in the x_i is the main theme of Chapter 5. Inverse terms or reciprocals (x_i^{-1}) of the component proportions are suggested for modeling extreme changes in the response behavior as the value of x_i becomes very close to zero. Extreme changes in the response behavior are very real occurrences in many areas of chemical experimentation, especially when certain component proportions approach boundary conditions. The modeling sequence of including inverse terms in the Scheffé polynomials starts with the simplest model form and grows with the addition of extra terms. This sequence is the reverse of that in Chapter 4.

When a component, say, i, serves as a diluent in the sense that as the proportion x_i increases the effect of the remaining components on the response diminishes in proportion, then component i is said to blend with the other components in an additive manner. To model such an effect of one component in an otherwise nonlinear blending system, models that are homogeneous of degree 1 are superior to the Scheffé polynomials. It is shown, using the analysis of a three-component special cubic model, how to discover when one of the components blends additively with the other two components.

In some experiments the ratios of the component proportions are more meaningful than the proportions alone. By defining $q - 1$ ratios as variables, standard polynomial models can be written in the ratio variables and the models can also be fitted to data collected at the points of standard factorial arrangements. This is illustrated by the construction of a 3^2 factorial arrangement in a constrained region of a three-component triangle, where the factorial arrangement uses equally spaced levels for the ratios $r_1 = x_3/x_1$ and $r_2 = x_3/x_2$.

When certain restrictions are placed on the parameter values in the standard polynomial models, the interpretations of the parameter estimates are closer in meaning to the treatment effects (or to the component effects defined in Section 4.6.) in an analysis of variance setting of a comparative-type experiment. Cox's mixture polynomials with restrictions imposed on the parameter values are compared to the Scheffé polynomials in terms of the meanings attached to the parameters in the respective models and some comments are made concerning the best designs to use for fitting Cox's polynomials.

The chapter ends with a discussion centered on fitting a slack-variable model where one of the components is designated as the slack variable and is left completely out of the model. Piepel and Cornell (1994) describe briefly a few situations where fitting a slack-variable model may be a very reasonable thing to do, whereas the question later addressed by Cornell (2000) is, "Does it matter which one of the q components is designated the slack variable if one chooses to fit the slack-variable model?" A three-component house paint example presented in the last section shows that because of the dependency that exists among the components in a mixture, the choice of component to serve as the slack variable can make a difference in the final model form if one tries to use model reduction techniques to arrive at a final model form.

REFERENCES AND RECOMMENDED READING

Aitchison, J., and J. Bacon-Shone (1984). Log contrast models for experiments with mixtures. *Biometrika*, **71** (2), 323–330.

Aitkin, M. A. (1974). Simultaneous inference and the choice of variable subsets. *Technometrics*, **16**, 221–227.

Becker, N. G. (1968). Models for the response of a mixture. *J. R. Stat. Soc. B*, **30**, 349–358.

Becker, N. G. (1978). Models and designs for experiments with mixtures. *Austral. J. Stat.*, **20** (3), 195–208.

Cain, M., and M. L. R. Price (1986). Optimal mixture design. *Appl. Stat.*, **35**, 1–7.

Chan, L. Y., J. H. Meng, Y. C. Jiang., and Y. N. Guan (1998). D-optimal axial designs for quadratic and cubic additive mixture models. *Aust., N. Z. J. Stat.*, **40**, 359–371.

Cornell, J. A. (1975). Some comments on designs for Cox's mixture polynomial. *Technometrics*, **17**, 25–35.

Cornell, J. A. (2000). Fitting a slack-variable model to mixture data: some questions raised. *J. Qual. Technol.*, **32**, 133–147.

Cornell, J. A. (2002). *Experiments with Mixtures: Design, Models, and the Analysis of Mixture Data*. Wiley New York.

Cornell, J. A., and J. W. Gorman (1978). On the detection of an additive blending component in multicomponent mixtures. *Biometrics*, **34** (2), 251–263.

Cox, D. R. (1971). A note on polynomial response functions for mixtures. *Biometrika*, **58** (1), 155–159.

Draper, N. R., and R. C. St. John (1977). A mixtures model with inverse terms. *Technometrics*, **19**, 37–46.

Kenworthy, O. O. (1963). Factorial experiments with mixtures using ratios. *Ind. Qual. Control*, **19**, 24–26.

Lim, Y. B. (1987). Symmetric D-optimal designs for log contrast models with mixtures. *J. Korean Stat. Soc.*, **16**, 71–79.

Marquardt, D. W., and R. D. Snee (1974). Test statistics for mixture models. *Technometrics*, **16**, 533–537.

Park, S. H., and J. I. Kim (1988). Slope-rotatable designs for estimating the slope of response surfaces in experiments with mixtures. *J. Korean Stat. Soc.*, **17**, 121–133.

Piepel, G. F., and J. A. Cornell (1994). Mixture experiment approaches: examples, discussion, and recommendations. *J. Qual. Technol.*, **26**, 177–196.

SAS Institute (1998). *SAS User's Guide*, Version 6.12. SAS Institute, Inc., Cary, NC.

Scheffé, H. (1958). Experiments with mixtures. *J. R. Stat. Soc. B*, **20** (2), 344–360.

Scheffé, H. (1963). Simplex-centroid design for experiments with mixtures. *J. R. Stat. Soc. B*, **25** (2), 235–263.

Smith W. F. (2005). *Experimental Design for Formulation*. ASA-SIAM Series on Statistics and Applied Probability. ASA, Alexandria, VA.

Smith, W. F. and T. A. Beverly (1997). Generating linear and quadratic Cox mixture models. *J. Qual. Technol*, **29**, 211–224.

Snee, R. D. (1971). Design and analysis of mixture experiments. *J. Qual. Technol.*, **3**, 159–169.

Snee, R. D. (1973). Techniques for the analysis of mixture data. *Technometrics*, **15** (3), 517–528.

Snee, R. D. (1974). Techniques for developing blending models. Paper presented at Annual Meeting of American Statistical Association, St. Louis, MO.

Snee, R. D., and D. W. Marquardt (1976). Screening concepts and designs for experiments with mixtures. *Technometrics*, **18**, 19–29.

Snee, R. D., and A. A. Rayner (1982). Assessing the accuracy of mixture model regression calculations. *J. Qual. Technol.*, **14**, 67–79.

Stat-Ease (1998). *DESIGN-EXPERT Software for Response Surface Methodology and Mixture Experiments*, Version 5.0. Stat-Ease, Inc., Minneapolis, MN.

QUESTIONS

5.1. The model in the ratios r_1 and r_2 shown by Eq. (5.25) was fitted to the octane data in Table 5.5 and produced a value of $R_A^2 = 0.72$. Fit a Scheffé-type linear-plus-inverse-terms model with the two terms (x_1^{-1}) and (x_3^{-1}) to the data and comment on the use of these terms in the model versus the use of second-degree terms. Calculate the C_p values using $s^2 = 2.50$.

5.2. In Table 5.3 are listed predicted mite numbers obtained with the special cubic Eq. (5.20) and the nonpolynomial Eq. (5.21). Using the observed mite numbers at the 15 blends, compute the residuals (r_u) where $r_u = y_u - \hat{y}_u$ with each of the models. Comment on the use of the residuals for testing the performance of each of the fitted model forms as well as for comparing the fitted model forms.

5.3. For the three-component system where data are collected at the points of a simplex-centroid design, the conditions listed in Eqs. (5.16) and (5.17) combined with a fitted model of the form in Eq. (5.15) imply the additive blending effect of component 3. A similar procedure would be to test the adequacy of the model

$$E[y_{lu}] = \mu + \beta_l(1 - x_{3u})$$

where μ is some overall mean value and β_l is the slope of the surface along the lth ray extending from $x_3 = 0$ to $x_3 = 1$, where we might have a set of k rays (i.e., $l = 1, 2, \ldots, k$). Refer to Figure 5.5 where data are positioned on three rays: the $x_1 - x_2$ edge, the x_3-axis, and the $x_2 - x_3$ edge. Fit a first-degree model of the form $y_u = \mu + \beta_l(1 - x_{3\mu}) + \varepsilon_\mu$ to the data on each of the three rays. Test the null hypothesis $H_0 : \beta_l = 0$ versus $H_A : \beta_l = 0$ at the $\alpha = 0.05$ level of significance for each model. Suppose you were asked to test the hypothesis $H_0 : \beta_i = \beta_j$, where $i < j = 2$ and 3 versus $H_A : \beta_i = \beta_j, ij = 1$ and 2, 1 and 3, and 2 and 3, only this time at the $\alpha = 0.01$ level of significance.

5.4. Cox's second-degree model is expressed in Eq. (5.27). If upon fitting the second-degree model to a set of data we choose to compare the response at **x** and **s** in Figure 5.9, for example, then the difference is expressed as

$$\Delta \hat{y}(\mathbf{x}) = \frac{\hat{\beta}_i \Delta_i}{1 - s_i} + \frac{\hat{\beta}_{ii} \Delta_i^2}{(1 - s_i)^2}$$

(a) Use the fitted model in Eq. (5.39) to illustrate the formula above, where the difference is desired between the point $\mathbf{x} = (0.82, 0.09, 0.09)$ and the point $\mathbf{s} = (0.654, 0.173, 0.173)$.

(b) Compare your result in (a) to that which is obtained by simply using the fitted Scheffé model in (5.40) and computing $\hat{y}(\mathbf{x}) - \hat{y}(\mathbf{s})$. Do the results differ? If so, why?

5.5. Shown are data values at the seven blends of a three-component simplex-centroid design. High values of the response are more desirable than low values. Indicate the type of blending, linear or nonlinear, synergistic or antagonistic, that is present between the components,

(a) A and B: _____ and is _____ .

(b) A and C: _____ and is _____ .

(c) B and C: _____ and is _____ .

(d) A and B and C: _____ and is _____ .

Hint: $\widehat{\text{var}}(b_i) = \text{MSE}/2$, $\widehat{\text{var}}(b_{ij}) = 24/2 \text{ MSE}$, $\text{var}(b_{\text{ABC}}) = 958.5 \text{ (MSE)}$. Also $\text{SST} = 63.79$, $\text{SSR} = 58.22$, and $\text{SSE} = 5.57$.

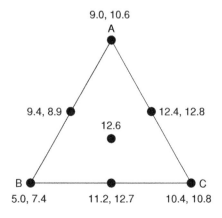

5.6. An extruded product was formed by blending wheat flour (W), peanut flour (P), and corn meal (C). The product was to be a candidate for a high-protein cereal or snack. The wheat flour resulted in a puffiness upon cooling after extrusion, but the wheat contained only approximately 10% protein. The peanut flour was high in protein, and the corn meal served mainly as a filler ingredient.

The data represent diameter measurements in millimeters of extruded and chopped 2-inch rods. Twelve combinations of W–P–C were performed. The pairs of diameter values represent data collected on each of two days.

(a) Fit a model to all 24 diameter readings, assuming that the difference in readings from day to day is random variation. The objective is to model the rod diameter surface over the triangle.

(b) Include a term in the model to represent the difference between the diameter values taken on the two days. Is the difference in the average diameter values for the two days significantly greater than zero?

(c) If only rods having a diameter of at least 7.5 mm are desirable, suggest mixtures that probably yield an acceptable product.

Data:

Ingredient Proportions			Diameter Measurements	
W	P	C	Day 1	Day 2
1.00	0	0	9.40	9.18
0	1.00	0	9.00	8.91
0.50	0.50	0	7.75	8.25
0	0	1.00	6.17	5.36
0	0.50	0.50	5.54	5.40
0.50	0	0.50	7.69	7.21
0.20	0.30	0.50	6.44	6.36
0.30	0.50	0.20	7.42	6.84
0.50	0.40	0.10	7.90	8.00
0.50	0.30	0.20	7.35	7.27
0.50	0.20	0.30	7.60	6.72
0.50	0.10	0.40	7.50	7.45

5.7. Displayed in Figure 5.5 are the average mite numbers along the three rays R1 ($x_1 - x_3$ edge), R2 ($x_2 - x_3$ edge) and R3 (x_3 axis). Listed below are the average mite numbers and the values of $(1 - x_{3u})$ and $(1 - x_{3u})^2$. The model to be fitted is

$$y_{lu} = \mu_l + \beta_l(1 - x_{3u}) + \gamma_l(1 - x_{3u})^2 + \varepsilon_u$$

Ray 1 ($x_1 - x_3$ edge) $(x_1 - x_3)^2$

$$y = \begin{bmatrix} 49.8 \\ 41.3 \\ 34.7 \\ 28.8 \\ 20.1 \end{bmatrix}, \quad X = \begin{bmatrix} 1 & 1.00 & \vdots & 1 \\ 1 & 0.75 & \vdots & 0.5625 \\ 1 & 0.50 & \vdots & 0.2500 \\ 1 & 0.25 & \vdots & 0.0625 \\ 1 & 0 & \vdots & 0 \end{bmatrix}$$

Ray 2 ($x_2 - x_3$ edge)

$$y = \begin{bmatrix} 84.2 \\ 66.0 \\ 52.4 \\ 39.4 \\ 20.1 \end{bmatrix}, \quad X = \begin{bmatrix} 1 & 1 & \vdots & 1 \\ 1 & 0.75 & \vdots & 0.5625 \\ 1 & 0.50 & \vdots & 0.2500 \\ 1 & 0.25 & \vdots & 0.0625 \\ 1 & 0 & \vdots & 0 \end{bmatrix}$$

Ray 3 (x_3 axis)

$$y = \begin{bmatrix} 35.8 \\ 32.7 \\ 29.6 \\ 27.9 \\ 26.1 \\ 23.3 \\ 20.1 \end{bmatrix}, \quad X = \begin{bmatrix} 1 & 1 & \vdots & 1 \\ 1 & 0.8 & \vdots & 0.64 \\ 1 & 0.6 & \vdots & 0.36 \\ 1 & 0.5 & \vdots & 0.25 \\ 1 & 0.4 & \vdots & 0.16 \\ 1 & 0.2 & \vdots & 0.04 \\ 1 & 0 & \vdots & 0 \end{bmatrix}$$

Along each of the three rays, fit the second-degree model and connected on whether or not the quadratic model is better than a first-degree model. Calculate SS Residual for each model and use the difference $SSE_{linear} - SSE_{quad}$ to support your claim.

5.8. On the axes of each component in a five-component blending system, the following response values were collected:

x_1	x_2	x_3	x_4	x_5	Response Values (y_u)
0.80	0.05	0.05	0.05	0.05	24.3, 28.5
0.05	0.80	0.05	0.05	0.05	14.0, 15.2
0.05	0.05	0.80	0.05	0.05	9.2, 10.6, 7.2
0.05	0.05	0.05	0.80	0.05	5.8, 4.8, 4.4
0.05	0.05	0.05	0.05	0.80	22.3, 24.5, 25.2
$s = (0.20, 0.20, 0.20, 0.20, 0.20)'$					28.0

(a) Obtain an estimate of the effect of each component using Eq. (5.29). Use the replicates to obtain an estimate of the error variance and test $H_0 : \beta_i = 0$ $H_a : \beta_i \neq 0$ for each $i = 1, 2, 3, 4, 5$.

(b) Estimate the parameters β_i, in the model $y = \beta_1 x_1 + \beta_2 x_2 + \beta_3 x_3 + \beta_4 x_4 + \beta_5 x_5 + \varepsilon$. How do these estimates b_i differ from the b_i in (a)? With the model parameter estimates, compute the component effects using Eq. (4.19) or Eq. (4.20) from Section 4.5. How do these estimates of the component effects differ from the estimates in (a)? Which set of estimates, the set from Section 4.5 or the set in Section 5.6 are most meaningful to you, and why?

CHAPTER 6

The Inclusion of Process Variables in Mixture Experiments

Often there are nonmixture variables in addition to the mixture components that are present in a mixture experiment. These nonmixture variables might represent external conditions whose settings or levels, if changed, can affect the values of the response or affect the blending properties of the mixture components. For example, in the fruit punch experiment of Section 2.17.1 the temperature of the fruit punch when served to a judge or panelist could have an effect on the judge's perception of overall or general acceptability. To control for temperature effects, we might want to consider two temperatures of the fruit punch—room temperature and a colder temperature—to see if the optimal juice blend is the same at both temperatures or changes with temperature. Another example might be a fertilizer experiment where three nutrient sources (N1, N2, N3) or (x_1, x_2, x_3) are applied as a liquid supplement with a sprayer at the beginning of the experiment. At the completion of the experiment, however, the following question was raised: Would we have observed the same effects of the nutrients on yield had the supplement been applied by drip irrigation over time rather than by a single spray application? Put differently, might the type or method of supplement application (spray versus drip) have an effect on yield or affect the blending properties of the three liquid nutrients? Such nonmixture variables as the temperature of the fruit punch and the method of supplement application are called **process variables**, and their inclusion in mixture experiments is addressed soon. Still a third application of process variables is in studying the effects of cooking temperature, cooking time, and deep fat frying time on the texture of fish patties made by blending three types of saltwater fish.

Including process variables in a mixture experiment can greatly increase the scope of the experiment by providing answers to questions such as; When changes are made in the settings of one or more process variables:

A Primer on Experiments with Mixtures, By John A. Cornell
Copyright © 2011 John Wiley & Sons, Inc. Published by John Wiley & Sons, Inc.

1. Does the value of the response change? Do the blending properties of the mixture components change? If yes to either question, what kinds of changes take place?

2. Does the optimal blend of the components change?

3. Are there specific blends among the components for which the value of the response *does not* change? In other words, are there specific blends that are insensitive (or robust) to changes made in the processing conditions?

An example of mixtures or blends that are robust to changes made in the levels of three process variables is illustrated in Figure 6.1, which displays the results of a fish patty experiment involving three types of saltwater fish and three process variables. In Figure 6.1, the triangles represent the composition space of three types of fish: mullet (x_1), sheepshead (x_2), and croaker (x_3). The triangles are positioned at the eight (2^3) vertices of a cube representing the different combinations of low and high settings of the cooking temperature (z_1), cooking time (z_2), and deep fat frying time (z_3). The shaded regions of the triangles (predominantly the upper four triangles in Figure 6.1 represent fish blends expected to produce patties with acceptable texture

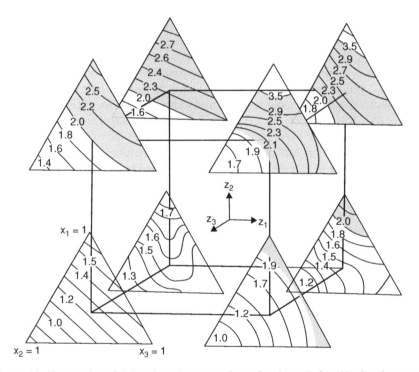

Figure 6.1 Contour plots of eight estimated texture surfaces of patties made from blending three types of fish. The eight three-fish triangles represent different combinations of cooking temperature (z_1), cooking time (z_2), and deep fat frying time (z_3). The shaded regions represent blends of the three fish estimated to produce texture values between 2.0 and 3.5 kilograms.

scores. Texture is assessed by a compression test that measures the force in grams required to puncture the surface of a patty where the firmer the patty, the higher the force value. Acceptable texture values range from 2000 to 3500 grams measured with an Instron Model TM. The locations of the upper four triangles in Figure 6.1 are at the high level of cooking time (z_2) and represent quite different processing conditions in terms of cooking temperature (z_1) and deep fat frying time (z_3). Yet there are some two- and three-fish blends within the shaded regions of the four upper triangles for which the estimated texture of the patties does not vary by more than 300 grams of force when taken across the four triangles. These blends are considered robust to changes in cooking temperature and deep fat frying time, at least to the changes that were made within their experimental ranges, when cooked at the high cooking time (high z_2).

6.1 DESIGNS CONSISTING OF SIMPLEX-LATTICES AND FACTORIAL ARRANGEMENTS

Including process variables in a mixture experiment entails setting up a design con- sisting of the different settings of the process variables in combination with the design that defines the blends of the mixture components. The most popular type of com- bined design is a mixture design set up at each point of a factorial arrangement in the levels of the process variables (or more truthfully coded variables). Let us take, for example, three components with the proportions x_1, x_2, and x_3 to which we wish to include two process variables, denoted by z_1 and z_2. If each process variable is to be studied at two levels, then a good choice of design for studying the effects of the two process variables is a $2 \times 2 = 2^2$ factorial arrangement. If the entire three- component triangular region is to be explored by using, for example, a six-point $\{3, 2\}$ simplex-lattice as shown in Figure 6.2, coupled with the 2^2 factorial

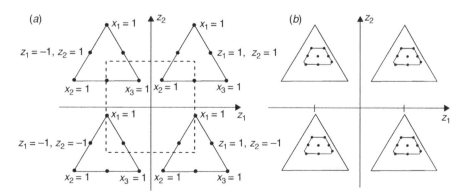

Figure 6.2 (*a*) A six-blend mixture design in x_1, x_2, and x_3 at each of four combinations of two process variables z_1 and z_2. (*b*) An eight-blend mixture design for a constrained region in x_1, x_2, and x_3 at the four combinations of z_1 and z_2.

arrangement in z_1 and z_2, then the combined $\{3, 2\}$ lattice $\times 2^2$ factorial arrangement, displayed in Figure 6.2a, can be used. However, if the mixture experimental region is constrained by placing bounds on the component proportions, then the combined design might be the mixture design consisting of the extreme vertices and/or midpoints of the edges, and so forth, depending on the form of the mixture model to be fitted, set up at each of the four combinations of z_1 and z_2, as shown in Figure 6.2b.

The most complete form the of combined mixture component–process variable model can be expressed as the product of the terms in the mixture component model and the terms in the process variable model,

$$y(\mathbf{x}, \mathbf{z}) = \{\text{mixture component model}\} \times \{\text{process variable model}\} + \varepsilon$$

To take an example, once again we use the $\{3, 2\}$ simplex-lattice in Figure 6.2a, first with the idea of fitting the six-term second-degree mixture model (omitting the error term),

$$y(\mathbf{x}) = \beta_1 x_1 + \beta_2 x_2 + \beta_3 x_3 + \beta_{12} x_1 x_2 + \beta_{13} x_1 x_3 + \beta_{23} x_2 x_3 \tag{6.1}$$

and then the 2^2 factorial arrangement in Figure 6.2a for fitting the four-term model in z_1 and z_2,

$$y(\mathbf{z}) = \alpha_0 + \alpha_1 z_1 + \alpha_2 z_2 + \alpha_{12} z_1 z_2. \tag{6.2}$$

So, if data are collected at the $6 \times 4 = 24$ points of the combined design in Figure 6.2a we can fit the combined 24-term model:

$$
\begin{aligned}
y(\mathbf{x}, \mathbf{z}) = & \{\text{six terms in (6.1)}\} \times \{\text{four terms in (6.2)}\} \\
= & \gamma_1^0 x_1 + \gamma_2^0 x_2 + \gamma_3^0 x_3 + \gamma_{12}^0 x_1 x_2 + \gamma_{13}^0 x_1 x_3 + \gamma_{23}^0 x_2 x_3 \\
& + \{\gamma_1^1 x_1 + \gamma_2^1 x_2 + \gamma_3^1 x_3 + \gamma_{12}^1 x_1 x_2 + \gamma_{13}^1 x_1 x_3 + \gamma_{23}^1 x_2 x_3\} z_1 \quad (6.3) \\
& + \{\gamma_1^2 x_1 + \gamma_2^2 x_2 + \gamma_3^2 x_3 + \gamma_{12}^2 x_1 x_2 + \gamma_{13}^2 x_1 x_3 + \gamma_{23}^2 x_2 x_3\} z_2 \\
& + \{\gamma_1^{12} x_1 + \gamma_2^{12} x_2 + \gamma_3^{12} x_3 + \gamma_{12}^{12} x_1 x_2 + \gamma_{13}^{12} x_1 x_3 \\
& + \gamma_{23}^{12} x_2 x_3\} z_1 z_2 + \varepsilon
\end{aligned}
$$

where $\gamma_i^\ell = \beta_i \alpha_\ell$, $\gamma_{ij}^\ell = \beta_{ij} \alpha_\ell$, $\ell = 0, 1, 2$, $\gamma_i^{12} = \beta_i \alpha_{12}$ and $\gamma_{ij}^{12} = \beta_{ij} \alpha_{12}$. In the model (6.3) the coefficients γ_i^0, $i = 1, 2, 3$ and γ_{ij}^0, $i < j$ are measures of the linear blending of component i and nonlinear blending of components i and j, respectively, when averaged over all four combinations of the two-level process variables z_1 and z_2. The coefficients γ_i^ℓ and γ_{ij}^ℓ, $\ell = 1$ and 2, represent the main effect of process variable z_ℓ on the linear blending of component i and on the nonlinear blending of components i and j. The meanings will be made clearer in the following numerical example.

6.2 MEASURING THE EFFECTS OF COOKING TEMPERATURE AND COOKING TIME ON THE TEXTURE OF PATTIES MADE FROM TWO TYPES OF FISH

The data in this example were taken from a larger data set involving three types of saltwater fish, mullet (x_1), sheepshead (x_2), and croaker (x_3), that were blended to form sandwich patties. Three process variables where also included. They were cooking temperature (375° and 425°F), cooking time (25 and 40 minutes), and deep fat frying time (25 and 40 seconds). Seven different formulations of the three fish species and two levels of each of the three processing factors were studied and analyzed for their effects on textural changes and sensory acceptability of the fish patties. The experiment was performed in the Department of Food Science and Human Nutrition at the University of Florida located in Gainesville.

In this example we focus on two of the three fish, which are mullet (x_1) and sheepshead (x_2), and on cooking temperature (375° and 425°F) and cooking time (25 and 40 minutes). At each of the four (2×2) cooking temperature–time combinations, three fish blends are considered: mullet only ($x_1 = 1$, $x_2 = 0$), sheepshead only ($x_1 = 0$, $x_2 = 1$), and the 50:50 blend of mullet and sheepshead ($x_1 = x_2 = \frac{1}{2}$). Listed in Table 6.1 are the average texture (firmness) readings in kilograms of force taken on two patties at each blend by cooking combination.

Table 6.1 Average Texture Readings of Force on Two Patties at Each Treatment (Cooking Temperature and Time) Combination

Cooking		Coded Variables		Fish Proportions		
Temperature (°F)	Time (min)	z_1	z_2	Mullet (x_1)	Sheepshead (x_2)	Texture Score (grams $\times 10^{-3}$)
375	25	-1	-1	1	0	1.84, 1.65
				0	1	0.67, 0.58
				1/2	1/2	1.29, 1.18
425	25	1	-1	1	0	2.86, 2.32
				0	1	1.10, 0.97
				1/2	1/2	1.53, 1.45
375	40	-1	1	1	0	3.01, 3.04
				0	1	1.21, 1.16
				1/2	1/2	1.93, 1.85
425	40	1	1	1	0	4.13, 4.13
				0	1	1.67, 1.30
				1/2	1/2	2.26, 2.06

The first step taken in analyzing the data of Table 6.1 is to investigate the effect of blending mullet with sheepshead at each combination of cooking temperature and time.

This entails fitting the second-degree (or quadratic) model

$$y(\mathbf{x}) = \beta_1 x_1 + \beta_2 x_2 + \beta_{12} x_1 x_2 + \varepsilon \tag{6.4}$$

to the six data values at each temperature–time combination in Table 6.1. Once this is done and we have determined the type of blending (linear or nonlinear) that is present, the next step is to investigate the effects of cooking temperature and cooking time on the blending properties of mullet and sheepshead by fitting the combined model:

$$
\begin{aligned}
y(\mathbf{x}, \mathbf{z}) ={}& \{\beta_1 x_1 + \beta_2 x_2 + \beta_{12} x_1 x_2\} + \{\beta_1 x_1 + \beta_2 x_2 + \beta_{12} x_1 x_2\} z_1 \\
&+ \{\beta_1 x_1 + \beta_2 x_2 + \beta_{12} x_1 x_2\} z_2 + \{\beta_1 x_1 + \beta_2 x_2 + \beta_{12} x_1 x_2\} z_1 z_2 + \varepsilon \\
={}& \gamma_1^0 x_1 + \gamma_2^0 x_2 + \gamma_{12}^0 x_1 x_2 + \gamma_1^1 x_1 z_1 + \gamma_2^1 x_2 z_2 + \gamma_{12}^1 x_1 x_2 z_1 \\
&+ \gamma_1^2 x_1 z_2 + \gamma_2^2 x_2 z_2 + \gamma_{12}^2 x_1 x_2 z_2 + \gamma_1^{12} x_1 z_1 z_2 \\
&+ \gamma_2^{12} x_2 z_1 z_2 + \gamma_{12}^{12} x_1 x_2 z_1 z_2 + \varepsilon
\end{aligned}
\tag{6.5}
$$

to the complete set of 24 texture values in Table 6.1. The variables z_1 and z_2 in Eq. (6.5) and in Table 6.1 are defined (coded) as

$$z_1 = \frac{\text{Temp} - 400°\text{F}}{25°\text{F}}, \qquad z_2 = \frac{\text{Time} - 32.5\,\text{min}}{7.5\,\text{min}}$$

These definitions produce -1 and $+1$ levels for z_1 and z_2 when cooking temperature and time are set at their low and high levels, respectively. Variables redefined in this way are referred to as coded and the use of coded variables in the model (6.5) simplifies the interpretation of the effects of cooking temperature and time on patty texture.

The fitted mixture models along with the mean square error (MSE) value from the analysis of variance of the six texture values at each of the four cooking temp–time combinations are

$$z_1 = -1,\ z_2 = -1 \qquad (\text{temp} = 375°\text{F},\ \text{time} = 25\,\text{min}) \tag{6.6}$$

$$\text{Texture} = 1.745 x_1 + 0.625 x_2 + 0.200 x_1 x_2, \qquad \text{MSE} = 0.0094$$

$$z_1 = 1,\ z_2 = -1 \qquad (\text{temp} = 425°\text{F},\ \text{time} = 25\,\text{min}) \tag{6.7}$$

$$\text{Texture} = 2.590 x_1 + 1.035 x_2 - 1.290 x_1 x_2, \qquad \text{MSE} = 0.0525$$

$$z_1 = -1,\ z_2 = 1 \qquad (\text{temp} = 375°\text{F},\ \text{time} = 40\,\text{min}) \tag{6.8}$$

$$\text{Texture} = 3.025 x_1 + 1.185 x_2 - 0.860 x_1 x_2, \qquad \text{MSE} = 0.0016$$

$$z_1 = 1, \ z_2 = 1 \qquad (\text{temp} = 425°\text{F, time} = 40\,\text{min}) \tag{6.9}$$

$$\text{Texture} = 4.130x_1 + 1.485x_2 - 2.590x_1x_2, \qquad \text{MSE} = 0.0295$$

Among the four fitted models (6.6) through (6.9), only the last two, (6.8) and (6.9), provide clear evidence of nonlinear blending between mullet and sheepshead as determined by a test on the magnitude of the coefficient estimate b_{12}. One could argue however that nonlinear blending is present with model (6.7) also, but that the lack of significance of the test on b_{12} results from the higher than average MSE value of 0.0525 with (6.7).

Fitted to the 24 texture values in Table 6.1, the combined model of Eq. (6.5) is

$$\hat{y}(\mathbf{x}, \ \mathbf{z}) = 2.873x_1 + 1.083x_2 - 1.135x_1x_2 + 0.488x_1z_1 + 0.178x_2z_1 - 0.805x_1x_2z_1$$

$$\quad (0.054) \qquad (0.054) \qquad (0.264) \qquad (0.054) \qquad (0.054) \qquad (0.264)$$

$$+ 0.705x_1z_2 + 0.253x_2z_2 - 0.590x_1x_2z_2 + 0.065x_1z_1z_2 - 0.028x_2z_1z_2$$

$$\quad (0.054) \qquad (0.054) \qquad (0.264) \qquad (0.054) \qquad (0.054)$$

$$- 0.060x_1x_2z_1z_2 \tag{6.10}$$

$$\quad (0.264)$$

with an $R^2 = 0.9869$ and MSE $= 0.0232$. The quantities in parentheses below the coefficient estimates are the estimated standard errors of the coefficient estimates. In the combined model above, several of the coefficient estimates, when divided by their estimated standard errors, are significantly ($P < 0.05$) different from zero. The specific estimates that are significant convey information about the blending properties of mullet with sheepshead as pertains to patty texture as well as the effects of changing cooking temperature and time on the blending properties of mullet with sheepshead.

To describe these blending properties and effects, we proceed as follows. We partition the model (6.10) into four parts: terms involving only x_1 and x_2, terms involving x_1 and x_2 with z_1, terms involving x_1 and x_2 with z_2, and finally terms with z_1z_2. Presented with each part of the model are the results of a statistical t-test performed on the coefficient estimate of each term. The coefficient estimates having a value of "t" appearing in the brackets below the estimate and of the magnitude $t > 2.179$, are significantly ($P < 0.05$) different from zero.

Shown in Figure 6.3 are plots of the average texture values:

a. For pure mullet, pure sheepshead, and the 50%–50% mullet–sheepshead patties where each average is based on eight patties.

b. For the same three blends at the low (solid) and high (dashed) cooking temperatures where each average is based on four patties.

c. For the same three blends at the low (solid) and high (dashed) cooking times where again each average is based on four patties.

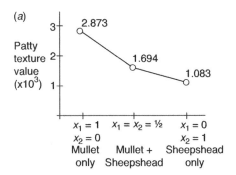

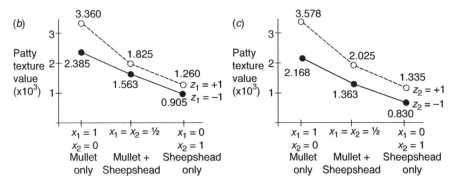

Figure 6.3 Plots of the average texture values of pure mullet, pure sheepshead, and 50%:50% mullet-sheepshead patties (*a*) averaged over the four cooking conditions, (*b*) at the low and high cooking temperatures, and (*c*) at the low and high cooking times.

The plots in Figures 6.3*a–c* are helpful when it comes to understanding the effects of cooking temperature and time. Returning to the first three terms in the combined model (6.10), we have

$$2.873x_1 + 1.083x_2 - 1.135x_1x_2$$
$$(t = -4.30)$$

Interpretation: The coefficient estimates 2.873 and 1.083 are the average texture values (grams $\times$ 10^{-3}) of the pure mullet and pure sheepshead patties, respectively, where averages are taken over the four cooking temperature and time combinations; see Figure 6.3*a*. The coefficient estimate, -1.135, of the binary blending term is highly significantly ($P < 0.01$) different from zero. The negative value of -1.135 implies that the average texture (1.694 g $\times$ 10^{-3}) of the patties made from the 50%–50% blend of mullet and sheepshead is highly significantly lower (meaning the patties are softer) than the average $(2.873 + 1.083)/2 = 1.978$ texture value of the pure mullet and pure sheepshead patties.

The second three terms in Eq. (6.10) are

$$0.488x_1z_1 + 0.178x_2z_1 - 0.805x_1x_2z_1$$
$$(t = 9.04) \quad (t = 3.29) \quad (t = -3.05)$$

Interpretation: The significance of the coefficient estimates of the first two terms implies that the average texture values of the pure mullet and pure sheepshead patties were highly significantly ($P < 0.01$) higher after being cooked at the high temperature ($425°F$) than after being cooked at the low temperature ($375°F$); see Figure 6.3b. The significance of the estimate -0.805 implies the magnitude of nonlinear blending of mullet with sheepshead is not the same at both cooking temperatures. We can refer to the plot of average texture values or fit a model to the average texture values at each temperature separately,

$$\text{Texture (at low temp)} = 2.385x_1 + 0.905x_2 - 0.328x_1x_2$$
$$\text{Texture (at high temp)} = 3.360x_1 + 1.260x_2 - 1.940x_1x_2$$

In this case it is clear that the magnitude of nonlinear blending or synergistic blending of mullet with sheepshead is significantly greater at the high cooking temperature since $|-1.940|$ is nearly six times greater than $|-0.328|$.

The third set of three terms in Eq. (6.10) is

$$0.705x_1z_2 + 0.253x_2z_2 - 0.590x_1x_2z_2$$
$$(t = 13.08) \quad (t = 4.68) \quad (t = -2.23)$$

Interpretation: The significance of the coefficient estimates of the first two terms implies increasing the cooking time from 25 to 40 minutes resulted in single-fish patties that had higher texture values than when cooked for the shorter time; see Figure 6.3c. The significance of the estimate -0.590 says the magnitude of nonlinear blending is not the same at the two cooking times. Modeling the average texture values in Figure 6.3c separately at the two times, we have

$$\text{Texture (at low temp)} = 2.168x_1 + 0.830x_2 - 0.544x_1x_2$$
$$\text{Texture (at high temp)} = 3.578x_1 + 1.335x_2 - 1.726x_1x_2$$

Again, it is clear that the magnitude of nonlinear blending is significantly greater at the higher (longer) cooking time.

Finally, the last three terms in Eq. (6.10) are

$$0.065x_1z_1z_2 - 0.028x_1z_1z_2 - 0.060x_1z_1z_2$$
$$(t = 1.21) \quad (t = -0.51) \quad (t = -0.23)$$

Interpretation: None of the coefficient estimates is significantly different from zero. This means there is no evidence of an interaction effect between cooking temperature and time on the average texture of the single fish patties or on the nonlinear blending of the two types of fish.

In summary, the results of the analysis of the texture values listed in Table 6.1, based on the fitted combined model of Eq. (6.10) are as follows:

1. Combining sheepshead with mullet in a 50%–50% blend produced softer textured patties than would be expected with linear blending when the patties were cooked over a temperature range of 375° to 425°F and over cooking times of 25 to 40 minutes; see Figure 6.3*a*.

2. Single-fish patties that were cooked at the higher temperature and/or for the longer time had significantly ($P < 0.05$) higher texture values (were firmer) than the single-fish patties that were cooked at the lower temperature and/or for the shorter cooking time; see Figures 6.3*b,c*.

3. The softening of the patty texture by blending sheepshead with mullet was more pronounced with patties cooked at the high temperature and/or high time than with patties cooked at the low temperature or at the low time; see Figures 6.3*b,c*.

4. If acceptable patties are defined as having a texture value in the interval 2000 g $\leq$ texture $\leq$ 3500 g then acceptable patty composition depends on the cooking temperature-time combination,

    a. At the low temperature–low time combination, none of the patties were acceptable.

    b. At the low temperature–high time combination, patties made with at least 56% mullet were acceptable.

    c. At the high temperature–low time combination, patties made with at least 77% mullet were acceptable.

    d. At the high temperature–high time combination, patties having between 44% and 87% mullet were acceptable.

These estimates of acceptable patty composition were obtained separately from the individual fitted models (6.6) through (6.9). Appendix 6A shows how to calculate the estimated combined mixture components–process variable model of Eq. (6.10) without using the computer.

6.3 MIXTURE-AMOUNT EXPERIMENTS

A mixture-amount (M-A) experiment is a mixture experiment that is performed at two or more levels of total amount. The response in a M-A experiment is assumed to depend not only on the relative proportions of the ingredients present in the mixture but on the amount of the mixture as well. As in the previous section on including process variables where the effects of the process variables either affected the value of the response or affected the blending properties of the mixture components or both, we are again thinking of factors or variables other than the mixture components that have an impact on the quality of the product being made.

The most popular type of experimental design for a M-A experiment is to use the same mixture design at each level of amount. Although it is not necessary to use the same mixture design at each level of amount, doing so allows for easier interpretation of the model's coefficient estimates with most of the M-A model forms.

The form of the combined model to be fitted to data collected in a M-A experiment depends on how the effect of amount is defined. Generally speaking, changing the amount of the mixture can affect the response in one of two ways. It can affect the response value only without affecting the blending properties of the mixture components, or, it can affect the value of the response by affecting the blending properties of the components. In the case where the value of the response only is affected, and not the blending properties of the components, the effect of amount is **additive** and the M-A model is expressed as

$$y(\mathbf{x}, A) = \{\text{mixture model}\} + \{\text{effect of amount}\} + \varepsilon \qquad (6.11)$$

where

$$
\begin{aligned}
[\text{effect of amount}] &= \alpha_1 A && \text{(linear)} \\
&= \alpha_1 A + \alpha_{11} A^2 && \text{(quadratic)}
\end{aligned}
$$

with A being a coded variable or a variable in the amount units. The number of terms in model (6.11) is the sum of the number in the mixture model plus the number of terms representing the effect of amount.

When changing the level of amount affects the blending properties of the components, the blending properties are said to interact with the amount effect. An **interactive M-A model** that expresses the blending properties of the mixture components and the effect of amount on the blending properties is

$$y(\mathbf{x}, A) = \{\text{mixture model}\} + \{\text{mixture model}\}\,[\text{effect of amount}] + \varepsilon \qquad (6.12)$$

The number of terms in the interactive model (6.12) is greater than the number of terms in the additive model (6.11), and therefore to fit the model (6.12) requires a larger number of experiments than to fit model (6.11).

To illustrate the model forms in (6.11) and (6.12) along with their respective M-A designs, let us consider having three mixture components whose proportions are denoted by x_1, x_2, and x_3. Suppose further that the mixture experimental region is the complete simplex or triangular region and that the mixture model to be fitted is the second-degree model

$$y(\mathbf{x}) = \beta_1 x_1 + \beta_2 x_2 + \beta_3 x_3 + \beta_{12} x_1 x_2 + \beta_{13} x_1 x_3 + \beta_{23} x_2 x_2 + \varepsilon \qquad (6.13)$$

Let the number of levels of amount be three: low, middle, and high. With three quantitative levels the effect of amount has two degrees of freedom that can be partitioned into a linear effect and a quadratic effect of the amount. In this case we let A represent a coded version of the total amount, such as $A = 2(\text{amount} - \text{middle level})/(\text{high level} - \text{low level})$, then the effect of amount in Eqs. (6.12) and (6.13) can be expressed as $\alpha_1 A + \alpha_{11} A^2$. The **additive** model (6.11) takes the form

$$y(\mathbf{x}, A) = \{(6.13)\} + \alpha_1 A + \alpha_{11} A^2 + \varepsilon \tag{6.14}$$

and model (6.14) contains eight terms plus the error term ε. The **interactive** model (6.12) takes the form,

$$y(\mathbf{x}, A) = \{(6.13)\} + \{(6.13)\}A + \{(6.13)\}A^2 + \varepsilon \tag{6.15}$$

and (6.15) contains 18 terms plus an error term. Notationally we write model (6.15) as

$$
\begin{aligned}
y(\mathbf{x}) = & \gamma_1^0 x_1 + \gamma_2^0 x_2 + \gamma_3^0 x_3 + \gamma_{12}^0 x_1 x_2 + \gamma_{13}^0 x_1 x_3 + \gamma_{23}^0 x_2 x_3 \\
& + \{\gamma_1^1 x_1 + \gamma_2^1 x_2 + \gamma_3^1 x_3 + \gamma_{12}^1 x_1 x_2 + \gamma_{13}^1 x_1 x_3 + \gamma_{23}^1 x_2 x_3\}A \\
& + \{\gamma_1^2 x_1 + \gamma_2^2 x_2 + \gamma_3^2 x_3 + \gamma_{12}^2 x_1 x_2 + \gamma_{13}^2 x_1 x_3 + \gamma_{23}^2 x_2 x_3\}A^2 + \varepsilon
\end{aligned}
\tag{6.16}
$$

Interpreting the coefficients of the terms in Eqs. (6.14) and (6.15) is straightforward, especially if the three levels of amount are equally spaced. Then the values that A in the model (6.16) takes are -1, 0, and $+1$ corresponding to the data taken at the low, middle, and high amounts, respectively, while the values of A^2 in (6.16) are 1, 0, and 1, respectively. Then the first six terms in Eq. (6.16) represent the response surface over the mixture region at $A = 0$ or at the middle level of amount. The second six terms in (6.16) represent the differences between the shapes of the mixture surfaces (in terms of the blending properties of x_1, x_2, and x_3) at $A = 1$, or at the high level of amount, and at $A = -1$, or at the low level of amount. The last six terms in Eq. (6.16) represent the differences between the mixture surface at $A = 0$ and the average of the mixture surfaces at $A = -1$ and $A = +1$. Coefficient estimates of any of the final 12 terms in Eq. (6.16) that are significantly different from zero represent the effect of changing the amount of the mixture on the specific blending properties of the mixture components.

An experimental design for supporting the additive model (6.14) differs, in terms of the number of design points, from an experimental design for supporting the interactive model of Eq. (6.15) or (6.16). To support the fit of the additive model of Eq. (6.14), the required number of distinct or different mixture blends has to equal or exceed the number of terms in the mixture model. If the mixture model is (6.13), then the six blends defined by the $\{3, 2\}$ simplex-lattice will suffice. In addition to these six blends at one amount level, we only need to replicate one of the six blends at each of the other two amounts in order to support the model of Eq. (6.14). Additional blends and/or replicates at the other two amounts would be a plus, however; see Figure 6.4a.

To support the fit of the 18-term interaction model (6.15), the six blends of the $\{3, 2\}$ simplex-lattice would be required at each of the three levels of amount as

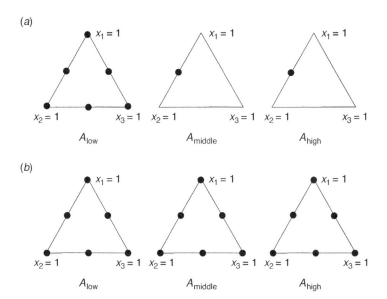

Figure 6.4 (*a*) An eight-point mixture-amount design for fitting the additive model (6.14). (*b*) An 18-point mixture-amount design for fitting the interaction model (6.15) or (6.16).

shown in Figure 6.4*b*. Replicates of some of the blends at one or more levels of amount would provide an estimate of the experimental error variance to be used for testing the significance of the coefficient estimates in the fitted model.

Upper and lower bound constraints placed on the component proportions of the form introduced in Section 3.7 of Chapter 3 do not change the design strategy. One simply specifies the form of the mixture model to be fitted and chooses either the additive model of (6.11) or the interactive model of (6.12) to represent the effect of changing the level of amount. Candidate design points for fitting the mixture model are once again the extreme vertices of the region, the midpoints of the edges of the region, and so on. The eventual mixture design is generally a subset of the candidate points as was discussed earlier in Section 3.7 on lower- and upper-bound constraints on the x_i.

Let us illustrate the use of candidate design points for a constrained region M-A design for a constrained mixture region where the number of levels of amount is two (a low and high amount). Suppose that the constraints on the component proportions are

$$0.2 \le x_1 \le 0.6, \quad 0.1 \le x_2 \le 0.8, \quad 0 \le x_3 \le 0.4 \tag{6.17}$$

The region defined by the constraints in (6.17) has five extreme vertices and five edges connecting the vertices and is shown in Figure 6.5. Listed in Table 6.2 are the coordinates of the extreme vertices, the midpoints of the edges, and the overall centroid of the region that make up the candidate list of points. A 17-point design

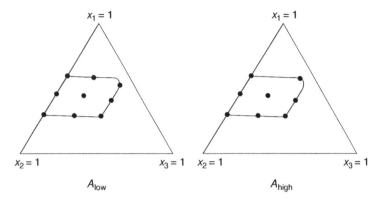

Figure 6.5 Seventeen-point design, with 9 points at A_{low} and 8 points at A_{high}, covering the constrained region $0.2 \leq x_1 \leq 0.6$, $0.1 \leq x_2 \leq 0.8$, $0 \leq x_3 \leq 0.4$.

consisting of 9 points at the low level of amount and 8 points at the high level of amount that could be used for fitting models (6.18a) to (6.18e) is shown in Figure 6.5.

The choice of design points at the two levels of amount depends on the form of the M-A model to be fitted. Listed below are five models that could be considered:

$$y(\mathbf{x}, \mathbf{A}) = \sum_{i=1}^{3} \gamma_i^0 x_i + \sum_{i<j}^{3} \gamma_{ij}^0 x_i x_j + \gamma_0^1 A + \varepsilon \qquad (6.18a)$$

$$y(\mathbf{x}, \mathbf{A}) = \sum_{i=1}^{3} \gamma_i^0 x_i + \sum_{i<j}^{3} \gamma_{ij}^0 x_i x_j + \sum_{i=1}^{3} \gamma_i^1 x_i A + \varepsilon \qquad (6.18b)$$

Table 6.2 Candidate Design Points of the Extreme Vertices, Midpoints of the Edges, and the Overall Centroid of the Constrained Regions at the Low and High Levels of Amount in Figure 6.5

Point	Region Boundary	x_1	x_2	x_3
1	Vertex	0.60	0.40	0
2	Mid-edge	0.60	0.25	0.15
3	Vertex	0.60	0.10	0.30
4	Mid-edge	0.55	0.10	0.35
5	Vertex	0.50	0.10	0.40
6	Mid-edge	0.35	0.25	0.40
7	Vertex	0.20	0.40	0.40
8	Mid-edge	0.20	0.60	0.20
9	Vertex	0.20	0.80	0
10	Mid-edge	0.40	0.60	0
11	Overall centroid	0.42	0.36	0.22

$$y(\mathbf{x}, \mathbf{A}) = \sum_{i=1}^{3} \gamma_i^0 x_i + \sum_{i<j}^{3} \gamma_{ij}^0 x_i x_j + \sum_{i=1}^{3} \gamma_i^1 x_i A + \sum_{i<j}^{3} \gamma_{ij}^1 x_i x_j A + \varepsilon$$

$$(6.18c)$$

$$y(\mathbf{x}, \mathbf{A}) = \sum_{i=1}^{3} \gamma_i^0 x_i + \sum_{i<j}^{3} \gamma_{ij}^0 x_i x_j + \gamma_{123}^0 x_1 x_2 x_3 + \sum_{i=1}^{3} \gamma_i^1 x_i A + \varepsilon \quad (6.18d)$$

$$y(\mathbf{x}, \mathbf{A}) = \sum_{i=1}^{3} \gamma_i^0 x_i + \sum_{i<j}^{3} \gamma_{ij}^0 x_i x_j + \gamma_{123}^0 x_1 x_2 x_3$$

$$+ \left\{ \sum_{i=1}^{3} \gamma_i^1 x_i + \sum_{i<j}^{3} \gamma_{ij}^1 x_i x_j + \gamma_{123}^1 x_1 x_2 x_3 \right\} A + \varepsilon \quad (6.18e)$$

All five models (6.18a) to (6.18e) take into account nonlinear blending of the components—quadratic in (6.18a) to (6.18c), special cubic in (6.18d) and (6.18e)), but the models differ with respect to the effect of changing the amount. The model (6.18a) is an additive model implying only the value of the response changes and not the blending properties of the components when going from the low to the high amount. Models (6.18b) and (6.18d) imply that the change in amount affects the linear blending properties of the components only, while models (6.18c) and (6.18e) imply both the linear and nonlinear blending properties of the components are affected when changing the mixture amount. The latter two models, (6.18c) and (6.18e), having 12 and 14 terms will require at least 6 and 7 distinct blends respectively, at each level of amount. Models (6.18b) and (6.18d), having 9 and 10 terms, will require at least 6 and 7 distinct blends, respectively, at one level of amount with at least three of the blends duplicated at the other level of amount.

Listed in Table 6.3. are some suggested designs for fitting models (6.18a) to (6.18e). The design point designations or numbers in Table 6.3 correspond to the design point number in Table 6.2. Each of the suggested designs contains a few more points at the low level of amount than there are distinct terms in the mixture-only part of the model. In addition to the suggested points, second replicates of two or more points at either amount should be performed in order to obtain an estimate of the experimental error variance.

Table 6.3 Suggested Designs of Subsets of Points in Table 6.2 for Fitting the Mixture-Amount Models (6.18a) to (6.18e)

Model	Number of Terms	Points at A_{low}	Points at A_{high}
6.18a	7	1, 3, 6, 7, 8, 9, 10, 11	1, 5, 7, 9
6.18b	9	1, 3, 6, 7, 8, 9, 10, 11	1, 5, 7, 9
6.18c, 6.18e	12, 14	1, 3, 6, 7, 8, 9, 10, 11	1, 2, 5, 6, 7, 8, 9, 10, 11
6.18d	10	1, 3, 6, 7, 8, 9, 10, 11	1, 5, 7, 9

6.4 DETERMINING THE OPTIMAL FERTILIZER BLEND AND RATE FOR YOUNG CITRUS TREES

In the spring of 1996 in Sao Paulo, Brazil, a large fertilization project involving 288 young ($\leq$ 3 years old) citrus trees was supported by the Brazilian Agricultural Commission to try and answer questions concerning certain physical characteristics of the trees such as tree height, trunk diameter, and canopy. In the past it was felt that the growers, under recommendations by the fertilizer producers, were over fertilizing the trees in an effort to accelerate the growth of the trees only to discover the trees were not responding as expected. As a result, the project involved studying the effects of fertilizer composition and application rate on the growth of the trees.

At the beginning of the experiment, 27 treatment by rate combinations were assigned completely at random to 39 groups of 8 trees each. Each tree in the group individually received the specific fertilizer composition–rate combination assigned to the group. The 39 combinations were 9 separate fertilizer blends plus second replicates of 4 of the blends at each of the three rates. The nine different fertilizer blends evolved from mixing the three fertilizer ingredients, nitrogen (N), phosphorus (P), and potash (K) whose percentages had to conform to the following lower and upper bounds:

$$30\% \leq N \leq 60\%, \quad 4\% \leq P \leq 33\%, \quad 27\% \leq K \leq 56\% \tag{6.19}$$

When expressed as proportions, the constraints (6.19) are

$$0.30 \leq x_1 \leq 0.60, \quad 0.04 \leq x_2 \leq 0.33, \quad 0.27 \leq x_3 \leq 0.56 \tag{6.20}$$

The 27 distinct fertilizer blend-rate combinations are listed in Table 6.4 where the rates were defined as low (200 g/tree/application), medium (400 g/tree/application) and high (600 g/tree/application). Also listed in Table 6.4 are the average trunk diameter

Table 6.4 Average Trunk Diameter Values (mm) of 3-Year-Old Citrus Trees at the Nine Fertilizer Blends

Blend	Original Components			L-Pseudocomponents			Fertilizer Rate		
	x_1(N)	x_2(P)	x_3(K)	x'_1	x'_2	x'_3	Low	Med	High
1	0.60	0.13	0.27	0.769	0.231	0	46.7, 45.3	66.4, 68.3	76.4, 78.3
2	0.60	0.04	0.36	0.769	0	0.231	42.1	72.1	82.1
3	0.40	0.04	0.56	0.256	0	0.744	50.4, 53.9	65.0, 67.4	77.4, 75.0
4	0.30	0.14	0.56	0	0.256	0.744	46.9	56.9	66.9
5	0.30	0.33	0.37	0	0.744	0.256	49.5, 52.4	54.9, 52.2	64.2, 64.9
6	0.40	0.33	0.27	0.256	0.744	0	55.9	59.2	59.2
7	0.50	0.04	0.46	0.513	0	0.487	47.7	71.7	85.7
8	0.50	0.23	0.27	0.513	0.487	0	62.5	65.5	71.5
9	0.43	0.17	0.40	0.342	0.329	0.329	51.8, 50.1	77.6, 75.2	83.6, 85.2

values (in mm) millimeters, taken across the eight trees in each group, for the 27 distinct blend-rate combinations plus the second replicates of 12 blend-rate combinations. The nine fertilizer blends were selected from a candidate list of 6 vertices, 6 mid-edge points, and the overall centroid plus input from a grower offering his opinion.

The types of questions to be answered from the analysis of the data listed in Table 6.4 are as follows:

Q1: What is the optimal blend of N-P-K for tree trunk diameter? By optimal blend is meant the blend or blends that produced trees having the largest diameter.

Q2: Is the optimal blend of N-P-K the same at each fertilizer rate? If yes, what is the effect of rate on tree diameter? Is the optimal rate one of the three that was selected?

Q3: If the optimal blend of N-P-K is not the same at each fertilizer rate, what effect if any, does fertilizer rate or amount have on the blending properties of N-P-K?

Q4: Were all 27 fertilizer blend–rate combinations considered worthwhile or were some combinations not productive enough? For next years applications, will you continue with the not worthwhile combinations? Explain.

To answer the questions above, let us consider the following two-step procedure. Step 1 consists of fitting separate mixture models to the average diameter values at each of the separate fertilizer rates. Once each of the three separate models is fitted, contour plots of the estimated average trunk diameter surfaces at the three fertilizer rates are generated from the three fitted models. These models and plots enable one to answer questions Q1 and Q2 above. Questions Q3 and Q4 are answered in step 2, which consists of a combined mixture-amount model to the total set of 39 average trunk diameter values listed in Table 6.4. As we will see, the particular form of the combined blend-rate model will depend on the forms of the separate fitted mixture models in step 1.

Question for the Reader   What degree model, planar or first, quadratic or second, special cubic, do you suppose produced the design in Figure 6.6? Would you use the original components N-P-K or would you use L-pseudocomponents? Please explain why?

In fitting the separate mixture models to the trunk diameter values at each of the separate amounts of fertilizer, the models could be fitted in the original component proportions or in L-pseudocomponent proportions. The L-pseudocomponents are

$$x_1' = \frac{x_1 - 0.30}{0.39}, \quad x_2' = \frac{x_2 - 0.04}{0.39}, \quad x_3' = \frac{x_3 - 0.27}{0.39} \tag{6.21}$$

and the coordinates (x_1', x_2', x_3') in the L-pseudocomponents corresponding to the nine mixture blends used in the experiment are listed in Table 6.4 also. The denominator, 0.39, in the definitions of the L-pseudocomponents in (6.21) is the height or altitude

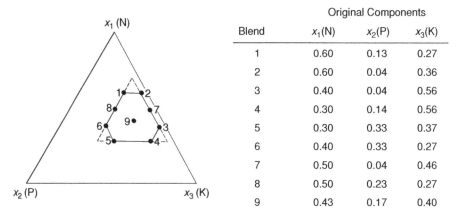

	Original Components		
Blend	$x_1(N)$	$x_2(P)$	$x_3(K)$
1	0.60	0.13	0.27
2	0.60	0.04	0.36
3	0.40	0.04	0.56
4	0.30	0.14	0.56
5	0.30	0.33	0.37
6	0.40	0.33	0.27
7	0.50	0.04	0.46
8	0.50	0.23	0.27
9	0.43	0.17	0.40

Figure 6.6 Nine N-P-K blends whose coordinates are defined in Table 6.4.

of the L-pseudocomponent simplex relative to the height of unity in the original components. Whenever the denominator is less than 0.50, it is probably advantageous to fit the L-pseudocomponent model rather than the model in the original component proportions.

At the low, medium, and high rates of fertilizer, the best-fitting models in the L-pseudocomponents are,

Low rate: *Quadratic model:*

$$\widehat{\text{Diameter}} = 32.79x_1' + 56.16x_2' + 52.01x_3' + 50.55x_1'x_2' + 23.48x_1'x_3' - 27.65x_2'x_3'$$
$$(t = 2.63) \quad (t = 1.22) \quad (t = -1.28)$$
$$\text{with } R^2 = 0.7821$$

$$(6.22)$$

Medium rate: *Special cubic model:*

$$\widehat{\text{Diameter}} = 66.27x_1' + 48.73x_2' + 54.78x_3' + 30.31x_1'x_2' + 44.97x_1'x_3'$$
$$(t = 1.65) \quad (t = 2.54)$$

$$+ 17.86x_2'x_3' + 250.75x_1'x_2'x_3'$$
$$(t = 0.83) \quad\quad (t = 3.38)$$
$$\text{with } R^2 = 0.9840$$

$$(6.23)$$

High rate: *Quadratic model:*

$$\widehat{\text{Diameter}} = 72.34x_1' + 40.89x_2' + 51.93x_3' + 61.72x_1'x_2' + 94.26x_1'x_3' + 104.47x_2'x_3'$$
$$(t = 5.23) \quad (t = 8.00) \quad\quad (t = 7.86)$$
$$\text{with } R^2 = 0.9707$$

$$(6.24)$$

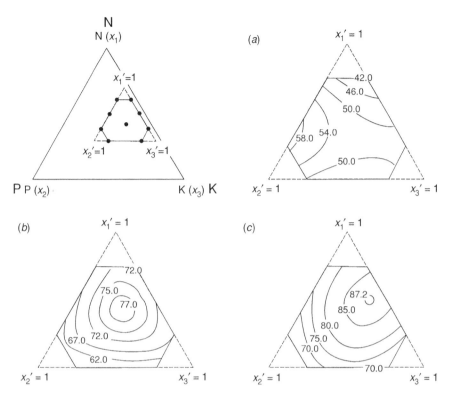

Figure 6.7 Contour plots of the estimated average trunk diameter surfaces at (a) low, (b) medium, and (c) high fertilizer rates. The triangular region (with dashed vertices) is the L-pseudocomponent triangle.

The values in parentheses below the coefficient estimates of the nonlinear blending terms are t-statistic values. Values of t greater than $|t| > 2.365$ are significantly ($P < 0.05$) different from zero. Contour plots of the estimated average trunk diameter surfaces at each of the three fertilizer rates are displayed in Figure 6.7. The triangles in Figures 6.7*a,c* are the smaller L-pseudocomponent simplex inside the solid boundary of the N-P-K simplex in the upper left-hand corner Figure 6.7.

At the low fertilizer rate, the shape of the estimated average trunk diameter surface in Figure 6.7*a* is very different from the shapes of the estimated average trunk diameter surfaces at the medium and high rates. At the low rate, the estimated surface rises in height as one approaches the neighborhood of blend 6 of Table 6.4 or $(x'_1, x'_2, x'_3) = (0.26, 0.74, 0)$, or (N, P, K) = (40%, 33%, 27%). The estimated average trunk diameter value in this neighborhood is slightly greater than 58 mm. At the medium rate, the estimated average diameter surface attains a maximum value of 77.8 mm at the blend $(x'_1, x'_2, x'_3) = (0.45, 0.22, 0.33)$, or (N, P, K) = (47.5%, 12.5%, 40%). At the high rate, the maximum estimated average diameter is 87.2 mm and occurs approximately at the blend $(x'_1, x'_2, x'_3) = (0.53, 0.10, 0.37)$ or (N, P, K) = (50.7%, 7.9%, 41.4%). The difference between the latter two estimated optimal blends is only a slight increase in N and K at the high rate of approximately 3.2% and 1.4% with a slight reduction in P of approximately 4.6% relative to the optimal blend at the medium rate. The difference in the maximum estimated average trunk diameters at the high and medium rates is 87.2 mm $-$ 77.8 mm $=$ 9.4 mm. The difference is of course a combination of the difference between the two blends plus the rate difference of $600 - 400 = 200$ g/tree/application.

The different shapes of the estimated average trunk diameter surfaces at the low, medium, and high fertilization rates in Figure 6.7 means the blending properties of N, P, and K are affected when changing the fertilization rate. To measure the effects of rate on the blending properties, we fit the combined model, (6.18e) plus the additional terms in {mixture model} A^2, to the total data set of 39 average trunk diameter values listed in Table 6.4. Since the highest degree or most complicated fitted model in the L-pseudocomponents is the special cubic model (6.23) at the medium rate, the special cubic model is used as the mixture model (MM) in the 21-term combined mixture-amount or mixture-rate model:

$$\text{Diameter} = \{(\text{MM}\} + \{(\text{MM})\}A + \{(\text{MM})A^2\} + \varepsilon \qquad (6.25)$$

where A is a coded variable for fertilizer rate. In this example, the low, medium, and high rates are equally spaced (200, 400, and 600 g/tree/application) so that the coded values assigned to A when estimating the coefficients in the 21-term model (6.25) are -1, 0, and $+1$ corresponding to the diameter values at the low, medium, and high rates, respectively. Consequently A^2 in (6.25) will take the values 1, 0, and 1 for the diameter values at the low, medium, and high rates, respectively.

The values of the coefficient estimates in the fitted combined mixture-amount model are listed in Table 6.5. Also listed in Table 6.5 are the t-test values for each of the coefficient estimates. The R^2 value for the fitted combined model is $R^2 = 0.9853$, and this means that approximately 98.53% (i.e., $R^2 \times 100\%$) of the total variation in the 39 average trunk diameter values is explained by the terms in the combined model. Of particular interest to us are the coefficient estimates in the fitted combined

Table 6.5 Coefficient Estimates of the Terms in the Combined Mixture-Amount Model

Mixture Model Term	Portion of Combined Model					
	Mixture Only	t-Value	Mixture $\times A$	t-value	Mixture $\times A^2$	t-Value
x_1'	66.27	—	25.51	6.57	−17.36	−2.58
x_2'	48.73	—	−0.20	−0.04	−4.94	−0.59
x_3'	54.78	—	6.66	1.49	−7.08	−0.91
$x_1' \, x_2'$	30.31	1.06	−27.24	−1.35	46.77	1.33
$x_1' \, x_3'$	44.97	1.63	4.00	0.20	33.93	1.00
$x_2' \, x_3'$	17.86	0.53	27.26	1.15	45.31	1.10
$x_1' \, x_2' \, x_3'$	250.76	2.17	149.71	1.83	−346.28	−2.45

model for which $|t_{.025,18}| > 2.101$. More specifically, from Table 6.5, the significant $(P < 0.05)$ terms and their approximate interpretations are:

Mixture-only portion of the model:

$$250.76x_1'x_2'x_3'$$

Interpretation: At the medium fertilizer rate, trees receiving fertilizer blend 9 (at the center of the region) have a larger average diameter than trees receiving blends defined along the edges of the L-pseudocomponent triangle.

{*mixture model*} × *A portion of the model* (This part of the model compares the shapes of the estimated diameter surfaces at the low and high rates.)

$$25.51x_1'A$$

Interpretation: At the high fertilization rate, trees receiving high percent N relative to P and K have significantly larger diameters than trees receiving the same blend but at the low fertilization rate. Although not one of the nine blends was studied, the blend referred to is the $x_1' = 1$, $x_2' = x_3' = 0$ blend or (N, P, K) = (69%, 4%, 27%).

{*mixture model*} × A^2 *portion of the model* (This part of the model compares the shape of the estimated diameter surface at the medium rate against the average of the shapes at the low and high rates.)

$$-17.36x_1'A^2,\ -346.28x_1'x_2'x_3'A^2$$

Interpretation: Trees receiving the high percent N relative to P and K blend at the middle rate have larger diameters than expected at this blend from a strictly linear effect of rate. Note the increase in the value of the estimated coefficient of the term $66.27x_1'$ in Eq. (6.23) compared to the value of the estimated coefficient of the same term, $32.79x_1'$, in Eq. (6.22).

At the middle rate, trees receiving fertilizer blend 9 (at the center of the region) have a larger average diameter than expected at blend 9 with a strictly linear effect of rate. This is because with a strictly linear effect of rate, the average diameter of the 16 trees at the low and high rates receiving blend 9 is, from Table 6.4, expected to be (50.95 + 84.40)/2 = 67.68 mm, where 50.95 and 84.40 mm are the average diameters of the trees receiving the low and high rates of blend 9, respectively. The trees receiving blend 9 at the medium rate had an average diameter of 76.40 mm, and this value is significantly $(P < 0.05)$ higher than 67.68 mm.

So what have we learned from the analysis of this fertilizer-blend-by-rate mixture-amount example? Although less than 25% of the data values used in Table 6.4 were artificial, we found that:

1. The optimal N-P-K blend is different at the three fertilizer rates. At the medium and high rates, the optimal N-P-K blends were close at (N, P, K) = (47.5%, 12.5%, 40%) for the medium rate and (50.7%, 7.9%, 41.4%) at the high rate.

At the low rate, the optimal blend was approximately $(N, P, K) = (40\%, 33\%, 27\%)$ where P was lower and N and K were higher than at the middle and high rates.

2. When the rate was increased from low to medium, blend 9 (at the center of the constrained region) became more influential in yielding trees with larger diameters. For example, trees receiving blend 9 at the medium rate had an average trunk diameter of 76.40 mm. This value represents an increase of $76.40 - 50.95 = 25.45$ mm over trees receiving blend 9 at the low rate. Furthermore trees receiving blend 9 at the high rate had an average trunk diameter of 84.40 mm, which represented an increase of 8.00 mm over trees receiving blend 9 at the medium rate. Similar increases were observed at the medium and high rates with trees receiving blend 7 having composition $(N, P, K) = (50\%, 4\%, 46\%)$. As a result one might conjecture that the optimal N-P-K blend, at the medium and high rates, is a compromise between blends 7 and 9, which would be $(N, P, K) = (48.75\%, 8.25\%, 43\%)$, and the contour plots of both estimated average trunk diameter surfaces shown in Figure 6.7 appear to support this conjecture.

Thus far in this chapter we have used Scheffé's canonical form of polynomial to model the blending properties of the components. The truth is, any of the different models in Chapter 5 such as including inverse terms or models that are homogeneous of degree one or models with ratios of the components as terms or Cox's polynomial, could serve as the mixture component model in a combined model. Similarly we could use designs other than the 2-level factorials such as a 3-level factorial, for example, in which to fit a standard polynomial in the process variables. Furthermore different design and analysis strategies are mentioned in Cornell (2002, ch. 7), the first being when the group of m different mixture blends are assigned to subplot units and are embedded in each of P processing conditions as depicted in Figure 6.2a and the processing conditions (level combinations of the process factors) are the treatments assigned to the main or whole plot units. The result is a split-plot scheme, and the full analysis requires two sources of error, both of which must be estimated in order to separate the main-plot and subplot error variances.

Next offered in Cornell (2002, ch. 7) is a numerical example of a three-component by two-process variables split-plot experiment, and the results are discussed in detail. The example is followed by a reparameterization of the combined model form for measuring the effects of the process variables. The idea of using fractional factorial designs in the process variables was considered for the first time by Cornell and Gorman (1984). In Figure 6.8 are two examples of the use of a half-fraction of a 2^3 factorial in three process variables (D, E, and F). In Figure 6.8a, the matched fraction design consists of the same 2^{3-1} fraction $(1 + DEF)$ in the process variables at each composition point or blend of the three-component simplex-centroid design. In Figure 6.8b, the mixed fraction design consists of the fraction $(I + DEF)$ at the vertices and centroid and the fraction $(I - DEF)$ at the midpoints of the edges (or binary blends) of the simplex-centroid design. Recommendations on when to use the

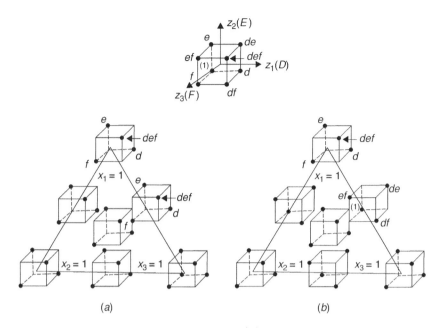

Figure 6.8 (*a*) Matched fraction design with the same 2^{3-1} fraction (I + DEF) in the process variables at each composition point of the three-component simplex-centroid design. (*b*) Mixed fraction design with the fraction (I + DEF) at the vertices and centroid and the fraction (I − DEF) at the binary blends of the simplex-centroid design.

matched or mixed fraction appear to be based on which interaction effects among the process variables are likely to be real as well as how many experimental runs are possible. A numerical example of the two-fish, mullet and sheepshead, along with cooking temperature (z_1), cooking time (z_2), and deep fat frying time (z_3) is considered next.

6.5   A NUMERICAL EXAMPLE OF THE FIT OF A COMBINED MODEL TO DATA COLLECTED ON FRACTIONS OF THE FISH PATTY EXPERIMENTAL DESIGN

The data in Table 6.6 represent average texture readings taken from twin patties prepared from seven blends of the three fish, mullet (x_1), sheepshead (x_2), and croaker (x_3), with the patties treated at eight different processing conditions. Suppose that the patties were prepared only from pure mullet, pure sheepshead, and the 50% : 50% blend of mullet and sheepshead, which restricts us to the three columns of the table headed by (1,0,0), (0,1,0), and ($\frac{1}{2}$, $\frac{1}{2}$, 0). Imagine we are interested only in the main effects of the three process variables and wish to study these effects by looking at only four (or a 2^{3-1} fraction) of the eight processing conditions.

Table 6.6 Average Texture Reading in Grams of Force $\times$ 10^{-3} Taken from Twin Patties

Scaled Settings of the Process Variables			Mixture Composition						
z_1	z_2	z_3	$(1,0,0)$	$(0,1,0)$	$(0,0,1)$	$(\frac{1}{2},\frac{1}{2},0)$	$(\frac{1}{2},0,\frac{1}{2})$	$(0,\frac{1}{2},\frac{1}{2})$	$(\frac{1}{3},\frac{1}{3},\frac{1}{3})$
-1	-1	-1	1.84	0.67	1.51	1.29	1.42	1.16	1.59
1	-1	-1	2.86	1.10	1.60	1.53	1.81	1.50	1.68
-1	1	-1	3.01	1.21	2.32	1.93	2.57	1.83	1.94
1	1	-1	4.13	1.67	2.57	2.26	3.15	2.22	2.60
-1	-1	1	1.65	0.58	1.21	1.18	1.45	1.07	1.41
1	-1	1	2.32	0.97	2.12	1.45	1.93	1.28	1.54
-1	1	1	3.04	1.16	2.00	1.85	2.39	1.60	2.05
1	1	1	4.13^a	1.30	2.75	2.06	2.82	2.10	2.32

^{a}The observed texture value was actually higher than 4.13, but it was thought that the patties were burnt and so the value 4.13 was substituted.

Two possible design plans for accomplishing our objective of fitting the model

$$y(\mathbf{x}, \mathbf{z}) = \sum_{i=1}^{2} \gamma_i^0 x_i + \gamma_{12}^0 x_1 x_2 + \sum_{l=1}^{3} \left[\sum_{i=1}^{2} \gamma_i^\ell x_i + \gamma_{12}^\ell x_1 x_2 \right] z_1 + \varepsilon \qquad (6.26)$$

are presented in Figure 6.8. Recall that the matched fraction design is Figure 6.8a while Figure 6.8b is the mixed fraction design. The texture data, taken from Table 6.6 and corresponding to each of the two designs in Figure 6.8, are listed in Table 6.7.

The data in Table 6.7 for design 6.8(a) were fitted using the model of Eq. (6.26).

Table 6.7 Patty Texture Values Taken from Table 6.6. Corresponding to the Matched Fraction Design in Figure 6.8a and Mixed Fraction Design of Figure 6.8b

			Design 6.8(a)			Design 6.8(b)		
			$I + DEF$	$I + DEF$	$I + DEF$	$I + DEF$	$I + DEF$	$I - DEF$
z_1	z_2	z_3	$(1,0,0)$	$(0,1,0)$	$(\frac{1}{2},\frac{1}{2},0)$	$(1,0,0)$	$(0,1,0)$	$(\frac{1}{2},\frac{1}{2},0)$
-1	-1	-1					1.29	
1	-1	-1	2.86	1.10	1.53	2.86	1.10	$(1.53)^a$
-1	1	-1	3.01	1.21	1.93	3.01	1.21	$(1.93)^a$
1	1	-1					2.26	
-1	-1	1	1.65	0.58	1.18	1.65	0.58	$(1.18)^a$
1	-1	1					1.45	
-1	1	1					1.85	
1	1	1	4.13	1.30	2.06	4.13	1.30	$(2.06)^a$

^{a}The numbers in parentheses under $(\frac{1}{2}, \frac{1}{2}, 0)$ of design 6.8(b) represent texture readings from the $I + DEF$ fraction and are used to fit Eq. (6.29).

The fitted model is

$$\hat{y}(\mathbf{x}, \mathbf{z}) = 2.91x_1 + 1.05x_2 - 1.22x_1x_2 + 0.58x_1z_1 + 0.15x_2z_1$$
$$\quad (0.07) \quad (0.07) \quad (0.33) \quad\quad (0.07) \quad\quad (0.07)$$
$$\quad - 0.99x_1x_2z_1 + 0.66x_1z_2 + 0.21x_2z_2 - 0.45x_1x_2z_2 - 0.02x_1z_3$$
$$\quad\quad (0.33) \quad\quad\quad (0.07) \quad\quad (0.07) \quad\quad (0.33) \quad\quad\quad (0.07)$$
$$\quad - 0.11x_2z_3 + 0.04x_1x_2z_3$$
$$\quad\quad (0.07) \quad\quad (0.33) \quad\quad\quad\quad\quad\quad\quad\quad\quad\quad\quad\quad (6.27)$$

where the estimated standard errors of the coefficient estimates were calculated using a pooled within- and between-patty variance of $s^2 = 0.0184$, that is, s.e.$(g_i^l) = (0.25s^2)^{1/2} = 0.07$ and s.e.$(g_{12}^l) = (6s^2)^{1/2} = 0.33$, for $l = 0, 1, 2, 3$. Corresponding to the data observed at the points of design 6.8(b) the fitted model is

$$\hat{y}(\mathbf{x}, \mathbf{z}) = 2.91x_1 + 1.05x_2 - 1.07x_1x_2 + 0.58x_1z_1 + 0.15x_2z_1$$
$$\quad (0.07) \quad (0.07) \quad (0.33) \quad\quad (0.07) \quad\quad (0.07)$$
$$\quad - 0.90x_1x_2z_1 + 0.66x_1z_2 + 0.21x_2z_2 - 0.36x_1x_2z_2 - 0.02x_1z_3$$
$$\quad\quad (0.33) \quad\quad\quad (0.07) \quad\quad (0.07) \quad\quad (0.33) \quad\quad\quad (0.07)$$
$$\quad -0.11x_2z_3 + 0.01x_1x_2z_3$$
$$\quad\quad (0.07) \quad\quad (0.33) \quad\quad\quad\quad\quad\quad\quad\quad\quad\quad\quad\quad (6.28)$$

Differences between the two fitted models (6.27) and (6.28) appear only in the estimates g_{12}^0, g_{12}^1, g_{12}^2, and g_{12}^3, since different treatments and thus different texture values were used at the $(\frac{1}{2}, \frac{1}{2}, 0)$ blend with the two designs. With both designs, there appears to be an overall synergistic effect (i.e., lowering of patty texture) of blending the two types of fish, since $g_{12}^0 < 0$ with both fitted models. Furthermore, with both designs, raising the level of cooking temperature (z_1) increased the amount of synergistic blending of mullet and sheepshead, since in both models (6.27) and (6.28), $g_{12}^1 < 0$. The effect of cooking time (z_2) is an increase in the texture of the pure fish patties, since with each of the models g_1^2 and g_2^1 are both greater than zero. Deep fat frying time (z_3) did not have an effect.

To check on the assumption of negligible or zero interaction effects among the three process variables, the design in 6.8(b) was augmented with the four process variable treatments d, e, f, and def at the $(\frac{1}{2}, \frac{1}{2}, 0)$ blend resulting in a complete 2^3-factorial experiment being performed at the $(\frac{1}{2}, \frac{1}{2}, 0)$ blend. The texture values corresponding to these treatments are enclosed in parentheses in Table 6.7. To the 16 texture values collected from the augmented design 6.8(b), the model

(6.26) with the four extra terms $\alpha_{12}z_1z_2$, $\alpha_{13}z_1z_3$, $\alpha_{23}z_2z_3$, and $\alpha_{123}z_1z_2z_3$ was fitted, producing

$$\hat{y}(\mathbf{x}, \mathbf{z}) = 2.93x_1 + 1.07x_2 - 1.22x_1x_2 + 0.59x_1z_1 + 0.16x_2z_1$$

$$(0.08) \quad (0.08) \quad (0.33) \quad (0.08) \quad (0.08)$$

$$- 0.99x_1x_2z_1 + 0.67x_1z_2 + 0.22x_2z_2 - 0.45x_1x_2z_2 - 0.03x_1z_3$$

$$(0.33) \quad\quad (0.08) \quad\quad (0.08) \quad\quad (0.33) \quad\quad (0.08)$$

$$- 0.11x_2z_3 + 0.04x_1x_2z_3 + 0.0004z_1z_2 - 0.011z_1z_3$$

$$(0.08) \quad\quad (0.33) \quad\quad (0.05) \quad\quad (0.05)$$

$$-0.011z_2z_3 - 0.019z_1z_2z_3 \quad\quad\quad\quad\quad\quad\quad\quad\quad (6.29)$$

$$(0.05) \quad\quad (0.05)$$

where the estimated standard errors for the last four terms are s.e.$(\hat{\alpha}_{ij})$ = s.e.$(\hat{\alpha}_{123})$ = $(0.125s^2)^{1/2}$ and $s^2 = 0.0184$. None of the estimates $\hat{\alpha}_{12}$, $\hat{\alpha}_{13}$, $\hat{\alpha}_{23}$, and $\hat{\alpha}_{123}$ appear to be different from zero, meaning that there is no evidence of interaction among the three process variables at the 50%: 50% blend of mullet and sheepshead. This suggests that the designs in Figures 6.8a,b along with the fitted models (6.28) and (6.29) are probably sufficient for studying the blending properties of mullet and sheepshead and the effects of cooking temperature, cooking time, and deep fat frying time on the blending properties of the two types of fish.

Definition. A *complete* model of degree d is one that contains all of the terms up to and including those of degree d.

Definition. A *reduced* model is one that contains only a subset of the terms of the complete model.

For example, the three-component *complete* Scheffé quadratic or second-degree model contains 6 terms (3 linear and 3 nonlinear terms) and the *complete* special cubic model contains 7 terms. A q-component *reduced* Scheffé quadratic model is one that contains all q linear terms but not all of the $q(q-1)/2$ binary cross-product terms.

6.6   QUESTIONS RAISED AND RECOMMENDATIONS MADE WHEN FITTING A COMBINED MODEL CONTAINING MIXTURE COMPONENTS AND OTHER VARIABLES

Many examples appear in the mixtures literature of experiments containing other factors. Two examples that caught the attention of Cornell (1995) are:

1. A printable coating material used for identification labels and tags consisted of two pigments (x_1 and x_2) and a polymeric binder (x_3). The response studied was opacity or absorbency of the coating plus the coating thickness (or amount of coating applied) denoted by Z. Constraints on the component proportions were set at

$$0.13 \leq x_1 \leq 0.45, \quad 0.21 \leq x_2 \leq 0.67, \quad 0.20 \leq x_3 \leq 0.34 \qquad (6.30)$$

and the thickness of the coating was set at three levels: 10, 19, and 28 μm. The example was introduced by Chau and Kelley (1993).

2. The fusion latitude of a toner ink is known to be related to weight fractions or relative proportions of three formulation components (monomers one (x_1), two (x_2), and three (x_3)) plus the settings of three process variables—cross-linker type (one vs. two), cross-linker amount (0.1% vs. 0.4%), and the amount of catalyst (1% vs. 3%). Constraints on the weight fractions of the monomers were set at

$$0.40 \leq x_1 \leq 0.90, \quad 0.10 \leq x_2 \leq 0.60, \quad 0 \leq x_3 \leq 0.20 \qquad (6.31)$$

and the effects of the three process variables on fusion latitude were to be studied by setting up a 2^3 factorial arrangement in the levels of cross-linker type (Z_1), cross-linker amount (Z_2), and amount of catalyst (Z_3). The example was introduced by Chitra and Ekong (1993).

The two examples in Figure 6.9 have several features that are common to mixture experiments containing other factors:

- The experimental region for the mixture components is constrained in shape by the imposition of lower and upper bounds on the component proportions in (6.30) and (6.31). The constrained regions defined in (6.30) and (6.31) are as shown in Figure 6.9.

- The data generated in the examples or experiments are fitted by a single model incorporating the blending properties of the mixture components and the effects of the other factors. The final fitted model form is used to generate contour plots of the estimated mixture surfaces at various amounts of coating or coating thickness (Figure 6.9a) or at the different settings of the process variables (Figure 6.9b).

There are also several features of the two examples that are different:

- In Figure 6.9a, a computer-generated 15-point D-optimal design was set up for fitting the combined 15-term second-degree equation consisting of Cox's (1971) mixture polynomial plus terms in the amount variable shown below as Eq. (6.32). Eight of the 15 points were replicated a second time. Eleven additional blend-amounts (different from the initial 15 blend-amounts) were performed, presumably to be used as check points whereby the data (opacity

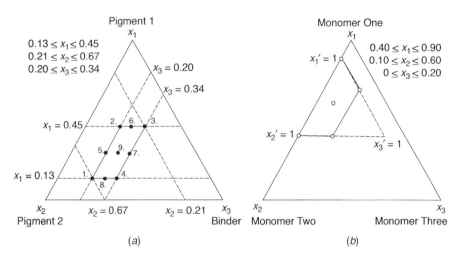

Figure 6.9 Constrained regions. (*a*) The nine points represent candidate blends of pigments 1 and 2 and binder for fitting a second-degree model. (*b*) The four extreme vertices and centroid of the region representing the design for fitting the five-term reduced Scheffé quadratic model in the L-pseudocomponents x_1', x_2', and x_3'.

values) at these check points were to be used to test the adequacy of the 15-term model that was fitted to the data from the initial 23 blend-amounts. The complete second-degree combined model is

$$\widehat{\text{Opac}} = b_0 + \sum_{i=1}^{3} b_i x_i + b_4 Z + \sum_{i=1}^{3} b_{ii} x_i^2 + b_{44} Z^2 \\ + \sum_{i<j}^{3} b_{ij} x_i x_j + \sum_{i=1}^{3} b_{i4} x_i Z$$

(6.32)

- In Figure 6.9*b*, a five-point mixture design consisting of the four extreme vertices plus the centroid of the constrained region was selected for the purpose of fitting a reduced quadratic model consisting of the three linear blending terms plus only two of the three nonlinear blending terms, of the six-term quadratic mixture model. The same design in the mixture blends was set up at each of the eight (2^3) combinations of the levels of the three process variables so that the entire data set could be fit by a combined Scheffé-type model in the mixture component proportions augmented with terms representing the effects of the process variables. The initial 35-term combined model was

$$F_1 = \{\gamma_1 x_1' + \gamma_2 x_2' + \gamma_3 x_3' + \gamma_{12} x_1' x_2' + \gamma_{23} x_2' x_3'\} + \sum_{j=1}^{3} \{\}z_j \\ + \sum_{j<k}^{3} \{\}z_j z_k + \varepsilon$$

(6.33)

where the same mixture model appears in the braces $\{\}$ z_j and $\{\}$ $z_j z_k$ that appears in the first set of braces on the right-hand side of the equal sign of (6.33), and z_1, z_2, and z_3 are coded variables taking the values -1 and $+1$ when the Z_p, $p = 1, 2, 3$, are at their low and high levels, respectively. The x_i' $i = 1, 2$, and 3, in (6.33) are L-pseudocomponents defined as $x_1' = (x_1 - 0.40)/0.50$, $x_2' = (x_2 - 0.10)/0.50$, and $x_3' = x_3/0.50$, where the L-pseudocomponent simplex is shown in Figure 6.9b.

Several questions were raised in Cornell (1995) regarding the designs used and the model-fitting strategies employed in the two examples. For Figure 6.9a two questions were:

QA1: What are the assumptions regarding the blending properties of the components and the effect of changing the amount of coating (or coating thickness) on the response when proposing a model of the form (6.32). Furthermore, does the 15-point design consisting of eight of the nine blends at $Z = 10$ μm, three of the nine blends at $Z = 19$ μm, and four of the nine blends at $Z = 28$ μm merely just support the proposed model, or does the design also allow for checking the assumptions made regarding the blending properties of the components?

QA2: Isn't it possible for a model of the form (6.32) to pick up nonlinear blending somewhere in the experimental region (such as at the low level $Z = 10$ μm, where data from eight of the nine blends were fit) and project the nonlinear blending, through predictions made with (6.32) to other locations of the experimental region (such as at the middle level $Z = 19$ μm or high level $Z = 28$ μm) where nonlinear blending is not present?

For Figure 6.9b, the two questions raised were:

QB1: Since the five-point mixture design in Figure 6.9b supports only a five-term reduced quadratic model, does it matter which two of the three nonlinear blending terms $\gamma_{12}x_1'x_2'$, $\gamma_{13}x_1'x_3'$, and $\gamma_{23}x_2'x_3'$ are included with the linear blending terms in the mixture-model contribution (in braces $\{\}$) of the combined model (6.33)? If yes, what effect might this have on the final form of the reduced combined model obtained using stepwise regression or any other variable selection procedure on the terms in the combined model of (6.33)? Furthermore, since the mixture experimental region, shown in Figure 6.9b, is irregularly shaped and the configuration of the design points is not symmetrical with respect to the axes of the L-pseudocomponents, might the significance (or importance) of a particular nonlinear blending term in the fitted five-term mixture model depend on which other nonlinear blending term is in the model?

QB2: If we reduce the form of the combined model of Eq. (6.33) by eliminating those terms that are tested and found to be insignificant, can the reduced combined model be used to generate estimated surface contour plots at each of the eight combinations of Z_1, Z_2, and Z_3, or should we fit eight separate mixture models, one at each of the combinations of Z_1, Z_2, and Z_3, and use these separate models to generate the individual contour plots?

The answers to these questions are left to the reader. The answers given in Cornell (1995) are listed at the end of this book, in the Answers to Selected Questions.

Several suggestions are made in Cornell (1995) for improving the mixture designs and model forms along with model-fitting strategies in Figures 6.9*a,b*. General recommendations are made as well as aimed specifically at combined mixture component-other factor experiments, where the primary objective is to generate predictions (or contour plots) of the mixture surface at the various other factor-level combinations. Generally, for experiments of these types to be successful, the number of components, q, and other factors, n, or levels of amount (≤ 3) must be reasonably manageable, say, $q \leq 4$ and $n \leq 3$. For design selection and modeling purposes, the recommendations listed in Cornell (1995) are:

R1. Propose a complete P-term mixture model that is capable of describing the most complicated-shaped mixture surface that is likely to exist at any of the combinations of the settings of the other factors.

R2. Select a mixture design consisting of $N \geq P$ points that will support the fit of the model in R1.

R3. Propose a complete Q-term model in the other factors that expresses all of the possible effects of interest among the other factors.

R4. Select a design in the $M \geq Q$ level-combinations of the other factors that supports the fit of the model in R3.

R5. Construct a complete combined design in the mixture components and other factors by crossing the points of the designs in R2 and R4, as was done earlier in this chapter in Figure 6.2. The combined design will consist of $N \times M$ points and will support the complete PQ-term combined model whose terms are defined by multiplying the terms in the models of R1 and R3.

R6. Select a subset of the effects in R3 and fractionate the design in R4 if the number NM of mixture blend-other-factor combinations in R5 is too large. See Cornell and Gorman (1984, sec. 7.6) for suggestions on choosing a fraction of a 2^n factorial in $n \geq 3$ process variables. Stay with the mixture design in R2.

R7. Cross the mixture design in R2 with the fraction selected in R6, and for the combined model, multiply the terms in the model in R1 and the reduced model in R6 used for selecting the fractional design.

For fitting models and data analysis:

R8. If mixture model reduction is one of the goals, proceed with reducing those particular individual mixture model forms where reduction is justified. Fit M separate complete P-term mixture models as defined in R1 to the N *data values at each of the M* level-combinations of the other factors.

R9. If the combined NM-point design is feasible, fit the complete PQ-term combined mixture components-other factors model defined in R5. If

$MN > PQ$, perform tests of significance on the coefficient estimates of the cross-product terms representing the effects of the other factors on the blending properties of the mixture components. Interpreting these effects is often aided by having compared and interpreted the coefficient estimates in the separate fitted mixture models in R8 initially.

R10. If reduction of the combined model is one of the goals, proceed by dropping the nonsignificant cross-product or higher degree terms involving the other factors and the mixture nonlinear blending properties whose coefficient estimates are not different from zero as determined from the tests in R9.

R11. Once the form of the reduced combined model is determined, fit the reduced model to the complete set of data and perform a lack-of-fit test on the reduced model. If the fitted reduced combined model is adequate, use this model for prediction purposes as defined in R13 below.

Recommendations for predicting the mixture surfaces at the factor-level combinations of the other factors are:

R12. If the factor-level combinations are some or all of the M design points in R4, use the fitted complete combined model in R9 for prediction purposes. Under certain design and model conditions (e.g., the same mixture design at each of the factor-level combinations of the other factors along with the use of coded variables for the other factors in the model) at the factor-level combinations of the other factors, the resulting prediction equation of the mixture surface and the complete P-term mixture model (obtained in R8) fitted to the N data values only at the particular factor-level combination are equivalent. This means simply that the contour plot of the predicted or estimated mixture surface generated with the now reduced combined model and the contour plot generated by the P-term mixture model fitted only to the N data values at the particular factor-level combination are identical.

R13. If the chosen factor-level settings or combinations of the other factors are not any of the design points in R4, use the fitted reduced combined model obtained in R11. The use of the reduced combined model for prediction purposes requires that certain assumptions be made about the blending properties of the components and the effects of the other factors. If the fitted reduced combined model is judged to be adequate, the prediction error with the reduced combined model is lower in magnitude than the prediction error associated with the complete combined model that contains one or more unimportant terms.

6.7 SUMMARY

In this chapter we focused on the inclusion of the process variable in mixture experiments from the standpoint of designing experiments and fitting combined model forms in the mixture components and process variables. The suggested designs consist of a factorial arrangement in the levels of the process variables positioned at each point or blend among the components in a mixture design. By the same token, we could have selected a design in the mixture components based on a proposed mixture model and position the mixture design at each point of a factorial arrangement in the levels of the process variables. The combined model form consists of crossing each of the terms in the mixture model with each and every term of a standard polynomial in the process variables. The terms of the combined model represent the blending properties of the components and the effects of changing the processing conditions on the component blending properties.

Varying the amount of the mixture in the form of a mixture-amount experiment is the next topic discussed. A mixture-amount experiment is a mixture experiment that is performed at two or more levels of amount. The strategy used in constructing designs and fitting model forms for mixture-amount experiments follows very closely the strategy employed when including process variables in mixture experiments. A fertilizer blend-fertilizer rate experiment that was performed in Sao Paulo, Brazil in 1996, served as the background for growing young (less 3 years old) citrus trees. Nine fertilizer N-P-K combinations were applied at each of three equally spaced rates (low, middle, and high), and it was discovered that the optimal N-P-K blend differed at the three rates even though the optimal blends at the middle and high rates were very close to one another. The final section of the chapter presents two examples taken from the literature of mixture experiments involving three mixture components along with other factors. The examples illustrate some of the potential pitfalls that await the unsuspecting practitioner who relies strictly on computer-generated designs while proposing impractical combined model forms. Listed at the end of the section are some general recommendations for designing experiments and fitting models to data collected from mixture experiments with other factors.

REFERENCES AND RECOMMENDED READING

Anderson-Cook, C. M., H. Goldfarb, C. M. Borror, D. C., Montgomery, K. Canter, and J. Twist (2004). Mixture and mixture-process variable experiments for pharmaceutical applications. *Pharmaceut. Stat., 3*, 247–260.

Chau, K. W., and W. R. Kelley (1993). Formulating printable coatings via *D*-optimality. *J. Coat. Tech.*, **65**, 71–78.

Chitra, S. P., and E. A. Ekong (1993). An alternate approach to product development. *ASQ Qual. Cong. Trans.*, 837–843.

Claringbold, P. J. (1995). *Use* of the simplex design in the study of the joint action of related hormones. *Biometrics*, **11**, 174–185.

Cornell, J. A. (1971). Process variables in the mixture problem for categorized components. *J. Am. Stat. Assoc.*, **66**, 42–48.

Cornell, J. A. (1988). Analyzing data from mixture experiments containing process variables: a split-plot approach. *J. Qual. Technol.*, **20**, 2–33.

Cornell, J. A., (1995). Fitting models to data from mixture experiments containing other factors. *J. Qual. Technol.*, **27**, 13–33.

Cornell, J. A. (2002). *Experiments with Mixtures: Designs, Models, and the Analysis of Mixture Data*, 3rd edition, Wiley, New York.

Cornell, J. A., and J. W. Gorman (1984). Fractional design plans for process variables in mixture experiments. *J. Qual. Technol.*, **16**, 20–38.

Cornell, J. A., and J. W. Gorman (2003). Two new mixture models: living with collinearity but removing its influence. *J. Qual. Technol.* **35**, 78–89.

Cox, D. R. (1971). A note on polynomial response functions for mixtures. *Biometrika*, **58**, 155–159.

Czitrom, V. (1989). Experimental design for four mixture components with process variables. *Commun. Stat.*, **18**, 4561–4581.

Czitrom, V. (1992). Note on a mixture experiment with process variables. *Commun. Stat.*, **21**, 493–198.

Donev, A. N. (1989). Design of experiments with both mixture and qualitative factors. *J. R. Stat. Soc.*, **51**, 297–302.

Goldfarb, H., C. M. Borror, and D. C. Montgomery (2003). Mixture-process variable experiments with noise variables. *J. Qual. Technol.*, **35**, 393–405.

Goldfarb, H., C. M Anderson-Cook, C. M, Borror, and D. C. Montgomery (2004). Fraction of design space plots for assessing mixture and mixture-process designs. *J. Qual. Technol.*, **36**, 69–79.

Goldfarb, H., C. M. Borror, D. C. Montgomery, and C. M. Anderson-Cook, (2004). Three-dimensional variance dispersion graphs for mixture-process experiments. *J. Qual. Technol.*, **36**, 109–124.

Goldfarb H., C. M. Borror, D. C. Montgomery, and C. M. Anderson-Cook, (2004). Evaluating mixture-process designs with control and noise variables. *J. Qual. Technol.*, **36** 245–262.

Goldfarb, H., C. M. Borror, D. C. Montgomery, and C. M. Anderson-Cook, (2005). Using genetic algorithms to generate mixture-process designs involving control and noise variables. *J. Qual. Technol.*, **37**, 60–74.

Gorman, J. W., and J. A. Cornell (1982). A note on model reduction for experiments with both mixture components and process variables. *Technometrics*, **24**, 243–247.

Hare, L. B. (1979). Designs for mixture experiments involving process variables. *Technometrics*, **21**, 159–173.

Kowalski, S., J. A. Cornell, and G. G. Vining (2000). A new model and class of designs for mixture experiments with process variables. *Commun. Stat.*, **29**, 312–341.

Kowalski, S., J. A. Cornell, and G. G. Vining (2002). Split-plot designs and estimation methods for mixture experiments with process variables. *Technometrics*, **44** 72–79.

Murthy, M. S. R., and P. L. Manga (1996). Restricted region simplex designs for mixture experiments in the presence of process variables. *Sankhya*, **58**, 231–239.

Piepel, G. F. (1988). A note on models for mixture-amount experiments when the total amount takes a zero value. *Technometrics*, **30**, 449–450.

Piepel, G. F., and J. A. Cornell (1985). Models for mixture experiments when the response depends on the total amount. *Technometrics*, **27**, 219–227.

Piepel, G. F., and J. A. Cornell (1987). Designs for mixture-amount experiments. *J. Qual. Technol.*, **19**, 11–28.

Piepel, G. F., and J. A. Cornell (1994). Mixture experiment approaches: examples, discussion, and recommendations. *J. Qual. Technol.*, **26**, 177–196.

Satterthwaite, (1946). An approximate distribution of estimates of variance components. *Biometrics Bull.*, **2**, 110–114.

Scheffé, H. (1963). The simplex-centroid design for experiments with mixtures. *J. R. Stat. Soc. B*, **25**, 235–263.

Smith, W. F. (2005). *Experimental Design for Formulation*. ASA-SIAM Series on Statistics and Applied Probability. ASA, Alexandria, VA.

Vuchkov, I. N., H. A. Yonchev, and D. L. Damgaliev (1983). Continuous D-optimal designs for experiments with mixture and process variables. *Math. Oper. Stat.*, **14**, 33–51.

QUESTIONS

6.1. A $\{2, 2\}$ simplex-lattice was set up at each of two levels of a single process variable (z). The data are:

$(x_1, x_2) =$	$(1, 0)$	$\left(\frac{1}{2}, \frac{1}{2}\right)$	$(0, 1)$
$z = +1$	2.6, 2.4	6.1, 7.1	4.0, 3.4
$z = -1$	4.6, 4.8	5.0, 4.0	5.4, 5.6

Fit a second-degree mixture model in x_1 and x_2 to the data at the low level of z and then at the high level of z. Fit the combined model to the entire set of data. Answer the following questions:

(a) Do the mixture components affect the response?

(b) Are the blending properties of the components different at the different levels of the process variable? What effect does changing (raising) the level of the process variable have on the blending properties of the mixture components?

(c) If a high value of the response is considered desirable, what settings of x_1, x_2, and z do you recommend?

6.2. Beef patties were cooked at three separate time–temperature conditions (c_1, c_2, and c_3). The patties were prepared by blending two types of ground beef (denoted by A and B). The blends were selected according to a $\{2, 3\}$ lattice arrangement. Three replications of each blend–cooking treatment combination were performed. The data represent texture test readings as measured in grams of force $\times 10^{-3}$.

$(x_A : x_B)$	Replication 1			Replication 2			Replication 3		
	c_1	c_2	c_3	c_1	c_2	c_3	c_1	c_2	c_2
$1:0$	1.5	2.0	3.0	1.4	1.8	2.1	1.4	2.0	3.0
$\frac{2}{3} : \frac{1}{3}$	2.8	3.4	2.6	3.2	3.5	3.2	2.6	3.4	2.6
$\frac{1}{3} : \frac{2}{3}$	2.4	2.6	2.6	2.0	1.8	1.5	2.2	2.1	1.8
$0:1$	1.2	1.2	1.8	1.2	1.4	2.0	0.8	1.8	2.0

Let the cooking combinations represent equally spaced levels of a quantitative factor, where c_1, c_2, and c_3 are the low, medium, and high levels, respectively.

(a) Use the data from Replication 1 only and fit the model (estimate the coefficients γ_A^i, γ_B^i, γ_{AB}^i, δ_{AB}^i, $i = 0, 1, 2$)

$$y_u = \sum_{i=0,1,2} [\gamma_A^i x_A + \gamma_B^i x_B + \gamma_{AB}^i x_A x_B + \delta_{AB}^i x_A x_B (x_A - x_B)]c^i + \varepsilon$$

where $c^0 = 1$, c^1 represents the linear term for the cooking treatments, and c^2 is the curvilinear term for cooking treatments.

(b) Analyze the data from all three replications, and set up the ANOVA table. Describe the blending behavior between beef types A and B, Does raising the cooking level ($c_1 < c_2 < c_3$) increase the firmness of the patties? Are combinations of A and B affected similarly to the pure beef patties by the cooking treatments? Fully explain your analyses. Assume that the 12 patties in each replication were prepared in a completely random order

6.3. Fit a combined model of the form

$$y = \sum_{i=1}^{2} \left[\gamma_i^0 + \sum_{l=1}^{2} \gamma_i^l z_l \right] x_i + \varepsilon$$

to the data in Table 6.1. Set up an analysis of variance table based on the $N = 24$ observations and comment on your findings.

6.4. Shown are data values at the seven blends of a three-component simplex-centroid design. High values of the response are more desirable than low values. Indicate the type of blending, linear or nonlinear, synergistic or antagonistic, that is present between the components:

(a) A and B: _____ and is _____ .

(b) A and C: _____ and is _____ .

(c) B and C: _____ and is _____ .

(d) A and B and C: _____ and is _____ .

Hint: $\text{var}(b_i) = \sigma^2/2$, $\text{var}(b_{ij}) = 24\sigma^2/2$, $\text{var}(b_{ijk}) = 958.5\sigma^2$.
Also SST $= 63.79$, SSR $= 58.22$, and SSE $= 5.57$.

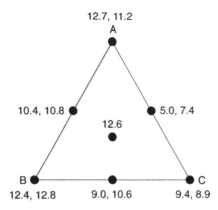

APPENDIX 6A CALCULATING THE ESTIMATED COMBINED MIXTURE COMPONENT–PROCESS VARIABLE MODEL OF EQ. (6.10) WITHOUT THE COMPUTER.

When the data set is balanced like the texture scores are in Table 6.1, it is not difficult to calculate the combined model of Eq. (6.5) to get Eq. (6.10). Initially, however, one must calculate the four mixture models of Eqs. (6.6) through (6.9). Each of the four models is calculated exactly the same way as we will show in calculating Eq. (6.6) using the six texture scores taken from the first three lines of Table 6.1; Eq. (6.7) is calculated using the texture scores from lines 4 to 6, and so on; Eq. (6.9) is calculated using the texture scores from the bottom three lines of Table 6.1.

To calculate Eq. (6.6), the estimates b_1, b_2, and b_{12} are obtained using the formulas shown earlier in Eq, (2.15):

$$b_1 = \frac{1.84 + 1.65}{2} = 1.745, \; b_2 = \frac{0.67 + 0.58}{2} = 0.625,$$

$$b_{12} = 4\left(\frac{1.29 + 1.18}{2}\right) - 2(b_1 + b_2) = 4(1.235) - 2(2.370)$$

$$= 4.940 - 4.740 = 0.200 \tag{6A.1}$$

The fitted model of Eq. (6.6) is

$$\hat{y}(x) = 1.745x_1 + 0.625x_2 + 0.200x_1x_2 \tag{6A.2}$$

The estimate of the error variance σ^2 is MSE and MSE is obtained in the case of two replicated texture scores for each blend, as

$$SSE = \frac{(1.84 - 1.65)^2 + (0.67 - 0.58)^2 + (1.29 - 1.18)^2}{2} = 0.02825$$

and since SSE, the sum of squares for pure error, has three degrees of freedom, then

$$MSE = \frac{SSE}{3} = 0.0094$$

Question for the Reader. To obtain an estimate of the variance and thus an estimate of the standard error for each of the estimates b_1, b_2 and b_{12}, so that we can test $H_0: \beta_{12} = 0$, versus $H_A: \beta_{12} \neq 0$ at the $\alpha = 0.05$ level of significance, how do we proceed?

Now let us continue in our goal of calculating the 12 coefficients in Eq. (6.10). We begin by listing the four Eqs. (6.6) to (6.9) along with values of z_1 and z_2 as

z_1	z_2	Model	MSE
-1	-1	$1.745x_1 + 0.625x_2 + 0.200x_1x_2$	0.0094
1	-1	$2.590x_1 + 1.035x_2 - 1.290x_1x_2$	0.0525
-1	1	$3.025x_1 + 1.185x_2 - 0.860x_1x_2$	0.0016
1	1	$4.130x_1 + 1.485x_2 - 2.590x_1x_2$	0.0295

The fitted combined mixture component-process variable model is

$$\hat{y}(x,\ z) = g_1^0 x_1 + g_2^0 x_2 + g_{12}^0 x_1 x_2 + \{g_1^1 x_1 + g_2^1 x_2 + g_{12}^1 x_1 x_2\} z_1$$
$$+ \{g_1^2 x_1 + g_2^2 x_2 + g_{12}^2 x_1 x_2\} z_2 + \{g_1^{12} x_1 + g_2^{12} x_2 + g_{12}^{12} x_1 x_2\} z_1 z_2 \tag{6A.3}$$

where

$$g_1^0 = (1.745 + 2.590 + 3.025 + 4.130)/4, \quad g_2^0 = (0.625 + 1.035 + 1.185 + 1.485)/4$$
$$= 2.873 \qquad\qquad = 1.083$$
$$g_{12}^0 = (0.200 - 1.290 - 0.860 - 2.590)/4, \quad g_1^1 = (-1.745 + 2.590 - 3.025 + 4.130)/4$$
$$= -1.135 \qquad\qquad = 0.488$$
$$g_2^1 = (-0.625 + 1.035 - 1.185 + 1.485)/4, \quad g_{12}^1 = (-0.200 - 1.290 + 0.860 - 2.590)/4$$
$$= 0.178 \qquad\qquad = -0.805$$
$$g_1^2 = (-1.745 - 2.590 + 3.025 + 4.130)/4, \quad g_2^2 = (-0.625 - 1.035 + 1.185 + 1.485)/4$$
$$= 0.705 \qquad\qquad = 0.253$$
$$g_{12}^2 = (-0.200 + 1.290 - 0.860 - 2.590)/4, \quad g_1^{12} = (1.745 - 2.590 - 3.025 + 4.130)/4$$
$$= -0.590 \qquad\qquad = 0.065$$
$$g_2^{12} = (0.625 - 1.035 - 1.185 + 1.485)/4, \quad g_{12}^{12} = (0.200 + 1.290 + 0.860 - 2.590)/4$$
$$= -0.028 \qquad\qquad = -0.060$$

$$\tag{6A.4}$$

Now, substituting the twelve estimates in (6A.4) into (6A.3), we obtain Eq. (6.10):

$$\hat{y}(\mathbf{x}, \mathbf{z}) = 2.873x_1 + 1.083x_2 - 1.135x_1x_2 + \{0.488x_1 + 0.178x_2 - 0.805x_1x_2\}z_1$$
$$+ \{0.705x_1 + 0.253x_2 - 0.590x_1x_2\}z_2 + \{0.065x_1 - 0.028x_2 - 0.060x_1x_2\}z_1z_2$$

and the new MSE $= (0.0094 + 0.0525 + 0.0016 + 0.0295)/4 = 0.0233$ with 12 degrees of freedom. The estimated standard errors for g_i^l and g_{12}^l are s.e.$(g_i^l) = (0.0233/8)^{1/2} = 0.054$ and s.e.$(g_{12}^l) = \sqrt{24}(0.054) = 0.264$ for $l = 0, 1, 2, 12$ and $i = 1, 2$.

CHAPTER 7

A Review of Least Squares and the Analysis of Variance

In this chapter we present some background ideas on the use of matrices and vectors. The use of matrices and vectors is fundamental for describing and constructing some of the mixture designs as well as necessary to some procedures for analyzing mixture data. We present only the basic concepts of matrix algebra that are pertinent to mixture experiments, and we refer the reader to the books listed in the end of this chapter for a more complete coverage.

7.1 A REVIEW OF LEAST SQUARES

Let us assume that provisionally N observations of the response are expressible by means of the linear first-degree equation

$$y_u = \beta_1 x_{u1} + \beta_2 x_{u2} + \cdots + \beta_q x_{uq} + \varepsilon_u \tag{7.1}$$

where y_u denotes the observed response for the uth trial, x_{ui} represents the proportion of component i at the uth trial, β_i represents an unknown parameter in the equation, and ε_u represents the random error, $u = 1, 2, \ldots, N$. The *method of least squares* selects as the estimates b_i, for the unknown parameters $\beta_i, i = 1, 2, \ldots, q$, those values of $b_1, b_2, \ldots, b_q$ that minimize the quantity

$$\sum_{u=1}^{N} (y_u - b_j x_{u1} - b_2 x_{u2} - \cdots - b_q x_{uq})^2$$

In *matrix notation* the first-degree model in Eq. (7.1), over the N observations, can be expressed as

$$\mathbf{y} = \mathbf{X}\boldsymbol{\beta} + \boldsymbol{\varepsilon} \tag{7.2}$$

A Primer on Experiments with Mixtures, By John A. Cornell
Copyright © 2011 John Wiley & Sons, Inc. Published by John Wiley & Sons, Inc.

where

$$
\mathbf{y} = \begin{bmatrix} y_1 \\ y_2 \\ \vdots \\ y_N \end{bmatrix}, \quad
\mathbf{X} = \begin{bmatrix} x_{11} & x_{12} & \cdots & x_{1q} \\ x_{21} & x_{22} & \cdots & x_{2q} \\ \vdots & \vdots & \vdots & \vdots \\ x_{N1} & x_{N2} & \cdots & x_{Nq} \end{bmatrix}, \quad
\boldsymbol{\beta} = \begin{bmatrix} \beta_1 \\ \beta_2 \\ \vdots \\ \beta_q \end{bmatrix}, \quad
\boldsymbol{\varepsilon} = \begin{bmatrix} \varepsilon_1 \\ \varepsilon_2 \\ \vdots \\ \varepsilon_N \end{bmatrix}
$$
$$
N \times 1 \qquad\qquad\qquad N \times q \qquad\qquad\qquad q \times 1 \qquad\qquad N \times 1
$$

7.1.1 Some Rules for Obtaining Means and Variances of Random Vectors

Let us assume the random variable y_u has expectation $E(y_u) = \mu_u$, where $\mu_u = \beta_1 x_{u1} + \ldots + \beta_q x_{uq}, u = 1, 2, \ldots, N$. Then

$$
E(\mathbf{y}) = E \begin{bmatrix} y_1 \\ y_2 \\ \vdots \\ y_N \end{bmatrix} = \begin{bmatrix} \mu_1 \\ \mu_2 \\ \vdots \\ \mu_N \end{bmatrix} = \boldsymbol{\mu}
$$

that is, the expectation of a vector is the vector of expectations. If the variances and the covariances of the y_u are given by $\mathrm{var}(y_u) = E(y_u - \mu_u)^2 = \sigma_u^2$, and cov $(y_u, y_{u'}) = E(y_u - \mu_u)(y_{u'} - \mu_{u'}) = \sigma_{uu'}, u \neq u'$, then the variance–covariance ($N \times N$ symmetric) matrix is denoted by

$$
\mathbf{V} = E(\mathbf{y} - \boldsymbol{\mu})(\mathbf{y} - \boldsymbol{\mu})' = \begin{bmatrix} \sigma_1^2 & \sigma_{12} & \sigma_{13} & \cdots & \sigma_{1N} \\ \sigma_{21} & \sigma_2^2 & \sigma_{23} & \cdots & \sigma_{2N} \\ \vdots & \vdots & \vdots & & \vdots \\ \sigma_{N1} & \sigma_{N2} & \sigma_{N3} & \cdots & \sigma_N^2 \end{bmatrix} \tag{7.3}
$$

If the variables $y_u, u = 1, 2, \ldots, N$, are jointly normally distributed with mean vector $\boldsymbol{\mu}$ and variance–covariance matrix $\mathbf{V}$, this is written as $\mathbf{y} \sim N(\boldsymbol{\mu}, \mathbf{V})$.

For the general case of the mathematical model in Eq. (7.2), $\mathbf{y}$ is an $N \times 1$ vector of observations, $\mathbf{X}$ is an $N \times p$ matrix whose elements are the mixture component proportions and functions of the component proportions, $\boldsymbol{\beta}$ is a $p \times 1$ vector of parameters, and $\boldsymbol{\varepsilon}$ is an $N \times 1$ vector of random errors. When the model $\mathbf{y} = \mathbf{X}\boldsymbol{\beta} + \boldsymbol{\varepsilon}$ is of the first-degree as in Eq. (7.1) then $p = q$. When the model is of the second-degree, then $p = q(q + 1)/2$.

The *normal equations* that are set up to estimate the elements of the parameter vector $\boldsymbol{\beta}$ in Eq. (7.2) are

$$
\mathbf{X}'\mathbf{Xb} = \mathbf{X}'\mathbf{y} \tag{7.4}
$$

where the $p \times p$ square matrix $\mathbf{X'X}$ consists of sums of squares and sums of cross-products of the mixture proportions and the $p \times 1$ vector $\mathbf{X'y}$ consists of sums of cross-products of the x_{ui} and y_u. The least-squares estimates of the elements of $\boldsymbol{\beta}$ are

$$\mathbf{b} = (\mathbf{X'X})^{-1}\mathbf{X'y} \tag{7.5}$$

where the $p \times p$ matrix $(\mathbf{X'X})^{-1}$ is the inverse of $\mathbf{X'X}$. Since $\mathbf{X'X}$ is symmetric, so is $(\mathbf{X'X})^{-1}$.

7.1.2 Properties of the Parameter Estimates

The statistical properties of the estimator $\mathbf{b}$ are easily verified once certain assumptions are made about the elements of $\boldsymbol{\varepsilon}$. We write the expectation of the elements of the random error vector in Eq. (7.2) as $E(\boldsymbol{\varepsilon}) = \mathbf{0}$ and $\text{var}(\boldsymbol{\varepsilon}) = E(\boldsymbol{\varepsilon\varepsilon'})\,\sigma^2\mathbf{I}_N$, where $\mathbf{I}_N$ is the identity matrix of order N. Then the expectation of $\mathbf{b}$ is

$$\begin{aligned}
E(\mathbf{b}) &= E[(\mathbf{X'X})^{-1}\mathbf{X'y}] \\
&= E[(\mathbf{X'X})^{-1}\mathbf{X'}(\mathbf{X}\boldsymbol{\beta} + \boldsymbol{\varepsilon})] \\
&= \boldsymbol{\beta} + E(\mathbf{X'X})^{-1}\mathbf{X'}\boldsymbol{\varepsilon} \\
&= \boldsymbol{\beta} \tag{7.6}
\end{aligned}$$

Thus, if the model $\mathbf{y} = \mathbf{X}\boldsymbol{\beta} + \boldsymbol{\varepsilon}$ is correct, $\mathbf{b}$ is an *unbiased estimator* of $\boldsymbol{\beta}$. The *variance-covariance* matrix of the elements of $\mathbf{b}$ is expressed as

$$\begin{aligned}
\text{var}(\mathbf{b}) &= \text{var}[(\mathbf{X'X})^{-1}\mathbf{X'y}] \\
&= (\mathbf{X'X})^{-1}\mathbf{X'}\text{var}(\mathbf{y})\mathbf{X}(\mathbf{X'X})^{-1} \\
&= (\mathbf{X'X})^{-1}\mathbf{X'}\mathbf{V}\mathbf{X}(\mathbf{X'X})^{-1}
\end{aligned}$$

Since $\text{var}(\varepsilon) = \sigma^2\mathbf{I}_N$, meaning the variances of the errors are the same σ^2 we have $\mathbf{V} = \sigma^2\mathbf{I}_N$ and

$$\text{var}(\mathbf{b}) = (\mathbf{X'X})^{-1}\sigma^2 \tag{7.7}$$

Along the main diagonal of the $p \times p$ matrix $(\mathbf{X'X})^{-1}\sigma^2$, the iith element is the variance of b_i, the ith element of $\mathbf{b}$. The ijth element of $(\mathbf{X'X})^{-1}\sigma^2$ is the covariance between the elements b_i and b_j of $\mathbf{b}$. Furthermore, if the errors $\boldsymbol{\varepsilon}$ are jointly normally distributed, then with the properties of $\mathbf{b}$ defined in Eqs. (7.6) and (7.7) the distribution of $\mathbf{b}$ is written as

$$\mathbf{b} \sim N(\boldsymbol{\beta},\, (\mathbf{X'X})^{-1}\sigma^2) \tag{7.8}$$

7.1.3 Predicted Response Values

Once the vector of estimates $\mathbf{b}$ is obtained using Eq. (7.5) prediction of the value of the response at some point $\mathbf{x} = (x_1, x_2, \ldots, x_q)'$ in the experimental region can be made and is expressed in matrix notation as

$$\hat{y}(\mathbf{x}) = \mathbf{x}'_p \mathbf{b} \tag{7.9}$$

where $\mathbf{x}'_p$ is a $1 \times p$ vector whose elements correspond to the elements in a row of the matrix $\mathbf{X}$ in Eq. (7.2). Specifically, if the predicted value of the response corresponding to the uth observation is desired, then $\hat{y}(\mathbf{x}) = \mathbf{x}'_u \mathbf{b}$, where x'_u is the uth row of $\mathbf{X}$. (We use the notation $\hat{y}(\mathbf{x})$ to denote the predicted value of $\hat{y}$ at the point $\mathbf{x}$.)

A measure of the precision of the estimate $\hat{y}(\mathbf{x})$ is defined as the *variance* of $\hat{y}(\mathbf{x})$ and is expressed as

$$
\begin{aligned}
\mathrm{var}[\hat{y}(\mathbf{x})] &= \mathrm{var}[\mathbf{x}'_p \mathbf{b}] \\
&= \mathbf{x}'_p \mathrm{var}(\mathbf{b})\mathbf{x}_p \\
&= \mathbf{x}'_p (\mathbf{X}'\mathbf{X})^{-1}\mathbf{x}_p \sigma^2
\end{aligned} \tag{7.10}
$$

The inverse matrix $(\mathbf{X}'\mathbf{X})^{-1}$ used for obtaining $\mathbf{b}$ in Eq. (7.5) also determines the variances and covariances of the elements of $\mathbf{b}$ in Eq. (7.7) as well as the variance of $\hat{y}(\mathbf{x})$.

7.2 THE ANALYSIS OF VARIANCE

The results of the analysis of a set of data from a mixture experiment can be displayed in table form. The table is called an *analysis of variance* table. The entries in the table represent measures of information about the separate sources of variation in the data.

The total variation in a set of data is called the "total sum of squares" and is abbreviated as SST. The quantity SST is computed by summing the squares of the observed y_u about their mean $\bar{y} = (y_1 + y_2 + \cdots + y_N)/N$,

$$\mathrm{SST} = \sum_{u=1}^{N} (y_u - \bar{y})^2 \tag{7.11}$$

The quantity SST has associated with it $N - 1$ degrees of freedom, since there are only $N - 1$ independent deviations $y_u - \bar{y}$ in the sum.

The sum of squares of the deviations of the observed y_u from their predicted values is

$$\mathrm{SSE} = \sum_{u=1}^{N} (y_u - \hat{y}_u)^2 \tag{7.12}$$

and SSE is called the "sum of squares of the residuals."

Table 7.1 Analysis of Variance Table

Source of Variation	Degrees of Freedom	Sum of Squares	Mean Square
Regression (fitted model)	$p - 1$	$\text{SSR} = \sum_{u=1}^{N} (\hat{y}_u - \bar{y})^2$	$\dfrac{\text{SSR}}{(p - 1)}$
Residual	$N - p$	$\text{SSE} = \sum_{u=1}^{N} (y_u - \hat{y}_u)^2$	$\dfrac{\text{SSE}}{(N - p)}$
Total	$N - 1$	$\text{SST} = \sum_{u=1}^{N} (y_u - \bar{y})^2$	

The difference between the sums of squares quantities is

$$\text{SSR} = \text{SST} - \text{SSE} = \sum_{u=1}^{N} (\hat{y}_u - \bar{y})^2$$

and SSR represents the portion of SST attributable on the fitted regression equation. The quantity SSR is called the *sum of squares due to regression*. In matrix notation, shortcut formulas for SST, SSE, and SSR are, letting $\mathbf{1}'$ be a $1 \times N$ vector of 1's,

$$\text{SST} = \mathbf{y}'\mathbf{y} - \frac{(\mathbf{1}'\mathbf{y})^2}{N}$$

$$\text{SSE} = \mathbf{y}'\mathbf{y} - \mathbf{b}'\mathbf{X}'\mathbf{y}$$

$$\text{SSR} = \mathbf{b}'\mathbf{X}'\mathbf{y} - \frac{(\mathbf{1}'\mathbf{y})^2}{N}$$

The partitioning of the total sum of squares is summarized with the familiar analysis of variance table, Table 7.1, where we assume that the fitted model in Eq. (7.9) contains p terms.

7.3 A NUMERICAL EXAMPLE: MODELING THE TEXTURE OF FISH PATTIES

Fish patties were formulated from three different types of saltwater fish. The fish were mullet (x_1), sheepshead (x_2), and croaker (x_3). The design chosen was a simplex-centroid design with the following fish percentages: mullet (100%), sheepshead (100%), croaker (100%), mullet : sheepshead (50% : 50%), mullet : croaker (50% : 50%), sheepshead : croaker (50% : 50%), and mullet : sheepshead : croaker (33% : 33% : 33%). The data in Table 7.2 represent average texture readings in grams of force ($\times 10^{-3}$) for replicate patties, where each average was computed from three readings taken on each patty. Initially a second-degree equation will be fitted to the data.

Table 7.2 Average Texture Readings Taken on Replicate Fish Patties

| Fish Percentages (%) | | | Component Proportions | | | Average Texture |
Mullet	Sheepshead	Croaker	x_1	x_2	x_3	(grams $\times 10^{-3}$)
100	0	0	1	0	0	2.02, 2.08
0	100	0	0	1	0	1.47, 1.37
50	50	0	$\frac{1}{2}$	$\frac{1}{2}$	0	1.91, 2.00
0	0	100	0	0	1	1.93, 1.83
50	0	50	$\frac{1}{2}$	0	$\frac{1}{2}$	1.98, 2.13
0	50	50	0	$\frac{1}{2}$	$\frac{1}{2}$	1.80, 1.71
33	33	33	$\frac{1}{3}$	$\frac{1}{3}$	$\frac{1}{3}$	1.46, 1.50

In matrix notation, the second-degree model is written as $\mathbf{y} = \mathbf{X}\boldsymbol{\beta} + \boldsymbol{\varepsilon}$ where

$$
\mathbf{y} = \begin{bmatrix} 2.02 \\ 2.08 \\ 1.47 \\ 1.37 \\ 1.91 \\ 2.00 \\ 1.93 \\ 1.83 \\ 1.98 \\ 2.13 \\ 1.80 \\ 1.71 \\ 1.46 \\ 1.50 \end{bmatrix}, \quad
\mathbf{X} = \begin{bmatrix}
x_1 & x_2 & x_3 & x_1x_2 & x_1x_3 & x_2x_3 \\
1 & 0 & 0 & 0 & 0 & 0 \\
1 & 0 & 0 & 0 & 0 & 0 \\
0 & 1 & 0 & 0 & 0 & 0 \\
0 & 1 & 0 & 0 & 0 & 0 \\
0.5 & 0.5 & 0 & 0.25 & 0 & 0 \\
0.5 & 0.5 & 0 & 0.25 & 0 & 0 \\
0 & 0 & 1 & 0 & 0 & 0 \\
0 & 0 & 1 & 0 & 0 & 0 \\
0.5 & 0 & 0.5 & 0 & 0.25 & 0 \\
0.5 & 0 & 0.5 & 0 & 0.25 & 0 \\
0 & 0.5 & 0.5 & 0 & 0 & 0.25 \\
0 & 0.5 & 0.5 & 0 & 0 & 0.25 \\
0.33 & 0.33 & 0.33 & 0.11 & 0.11 & 0.11 \\
0.33 & 0.33 & 0.33 & 0.11 & 0.11 & 0.11
\end{bmatrix}, \quad
\boldsymbol{\beta} = \begin{bmatrix} \beta_1 \\ \beta_2 \\ \beta_2 \\ \beta_{12} \\ \beta_{13} \\ \beta_{23} \end{bmatrix}
$$

The normal equations (7.4) are

$$
\underset{\mathbf{X'X}}{\begin{bmatrix}
3.222 & 0.722 & 0.722 & 0.324 & 0.324 & 0.074 \\
 & 3.222 & 0.722 & 0.324 & 0.074 & 0.324 \\
 & & 3.222 & 0.074 & 0.324 & 0.324 \\
 & & & 0.150 & 0.025 & 0.025 \\
 \text{symmetric} & & & & 0.150 & 0.025 \\
 & & & & & 0.150
\end{bmatrix}}
\underset{\mathbf{b}}{\begin{bmatrix} b_1 \\ b_2 \\ b_3 \\ b_{12} \\ b_{13} \\ b_{23} \end{bmatrix}}
=
\underset{\mathbf{X'y}}{\begin{bmatrix} 9.10 \\ 7.54 \\ 8.56 \\ 1.31 \\ 1.36 \\ 1.21 \end{bmatrix}}
$$

The solutions to the normal equations are

$$
\mathbf{b} \qquad = \qquad (\mathbf{X'X})^{-1} \qquad\qquad \mathbf{x'y}
$$

$$
\begin{bmatrix} 2.08 \\ 1.45 \\ 1.91 \\ 0.21 \\ -0.31 \\ -0.25 \end{bmatrix} = \begin{bmatrix} 0.496 & -0.004 & -0.004 & -0.924 & -0.924 & 0.076 \\ & 0.496 & -0.004 & -0.924 & 0.076 & -0.924 \\ & & 0.496 & 0.076 & -0.924 & -0.924 \\ & & & 10.485 & 0.485 & 0.485 \\ \text{symmetric} & & & & 10.485 & 0.485 \\ & & & & & 10.485 \end{bmatrix} \begin{bmatrix} 9.10 \\ 7.54 \\ 8.56 \\ 1.31 \\ 1.36 \\ 1.21 \end{bmatrix}
$$

and the fitted model (7.9) is

$$
\hat{y}(\mathbf{x}) = \mathbf{x}_p'\mathbf{b}
$$
$$
= 2.08x_1 + 1.45x_2 + 1.91x_3 + 0.21x_1x_2 - 0.31x_1x_3 - 0.25x_2x_3 \qquad (7.13)
$$
$$
(0.14) \quad (0.14) \quad (0.14) \quad (0.65) \qquad (0.65) \qquad (0.65)
$$

where the number in parentheses below each parameter estimate is the estimated standard error (est, s.e.) of the parameter of coefficient estimate.

The analysis of variance calculations are

$$
\mathrm{SST} = \mathbf{y'y} - \frac{(\mathbf{1'y})^2}{14} = 46.170 - 45.325 = 0.845
$$
$$
\mathrm{SSE} = \mathbf{y'y} - \mathbf{b'X'y} = 46.170 - 45.847 = 0.323
$$
$$
\mathrm{SSR} = \mathrm{SST} - \mathrm{SSE} = 0.845 - 0.323 = 0.522
$$

and the analysis of variance table for the fitted second-degree model (7.13) is

Source of Variation	Degrees of Freedom	Sum of Squares	Mean Square
Regression	5	0.522	0.104
Residual	8	0.323	0.04
Total	13	0.845	

The estimated standard errors of the parameter estimates in Eq. (7.13) are the square roots of the diagonal elements of $(\mathbf{X'X})^{-1}s^2$, where $s^2 = 0.04$. In other words,

$$
\widehat{s.e.}(b_1) = \sqrt{0.496(0.04)} = 0.14
$$
$$
\widehat{s.e.}(b_{12}) = \sqrt{10.485(0.04)} = 0.65
$$

The data in Table 7.2 were collected from a simplex-centroid design that actually supports the fitting of the special cubic model

$$y = \beta_1 x_1 + \beta_2 x_2 + \beta_3 x_3 + \beta_{12} x_1 x_2 + \beta_{13} x_1 x_3 + \beta_{23} x_2 x_3 + \beta_{123} x_1 x_2 x_3 + \varepsilon$$

$$(7.14)$$

Furthermore, since there are two replicate observations at each point of the design, an estimate of the experimental error variance (owing to the replicates) with seven degrees of freedom can be calculated and compared to the mean square for lack of fit of the model (7.13). The mean square for lack of fit of the model (7.13) has one degree of freedom as we show now.

Using the replicated observations at the seven design blends, the sum of squares for *pure error* is calculated to be

SS pure error

$$= \frac{(2.02 - 2.08)^2 + (1.47 - 1.37)^2 + (1.91 - 2.00)^2 + \cdots + (1.46 - 1.50)^2}{2}$$

$$= \frac{0.0639}{2} = 0.032 \quad \text{with 7 degrees of freedom}$$

The sum of squares for lack of fit of the model (7.13) is obtained by subtracting SS pure error from the residual sum of squares (SSE $= 0.323$) taken from the analysis of variance table associated with the fitted model (7.13). The sum of squares for lack of fit is

$$SS_{LOF} = SSE - SS \text{ pure error}$$

$$= 0.323 - 0.032 = 0.291$$

with one degree of freedom $(8 - 7 = 1)$. Finally, the value of the F-statistic for testing the null hypothesis, H_0: Lack of fit of the second-degree model $= 0$ versus the alternative hypothesis, H_A: Lack of fit $\neq 0$ is

$$F = \frac{0.291/1}{0.032/7} = 63.66 \qquad (7.15)$$

Since the value of F in Eq. (7.15) exceeds the table value $F_{(1, 7, 0.01)} = 12.25$, we reject zero lack of fit of the model (7.13) owing to the fact that something must be contributing to the large F-value of 63.66 and we believe that the something is the absence of one or more terms in the model. Thus we upgrade the form of the model by adding the term $\beta_{123} x_1 x_2 x_3$ to it as in Eq. (7.14).

The fitted special cubic model is

$$\hat{y}(\mathbf{x}) = 2.05 x_1 + 1.42 x_2 + 1.88 x_3 + 0.88 x_1 x_2 + 0.36 x_1 x_3$$

$$(0.05) \quad (0.05) \quad (0.05) \quad (0.23) \quad (0.23)$$

$$+ 0.42 x_2 x_3 - 13.17 x_1 x_2 x_3 \qquad (7.16)$$

$$(0.23) \qquad (1.65)$$

The estimate of the error variance is now the mean square pure error $= 0.0046$ with 7 degrees of freedom.

7.4 THE ADJUSTED MULTIPLE CORRELATION COEFFICIENT

A measure of goodness of fit of a standard regression equation is the multiple correlation coefficient, denoted by R. The proportion of the total sum of squares that is explained by the fitted model is denoted by $R^2 = \text{SSR}/\text{SST}$. There are many characteristics of the data collection scheme that can influence the value of R^2 (see Cornell and Berger, 1987).

Although the Scheffé-type canonical polynomials do not contain a constant term (β_0), during the analysis of the fitted model, the sum of squares for regression (SSR) and the total sum of squares (SST) are both corrected for the overall average. The overall average of the observations is an estimate of the mean response even when the mixture components do not affect the response; see Section 4.1. However, when SSR and SST are not corrected for the overall average, the value of $R^2 = \text{SSR}/\text{SST}$ associated with the fit of the canonical polynomials is inflated. As a result an adjusted R^2 statistic has been suggested when fitting the canonical polynomials, and it is calculated as

$$R_A^2 = 1 - \frac{\text{SSE}/(N - p)}{\text{SST}/(N - 1)} \tag{7.17}$$

From the formula for R_A^2 in Eq. (7.17), R_A^2 is a measure of the reduction in the estimate of the error variance due to the fitting of the model, $\text{SSE}/(N - p)$, relative to the estimate of the error variance based on fitting the simple model $y = \beta_0 + \varepsilon$. For the fish-patty data of the previous section, the value of R_A^2 associated with the fitted second-degree model in Eq. (7.13) is $1 - [(0.323/8)/(0.845/13)] = 0.379$, while the value of R_A^2 for the fitted special cubic model (7.16) is $R_A^2 = 0.929$. The significance of the special cubic term in Eq. (7.16) is reflected in the increase in the value of $R_A^2 = 0.379$ with the second degree model to $R_A^2 = 0.929$ with the special cubic model of Eq. (7.16).

7.5 THE PRESS STATISTIC AND STUDENTIZED RESIDUALS

Most software packages today offer the PRESS statistic as a measure of the capability of the fitted model to predict new data as well as a measure of the stability of the regression equation. When fitting the proposed model, imagine the uth observed value of the response y_u removed from the data set; that is, imagine the model is fit to the remaining $N - 1$ observations. With the fitted model, one can predict the uth observation; call it $\hat{y}_{(u)}$. Now the difference $y_u - \hat{y}_{(u)} = e_{(u)}$ is called the prediction error of the uth observation or uth PRESS residual. If all N observations are treated this way, producing $e_{(1)}, e_{(2)}, \ldots, e_{(N)}$, then the PRESS statistic is defined as the sum of squares of the N PRESS residuals as in

$$\text{PRESS} = \sum_{u=1}^{N} e_{(u)}^2 = \sum_{u=1}^{N} [y_u - \hat{y}_{(u)}]^2 \tag{7.18}$$

Regression models with small (to be defined shortly) values of PRESS are generally considered good prediction equations.

In Montgomery, Peck, and Vining (2001, app. C.3), it is shown that $e(u) = e_u/(1 - h_{uu})$; that is, the uth PRESS residual is equal to the ordinary residual $e_u = y_u - \hat{y}_u$ divided by the $1 - u$th diagonal element of the hat matrix $\mathbf{H}$. Thus a simpler computing formula than (7.18) is

$$\text{PRESS} = \sum_{u=1}^{N} \left(\frac{e_u}{1 - h_{uu}} \right)^2$$

In judging the relative size of PRESS it is often helpful to compute an R^2-like statistic based on PRESS of the form

$$R^2_{\text{PRESS}} = 1 - \frac{\text{PRESS}}{\sum_{u=1}^{N} (y_u - \bar{y})^2} \tag{7.19}$$

If the value of R^2_{PRESS} is reasonably close to the value of R^2 (which means the value of PRESS is approximately equal to SSE) and the value of R^2_A, then we infer the value of PRESS is of reasonable magnitude. Some authors call $R^2\text{PRESS}$ the $R^2(\text{PREDICTION})$ statistic. With the fitted special cubic model of Eq. (7.16) PRESS $= 0.127$ so that

$$R^2_{\text{PRESS}} = 1 - \frac{0.127}{0.843}$$

$$= 0.8488$$

From $R^2_{\text{PRESS}} = 0.8488$, we would say, "We expect the special cubic fitted model to explain about 85% of the variability in predicting new observations."

The measure of leverage of the design points is discussed in Section 4.7 where the $N \times 1$ vector of residuals at the design points is shown in Eq. (4.42) to be $\mathbf{e} = (\mathbf{I} - \mathbf{H})\mathbf{y}$. Since from Eq. (7.2), the $N \times 1$ vector of observations has variance, $\text{var}(\mathbf{y}) = \text{var}(\boldsymbol{\varepsilon}) = \sigma^2 \mathbf{I}_N$, we can write the variance of the residuals as

$$\text{var}(\mathbf{e}) = (\mathbf{I} - \mathbf{H})\text{var}(\varepsilon)(\mathbf{I} - \mathbf{H})$$

$$= (\mathbf{I} - \mathbf{H})\sigma^2 \tag{7.20}$$

because the matrix $\mathbf{I} - \mathbf{H}$ is symmetric and idempotent. Thus the variance of the uth residual e_u is

$$\text{var}(e_u) = \sigma^2(1 - h_{uu})$$

that is, the variance of the uth residual depends on the location of the uth design point in the simplex relative to the point $\bar{x}_1, \bar{x}_2, \ldots, \bar{x}_q$, which is the data centroid.

Plotting the residuals to check on the assumption of $\text{var}(\boldsymbol{\varepsilon}) = \sigma^2 \mathbf{I_N}$ is common. However, rather than plot the ordinary residuals (the e_u), most authors recommend plotting the *studentized residuals*

$$r_u = \frac{e_u}{\sqrt{\text{MSE}(1 - h_{uu})}}, \qquad u = 1, 2, \ldots, N \tag{7.21}$$

This is because the studentized residuals have a constant variance equal to unity after the standardization eliminates the effect of the location of the data point in the simplex when the form of the model is correct. Hence we recommend plotting studentized residuals also when checking assumptions made about the errors.

7.6 TESTING HYPOTHESES ABOUT THE FORM OF THE MODEL: TESTS OF SIGNIFICANCE

The properties of the parameter estimates shown in Eqs. (7.6) and (7.7) are based on the assumption that the fitted model is correct (i.e., that $E(\boldsymbol{\varepsilon}) = \mathbf{0}$, $E(\boldsymbol{\varepsilon\varepsilon}') = \sigma^2 \mathbf{I}_N$) and on the additional assumption that the errors are jointly normally distributed was made, which enabled us to say the distribution of the parameter estimates is normal. The normality assumption on the errors allows us additional flexibility: a significance test can be made of the goodness of fit of the fitted model using the entries from the analysis of variance table, Table 7.1.

Let us consider the general mixture model $\mathbf{y} = \mathbf{X}\boldsymbol{\beta} + \boldsymbol{\varepsilon}$, where the matrix $\mathbf{X}$ is $N \times p$ and where the errors are normally distributed. Then SSE in Eq. (7.12) can be shown to be distributed as

$$\frac{\text{SSE}}{\sigma^2} \sim \chi^2_{(N-p)} \tag{7.22}$$

where the symbol $\sim$ means "is distributed as" and $\chi^2_{(N-p)}$ is a chi-square random variable with $N - p$ degrees of freedom. Under the null hypotheses that the mixture components do not affect the response, that is, for $q = 3$, the surface is a level plane above the triangle whose height is constant (equal to $\boldsymbol{\beta}$) at all points,

$$H_0 : \text{ all } \beta_i = \beta \text{ and all } \beta_{ij}, \beta_{ijk} \text{ other than } \beta_i = 0$$

$$i = 1, 2, \ldots, q, \quad i \neq j \neq k$$

then

$$\frac{\text{SSR}}{\sigma^2} \sim \chi^2_{(p-1)} \tag{7.23}$$

and SSR/σ^2 is independent of SSE/σ^2. Hence the ratio

$$F = \frac{\text{SSR}/(p-1)}{\text{SSE}(N-p)} \tag{7.24}$$

has an F-distribution with $p - 1$ and $N - p$ degrees of freedom in the numerator and denominator, respectively. Computed values of the F-ratio in Eq. (7.24) can be compared with the tabled values of $F_{(p-1, N-p, \alpha)}$ to test the significance of the regression on $x_1, x_2, \ldots, x_q$ at the α level.

Occasionally we will want to test a hypothesis of the form $\mathbf{C}\boldsymbol{\beta} = \mathbf{m}$, where $\mathbf{C}$ is an $r \times p$ matrix and $\mathbf{m}$ is a $r \times 1$ vector of constants. The rows of $\mathbf{C}\boldsymbol{\beta}$ are linearly independent estimable functions, so that $\mathbf{C}$ has rank $r (r \leq p)$. For example, if the elements of the parameter vector are $\boldsymbol{\beta} = (\beta_1, \beta_2, \beta_3, \beta_{12}, \beta_{13}, \beta_{23})'$ and we want to test the hypothesis that $\beta_1 = \beta_2$ and $\beta_{12} = \beta_{23}$, we might have for the matrix $\mathbf{C}$ and the vector $\mathbf{m}$:

$$\mathbf{C}\boldsymbol{\beta} = \mathbf{m}$$

$$\begin{bmatrix} 1 & -1 & 0 & 0 & 0 & 0 \\ 0 & 0 & 0 & 1 & 0 & -1 \end{bmatrix} \boldsymbol{\beta} = \begin{bmatrix} 0 \\ 0 \end{bmatrix}$$

The test of the hypothesis $\mathbf{C}\boldsymbol{\beta} = \mathbf{m}$ requires first defining the complete model as $\mathbf{y} = \mathbf{X}\boldsymbol{\beta} + \boldsymbol{\varepsilon}$ and a reduced model $\mathbf{y} = \tilde{\mathbf{X}}\tilde{\boldsymbol{\beta}} + \boldsymbol{\varepsilon}$ where $\boldsymbol{\beta}$ is changed into $\tilde{\boldsymbol{\beta}}$ by the conditions of the hypothesis and $\tilde{\mathbf{X}}$ is the form of $\mathbf{X}$ corresponding to $\boldsymbol{\beta}$. Next the sum of squares due to fitting the complete model and that due to fitting the reduced model are obtained. These are

$$\mathrm{SSR}_{\mathrm{complete}} = \mathbf{b}'\mathbf{X}'\mathbf{y} - \frac{(\mathbf{1}'\mathbf{y})^2}{N} \quad \text{with } p - 1 \text{ degrees of freedom}$$

$$\mathrm{SSR}_{\mathrm{reduced}} = \tilde{\mathbf{b}}'\tilde{\mathbf{X}}'\mathbf{y} - \frac{(\mathbf{1}'\mathbf{y})^2}{N} \quad \text{with } p - 1 \text{ degrees of freedom} \tag{7.25}$$

where $\mathbf{b}$ and $\tilde{\mathbf{b}}$ are the estimators of $\boldsymbol{\beta}$ and $\tilde{\boldsymbol{\beta}}$, respectively. The test of $\mathbf{C}\boldsymbol{\beta} = \mathbf{m}$ involves the difference

$$F = \frac{[\mathrm{SSR}_{\mathrm{complete}} - \mathrm{SSR}_{\mathrm{reduced}}]/r}{\mathrm{SSE}_{\mathrm{complete}}/(N - p)} \tag{7.26}$$

or

$$F = \frac{[\mathrm{SSE}_{\mathrm{reduced}} - \mathrm{SSE}_{\mathrm{complete}}]/r}{\mathrm{SSE}_{\mathrm{complete}}/(N - p)} \tag{7.27}$$

where $\mathrm{SSE}_{\mathrm{complete}} = \mathbf{y}'\mathbf{y} - \mathbf{b}'\mathbf{X}'\mathbf{y}$ and $\mathrm{SSE}_{\mathrm{reduced}} = \mathbf{y}'\mathbf{y} - \tilde{\mathbf{b}}'\tilde{\mathbf{X}}'\mathbf{y}$. Comparing the computed F-ratio with tabulated values of $F_{(r, N-p, \alpha)}$ provides the test

A more readily obtainable numerator for the F-ratio in (7.26) for testing $\mathbf{C}\boldsymbol{\beta} = \mathbf{m}$ is possible. A proof of a theorem in Searle (1966, ch. 10), reveals that the F-ratio can be written as

$$F = \frac{(\mathbf{C}\mathbf{b} - \mathbf{m})'[\mathbf{C}(\mathbf{X}'\mathbf{X})^{-1}\mathbf{C}']^{-1}(\mathbf{C}\mathbf{b} - \mathbf{m})}{r\{\mathrm{SSE}_{\mathrm{complete}}/(N - p)\}} \tag{7.28}$$

where $\mathbf{Cb}$ is the estimator of $\mathbf{C\beta}$. Hence, once $\mathbf{C}$ and $\mathbf{m}$ are specified, only the complete model estimates are necessary for the calculations of F in Eq. (7.28).

Let us test the hypothesis $H_0: \beta_{12} = \beta_{13} = \beta_{23} = 0$ versus H_A: *One or more equality signs is false*, using the fish patty data in Table 7.2, having fitted the second-degree model (7.13). The matrix $\mathbf{C}$ and vector $\mathbf{m}$ are

$$\mathbf{C} = \begin{bmatrix} 0 & 0 & 0 & 1 & 0 & 0 \\ 0 & 0 & 0 & 0 & 1 & 0 \\ 0 & 0 & 0 & 0 & 0 & 1 \end{bmatrix}, \qquad \mathbf{m} = \begin{bmatrix} 0 \\ 0 \\ 0 \end{bmatrix}$$

The complete model is the second-degree equation (7.13) and $\text{SSE}_{\text{complete}} = 0.323$ with $N - p = 8$ degrees of freedom.

The numerator of the F-statistic in Eq. (7.28) is

$$(\mathbf{Cb} - \mathbf{m})'[\mathbf{C}(\mathbf{X'X})^{-1}\mathbf{C'}]^{-1}[\mathbf{Cb} - \mathbf{m}]$$

$$= \begin{bmatrix} b_{12} \\ b_{13} \\ b_{23} \end{bmatrix}' [10\mathbf{I} + 0.485\mathbf{J}]^{-1} \begin{bmatrix} b_{12} \\ b_{13} \\ b_{23} \end{bmatrix}$$

$$= (0.21, -0.31, -0.25)\left[\frac{1}{10}\mathbf{I} - \frac{0.485}{114.55}\mathbf{J}\right]\begin{bmatrix} 0.21 \\ -0.31 \\ -0.25 \end{bmatrix} = 0.02$$

and therefore the value of the F-statistic is

$$F = \frac{0.02}{3(0.04)} = 0.16$$

Since the tabled value $F_{(3,8,\alpha=0.05)} = 4.07$, we do not reject $H_0: \beta_{12} = \beta_{13} = \beta_{23} = 0$. The use of the residual sums-of-squares quantities for the complete and reduced models, where the reduced model is the three-term first-degree model, according to Eq. (7.27), produces

$$F = \frac{[\text{SSE}_{\text{reduced}} - \text{SSE}_{\text{complete}}]/r}{\text{SSE}_{\text{complete}}/(N - p)}$$

$$= \frac{[0.343 - 0.323]/3}{0.323/8} = 0.16$$

This value must be identical to the value of the previously calculated F.

REFERENCES AND RECOMMENDED READING

Cornell, J. A., and R. D. Berger (1987). Factors that influence the value of the coefficient of determination in simple linear and nonlinear regression models. *Phytopathology*, **77**, 63–70.

Draper, N. R., and H. Smith (1981). *Applied Regression Analysis*, 2nd ed. Wiley, New York.

Bibliography

Adeyeye, A. D., and F. A. Oyawale (2008). Mixture experiments and their applications in welding flux design. *J. Brazilian Soc. Mech. Sci. Eng.*, **30**, 319–326.

Aggarwal, M. L., V. Sarin, and P. Singh (2002). Optimal designs in two blocks for Becker's mixture models in three and four components. *Stat. Prob. Lett.*, **59**, 385–396.

Agreda, C. L., and V. H. Agreda (1989). Designing mixture experiments. *CHEMTECH*, **19**, 573–575.

Akay, K. U. (2007). A note on model selection in mixture experiments. *J. Math. Stat.*, **3**, 93–99.

Aitchison, J. (1982). The statistical analysis of compositional data (with discussion). *J. Roy. Stat. Soc.*, **B44**, 139–147.

Aitchison, J. (1985). A general class of distributions on the simplex. *J. Roy. Stat. Soc.*, **B47**, 136–146.

Aitchison, J. (1986). *The Statistical Analysis of Compositional Data.* Chapman and Hall, London.

Aitchison, J., and J. Bacon-Shone (1984). Log contrast models for experiments with mixtures. *Biometrika*, **71**, 323–330.

Aitkin, M. A. (1974). Simultaneous inference and the choice of variable subsets in multiple regression. *Technometrics*, **16**, 221–227.

Andere-Rendon, J., D. C. Montgomery, and D. A. Rollier (1997). Design of mixture experiments using Bayesian *D*-optimality. *J. Qual. Technol.*, **29**, 451–463.

Anderson, M. J., and P. J. Whitcomb (1996). Optimization of paint formulations made easy with computer-aided design of experiments for mixtures. *J. Coat. Technol.*, **68**, 71–75.

Anderson, M. J., and P. J. Whitcomb (2009). Making use of mixture design to optimize olive oil—a case study. *Chem. Process Indus. Divi. Newslett.*, Fall, pp. 5–10.

Anderson, V. L., and R. A. McLean (1974). *Design of Experiments: A Realistic Approach.* Dekker, New York.

A Primer on Experiments with Mixtures, By John A. Cornell
Copyright © 2011 John Wiley & Sons, Inc. Published by John Wiley & Sons, Inc.

Anderson-Cook, C. M., H. Goldfarb, C. M., Borror, D. C. Montgomery, K Canter, and J. Twist (2004). Mixture and mixture-process variable experiments for pharmaceutical applications. *Pharmaceut. Stat.*, **3**, 247–260.

Anik, S. T., and L. Sukumar (1981). Extreme vertices design in formulation development: solubility of butoconazole nitrate in a multicomponent system. *J. Pharmaceut. Sci.*, **70**, 897–900.

Atkinson, A. C., and A. N. Donev (1996). *Optimum Experimental Designs*. Clarendon Press, Oxford.

Bayne, C. K., and C. Y. Ma (1987). Optimization of solvent composition for high performance thin-layer chromatography. *J. Liq. Chromatogr.*, **10**, 3529–3546.

Becker, N. G. (1968). Models for the response of a mixture. *J. Roy. Stat. Soc.*, **B30**, 349–358.

Becker, N. G. (1969). Regression problems when the predictor variables are proportions. *J. Roy. Stat. Soc.*, **B31**, 107–112.

Becker, N. G. (1970). Mixture designs for a model linear in the proportions. *Biometrika*, **57**, 329–338.

Becker, N. G. (1978). Models and designs for experiments with mixtures. *Austral. J. Stat.*, **20**, 195–208.

Belloto, R. J. Jr., A. M. Dean, M. A. Moustafa, A. M. Molokhia, M. W. Gouda, and T. D. Sokoloski (1985). Statistical techniques applied to solubility predictions and pharmaceutical formulations: an approach to problem solving using mixture response surface methodology. *Int. J. Pharmaceut.*, **23**, 195–207.

Berliner, L. M. (1987). Bayesian control in mixture models. *Technometrics*, **29**, 455–460.

Bjorkestol, K., and T. Naes (2005). A discussion of alternative ways of modeling and interpreting mixture data. *Qual. Eng.*, **17**, 509–533.

Borkowski, J. J., and G. F. Piepel (2009). Uniform designs for highly-constrained mixture experiments. *J. Qual. Technol.*, **41**, 35–47.

Bohl, A. H. (1988). A formulation tool. *CHEMTECH*, **18**, 158–163.

Bownds, J. M., I. S. Kurotori, and D. R. Cruise (1965). Notes on simplex models in the study of multi-component mixtures, *U.S. Naval Ordinance Test Station, NAVWEPS* Report 8670, NOTS TP 3719, Copy 197, China Lake, CA, January, pp. 1–40.

Box, G. E. P. (1952). Multi-factor designs of first order. *Biometrika*, **39**, pp. 49–57.

Box, G. E. P. (1982). Choice of response surface design and alphabetic optimality. *Utilitas Math.*, **B21**, 11–55.

Box, G. E. P., and N. R. Draper (1959). A basis for the selection of a response surface design. *J. Am. Stat. Assoc.*, **54**, 622–654.

Box, G. E. P., and C. J. Gardiner (1966). Constrained designs — Part 1, First order designs. Technical Report 89, Department of Statistics, University of Wisconsin, Madison, pp. 1–14.

Box, G. E. P., and J. S. Hunter (1957). Multi-factor experimental designs for exploring response surfaces. *Ann. Math. Stat.*, **28**, 195–242.

Box, G. E. P., W. G. Hunter, and J. S. Hunter (1978). *Statistics for Experimenters: An Introduction to Design, Data Analysis, and Model Building*. Wiley, New York.

Cain, M., and M. L. R. Price (1986). Optimal mixture choice. *App. Stat.*, **35**, 1–17.

Chan, L. Y. (1988). Optimal design for a linear log contrast model for experiments with mixtures. *J. Stat. Plan. Infer.*, **20**, 105–113.

Chan, L. Y. (1992). *D*-optimal design for a quadratic log contrast model for experiments with mixtures. *Comm. Stat. Theory Meth.*, **21**, 2909–2930.

Chan, L. Y. (2000). Optimal designs for experiments with mixtures: a survey. *Comm. Stat. Theory Meth.*, **29**, 2281–2312.

Chan, L. Y., Y. N. Guan, and C. Q. Zhang (1998). *A*-Optimal designs for an additive quadratic mixture model. *Stat. Sinica*, **8**, 979–990.

Chan, L. Y., J. H. Meng, Y. C. Jiang, and Y. N. Guan (1998). *D*-Optimal axial designs for quadratic and cubic additive mixture models. *Aust. N. Z. J. Stat.,* **40**, 359–371.

Chan, L. Y., and M. K. Sandhu (1999). Optimal orthogonal block designs for a quadratic mixture model for three components. *J. Appl. Stat.*, **26**, 19–34.

Chan, L. Y., and Y. N. Guan (2001). *A*- and *D*-optimal designs for a log contrast model for experiments with mixtures. *J. Appl. Stat.*, **28**, 537–546.

Chantarat, N., T. T. Allen, N. Ferhatosmanoglu, and M. Bernshteyn (2006). A combined array approach to minimize expected prediction errors in experimentation involving mixture and process variables. *Int. J. Indust. Syst. Eng.*, **1**, 129–147.

Charnet, R., and R. J. Beaver (1988). Design and analysis of paired comparison experiments involving mixtures. *J. Stat. Plan. Infer.*, **20**, 91–103.

Chen, J. J., L.-A. Li, and C. D. Jackson (1996). Analysis of quantal response data from mixture experiments. *Environmetrics*, **7**, 503–512.

Chick, L. A., and G. F. Piepel (1984). Statistically designed optimization of a glass composition. *J. Am. Ceramic Soc.*, **67**, 763–768.

Chick, L. A., G. F. Piepel, W. J. Gray, G. B. Mellinger, and B. O. Barnes (1980). Statistically designed study of a nuclear waste glass system. In D. J. M. Northrup Jr., ed., *Scientific, Basis for Nuclear Waste Management*, Vol. **2**. Plenum Press, New York, pp. 175–181.

Claringbold, P. J. (1955). Use of the simplex design in the study of the joint action of related hormones. *Biometrics*, **11**, 174–185.

Cornell, J. A. (1971). Process variables in the mixture problem for categorized components. *J. Am. Stat. Assoc.*, **66**, 42–48.

Cornell, J. A. (1973). Experiments with mixtures: a review. *Technometrics*, **15**, 437–455.

Cornell, J. A. (1975). Some comments on designs for Cox's mixture polynomial. *Technometrics*, **17**, 25–35.

Cornell, J. A. (1977). Weighted versus unweighted estimates using Scheffé's mixture model for symmetrical error variances patterns. *Technometrics*, **19**, 237–247.

Cornell, J. A. (1979). Experiments with mixtures: an update and bibliography. *Technometrics*, **21**, 95–106.

Cornell, J. A. (1981, 1992, 2002). *Experiments with Mixtures: Designs, Models, and the Analysis of Mixture Data*, 1st ed., 2nd ed., and 3rd ed. Wiley, New York.

Cornell, J. A. (1985). Mixture experiments. In *Encyclopedia of Statistical Sciences*, Vol. **5**. Wiley, New York, pp. 569–579.

Cornell, J. A. (1986). A comparison between two ten-point designs for studying three-component mixture systems. *J. Qual. Technol.*, **18**, 1–15.

Cornell, J. A. (1988). Analyzing data from mixture experiments containing process variables: a split-plot approach. *J. Qual. Technol.*, **20**, 2–23.

Cornell, J. A. (1990a). The ASQC basic references in quality control: statistical techniques. In *How to Run Mixture Experiments for Product Quality*, 2nd ed., Vol. **5**. American Society for Quality Control, Milwaukee, WI, pp. 1–71.

Cornell, J. A. (1990b). Embedding mixture experiments inside factorial experiments. *J. Qual. Technol.*, **22**, 265–276.

Cornell, J. A. (1990c). Mixture experiments. In S. Ghosh, ed. *Statistical Design and Analysis of Industrial Experiments*. Dekker, New York, pp. 175–209.

Cornell, J. A. (1991a). Mixture designs for product improvement. *Comm. Stat. Theory Meth.*, **20**, 391–416.

Cornell, J. A. (1991b). The fitting of Scheffé-type models for estimating solubilities of multisolvent systems. *J. Biopharmaceut. Stat.*, **1**, 303–329.

Cornell, J. A. (1993). Saving money with a mixture experiment. *ASQC Stat. Div. Newslet.*, **14**, 11–12.

Cornell, J. A. (1995). Fitting models to data from mixture experiments containing other factors. *J. Qual. Technol.*, **27**, 13–33.

Cornell, J. A. (2000). Fitting a slack-variable model to mixture data: some questions raised. *J. Qual. Technol.*, **32**, 133–147.

Cornell, J. A., and I. J. Good (1970). The mixture problem for categorized components. *J. Am. Stat. Assoc.*, **65**, 339–355.

Cornell, J. A., and L. Ott (1975). The use of gradients to aid in the interpretation of mixture response surfaces. *Technometrics*, **17**, 409–424.

Cornell, J. A., and J. W. Gorman (1978). On the detection of an additive blending component in multicomponent mixtures. *Biometrics*, **34**, 251–263.

Cornell, J. A., and A. I. Khuri (1979). Obtaining constant prediction variance on concentric triangles for ternary mixture systems. *Technometrics*, **21**, 147–157.

Cornell, J. A., and J. C. Deng (1982). Combining process variables and ingredient components in mixing experiments. *J. Food Sci.*, **47**, 836–843, 848.

Cornell, J. A., J. T. Shelton, R. Lynch, and G. F. Piepel (1983). Plotting three-dimensional response surfaces for three-component mixtures or two-factor systems. *Bulletin 836, Agricultural Experiment Stations*. Institute of Food and Agricultural Sciences, University of Florida, Gainesville, pp. 1–31.

Cornell, J. A., and J. W. Gorman (1984). Fractional design plans for process variables in mixture experiments. *J. Qual. Technol.*, **16**, 20–38.

Cornell, J. A., and R. D. Berger (1987). Factors that influence the value of the coefficient of determination in simple linear and nonlinear regression models. *Phytopathology*, **77**, 63–70.

Cornell, J. A., and S. B. Linda (1991). Models and designs for experiments with mixtures. Part I: Exploring the whole simplex region. *Bulletin 879, Agricultural Experiment Station*. Institute of Food and Agricultural Sciences, University of Florida, Gainesville, pp. 1–49.

Cornell, J. A., and D. C. Montgomery (1996). Interaction models as alternatives to low-order polynomials. *J. Qual. Technol.*, **28**, 163–176.

Cornell, J. A., and J. M. Harrison (1997). Models and designs for experiments with mixtures. Part II: Exploring a subregion of the simplex and the inclusion of other factors in mixture experiments. Bulletin 899, Agricultural Experiment Station. Institute of Food and Agricultural Sciences, University of Florida, Gainesville, pp. 1–78.

Cornell, J. A., and P. J. Ramsey (1998). A generalized mixture model for categorized-component problems with an application to a photoresist-coating experiment. *Technometrics*, **40**, 48–61.

Cornell, J. A., and J. W. Gorman (2003). Two new mixture models: living with collinearity but removing its influence. *J. Qual. Technol.*, **35**, 78–88.

Cox, D. R. (1971). A note on polynomial response functions for mixtures. *Biometrika*, **58**, 155–159.

Crosier, R. B. (1984). Mixture experiments: geometry and pseudo-components. *Technometrics*, **26**, 209–216.

Crosier, R. B. (1986). The geometry of constrained mixture experiments. *Technometrics*, **28**, 95–102.

Crosier, R. B. (1991). Symmetry in mixture experiments. *Comm. Stat. Theory Meth.*, **20**, 1911–1935.

Cruise, D. R. (1966). Plotting the composition of mixtures on simplex coordinates. *J. Chem. Educ.*, **43**, 30–33.

Czitrom, V. (1988). Mixture experiments with process variables: *D*-optimal orthogonal experimental designs. *Comm. Stat. Theory Meth.*, **17**, 105–121.

Czitrom, V. (1989). Experimental design for four mixture components with process variables. *Comm. Stat. Theory Meth.*, **18**, 4561–4581.

Czitrom, V. (1992). Note on a mixture experiment with process variables. *Comm. Stat., Pt B: Simul. Comput.*, **21**, 493–498.

Daniel, C. (1963). Discussion on Professor Scheffé's paper. *J. Roy. Stat. Soc.*, **B25**, 256–257.

Daniel, C., and F. S. Wood (1971). *Fitting Equations to Data. Computer Analysis of Multifactor Data for Scientists and Engineers*. Wiley, New York.

Darroch, J. N., and J. Waller (1985). Additivity and interaction in three-component experiments with mixtures. *Biometrika*, **72**, 153–163.

Debets, H. J. G. (1985). Optimization methods for HPLC. *J. Liq. Chromatogr.*, **8**, 2725–2780.

Del Castillo, E., and D. C. Montgomery (1993). A nonlinear programming solution to the dual response problem. *J. Qual. Technol.*, **25**, 199–204.

Derringer, G. C. (1994). A balancing act: optimizing a product's properties. *Qual. Progr.*, **27**, 51–58.

Derringer, G. C., and R. Suich (1980). Simultaneous optimization of several response variables. *J. Qual. Technol.*, **12**, 214–219.

Diamond, W. J. (1967). Three dimensional models of extreme vertices designs for four component mixtures. *Technometrics*, **9**, 472–475.

Diamond, W. J. (1981). *Practical Experiment Designs for Engineers and Scientists*. Lifetime Learning Publications, Belmont, CA, pp. 263–274.

Donelson, D. H., and J. T. Wilson (1960). Effect of the relative quantity of flour fractions on cake quality. *Cereal Chem.*, **37**, 241–262.

Donev, A. N. (1989). Design of experiments with both mixture and qualitative factors. *J. Roy. Stat. Soc.*, **B51**, 297–302.

Draper, N. R., and W. E. Lawrence (1965a). Mixture designs for three factors. *J. Roy. Stat. Soc.*, **B27**, 450–465.

Draper, N. R., and W. E. Lawrence (1965b). Mixture designs for four factors. *J. Roy. Stat. Soc.*, **B27**, 473–478.

Draper, N. R., and A. M. Herzberg (1971). On lack of fit. *Technometrics*, **13**, 231–241.

Draper, N. R., and R. C. St. John (1977a). A mixtures model with inverse terms. *Technometrics*, **19**, 37–46.

Draper, N. R., and R. C. St. John (1977b). Designs in three and four components for mixtures models with inverse terms. *Technometrics*, **19**, 117–130.

Draper, N. R., and H. Smith (1981). *Applied Regression Analysis*, 2nd ed. Wiley, New York.

Draper, N. R., P. Prescott, S. M. Lewis, A. M. Dean, P. W. M. John, and M. G. Tuck (1993). Mixture designs for four components in orthogonal blocks. *Technometrics*, **35**, 268–276. Correction in *Technometrics*, 1994, **36**, 234.

Draper, N. R., and F. Pukelsheim (1997). Mixture models based on homogeneous polynomials. *J. Stat. Plan. Infer.*, **71**, 303–311.

Draper, N. R., and F. Pukelsheim (1999). Kiefer ordering of simplex designs for first- and second-degree mixture models. *J. Stat. Plan. Infer.*, **79**, 325–348.

Draper, N. R., and F. Pukelsheim (2002). Generalized ridge analysis under linear restrictions, with particular applications to mixture experiments problems. *Technometrics*, **44**, 250–259.

Dingstad, G., B. Egelandsdal, and T. Naes (2003). Modeling methods for crossed mixtures experiments—a case study from sausage production. *Chemometr. Intell. Lab. Syst.*, **66**, 175–190.

Drew, B. A. (1967). Experiments with mixtures. *Minnesota Chem.*, **19**, 4–9.

Duineveld, C. A. A., A. K. Smilde, and D. A. Doornbos (1993). Comparison of experimental designs combining process and mixture variables. Parts I and II. Design construction and theoretical evaluation. *Chemometr. Intell. Lab. Syst.*, **19**, 295–308.

Farrand, E. A. (1969). Starch damage and alpha-amylase as bases for mathematical models relating to flour-water absorption. *Cereal Chem.*, **46**, 103–116.

Galil, Z., and J. Kiefer (1977a). Comparison of Box–Draper and *D*-optimum designs for experiments with mixtures. *Technometrics*, **19**, 441–444.

Galil, Z., and J. Kiefer (1977b). Comparison of simplex designs for quadratic mixture models. *Technometrics*, **19**, 445–453.

Glajch, J. L., J. J. Kirkland, K. M. Squire, and J. M. Minor (1980). Optimization of solvent strength and selectivity for reversed-phase liquid chromatography using an interactive mixture-design statistical technique. *J. Chromatogr.*, **199**, 57–79.

Glajch, J. L., J. J. Kirkland, and L. R. Snyder (1982). Practical optimization of solvent selectivity in liquid-solid chromatography using a mixture-design statistical technique. *J. Chromatogr.*, **238**, 269–280.

Goel, B. S. (1980a). Systematic designs for experiments with mixtures. *Biometr. J.*, **22**, 345–194.

Goel, B. S. (1980b). Designs for restricted exploration in mixture experiments. *Biometr. J.*, **22**, 351–358.

Goel, B. S., and A. K. Nigam (1979). Sequential exploration in mixture experiments. *Biometr. J.*, **21**, 277–285.

Goldfarb, H. B., C. M. Borror, and D. C. Montgomery (2003). Mixture-process variable experiments with noise variables. *J. Qual. Technol.*, **36**, 393–405.

Goldfarb, H. B., C. M. Borror, D. C. Montgomery, and C. M. Anderson-Cook (2004). Three-dimensional variance dispersion graphs for mixture-process Experiments. *J. Qual. Technol.*, **36**, 109–124.

Goldfarb, H. B., C. M. Anderson-Cook, C. M. Borror, and D. C. Montgomery (2004). Fraction of design space plots for assessing mixture and mixture-process designs. *J. Qual. Technol.*, **36**, 169–179.

Goldfarb, H. B., C. M. Borror, D. C. Montgomery, and C. M. Anderson-Cook (2004). Evaluating mixture-process designs with control and noise variables. *J. Qual. Technol.*, **36**, 245–262.

Goldfarb, H. B., C. M. Borror, D. C. Montgomery, and C. M. Anderson-Cook (2005). Using genetic algorithms to generate mixture-process experimental designs involving control and noise variables. *J. Qual. Technol.*, **37**, 60–74.

Goldfarb, H. B., and D. C. Montgomery (2006). Graphical methods for comparing response surface designs for experiments with mixture components. In A. I. Khuri, ed., *Response Surface Methodology and Related Topics*. World Scientific Press, Singapore, pp. 329–348.

Goos, P., and M. Vandebroek (2001). How to relax inconsistent constraints in a mixture experiment. *Chemometr. Intell. Lab. Syst.*, **55**, 147–149.

Goos, P., and A. N. Donev (2006). The *D*-optimal design of blocked experiments with mixture components. *J. Qual. Technol.*, **38**, 319–332.

Goos, P., and A. N. Donev (2007). *D*-optimal minimum support mixture designs in blocks. *Metrika*, **65**, 53–68.

Gorman, J. W. (1966). Discussion of extreme vertices designs of mixture experiments by R. A. McLean and V. L. Anderson. *Technometrics*, **8**, 455–456.

Gorman, J. W. (1970). Fitting equations to mixture data with restraints on compositions. *J. Qual. Technol.*, **2**, 186–194.

Gorman, J. W., and J. E. Hinman (1962). Simplex-lattice designs for multicomponent systems. *Technometrics*, **4**, 463–487.

Gorman, J. W., and J. A. Cornell (1982). A note on model reduction for experiments with both mixture components and process variables. *Technometrics*, **24**, 243–247.

Gorman, J. W., and J. A. Cornell (1985). A note on fitting equations to freezing point data exhibiting eutectics for binary and ternary mixture systems. *Technometrics*, **27**, 229–239.

Gous, R. M., and H. K. Swatson (2000). Mixture experiments: a severe test of the ability of a broiler chicken to make the right choice. *Brit. Poultry Sci.*, **41**, 136–140.

Graybill, F. A. (1961). *An Introduction to Linear Statistical Models*. McGraw-Hill, New York.

Green, J. R. (1971). Testing departure from a regression, without using replication. *Technometrics*, **13**, 609–615.

Hackler, W. C., W. W. Kriegel, and R. J. Hader (1956). Effect of raw-material ratios on absorption of whiteware compositions. *J. Am. Ceramic Soc.*, **39**, 20–25.

Hahn, G. J., W. Q. Meeker, and P. I. Feder (1976). The evaluation and comparison of experimental designs for fitting regression relationships. *J. Qual. Technol.*, **8**, 140–157.

Hamada, M., H. F. Martz, and S. Steiner (2005). Accounting for mixture errors in analyzing mixture experiments. *J. Qual. Technol.*, **37**, 139–148.

Hare, L. B. (1974). Mixture designs applied to food formulation. *Food Technol.*, **28**, 50–62.

Hare, L. B. (1979). Designs for mixture experiments involving process variables. *Technometrics*, **21**, 159–173.

Hare, L. B. (1985). Graphical display of the results of mixture experiments. In R. D. Snee, L. B. Hare, and R. Trout, eds., *Experiments in Industry: Design, Analysis, and Interpretation of Results*. ASQC Chemical and Process Industries Division, ASQC, Milwaukee, WI, pp. 99–109.

Hare, L. B., and P. L. Brown (1977). Plotting response surface contours for three-component mixtures. *J. Qual. Technol.*, **9**, 193–197.

Harris, I. A., and P. J. Bray (1984). B-nuclear magnetic resonance studies of borate glasses containing tellurium dioxide. *Phys. Chem. Glasses*, **25**, 44–51.

Heinsman, J. A., and D. C. Montgomery (1995). Optimization of a household product formulation using a mixture experiment. *Qual. Eng.*, **7**, 583–599.

Heiligers, B., and R. D. Hilgers (2003). A note on optimal mixture and mixture amount designs. *Stat. Sinica*, **13**, 709–725.

Henika, R. G., and M. R. Henselman (1969). Stable brew concentrate for whey cysteine batch breadmaking and methods. *Cereal Sci. Today*, **14**, 248–253.

Hesler K. K., and J. R. Lofstrom (1981). Application of simplex lattice design experimentation to coatings research. *J. Coat. Technol.*, **53**, 33–40.

Hewlett, P. S., and R. L. Plackett (1961). Models for quantal responses to mixtures of two drugs. In H. de Jonge, ed., *Quantitative Methods in Pharmacology*. North-Holland, Amsterdam, pp. 328–336

Hilgers, R. D., and P. Bauer (1995). Optimal designs for mixture amount experiments. *J. Stat. Plan. Infer.*, **48**, 241–246.

Hoerl, R. W. (1987). The application of ridge techniques to mixture data: ridge analysis. *Technometrics*, **29**, 161–172.

Huor, S. S., E. M. Ahmed, P. V. Rao, and J. A. Cornell (1980). Formulation and sensory evaluation of a fruit punch containing watermelon juice. *J. Food Sci.*, **45**, 809–813.

Jang, D. H., and G. J. Na (1996). A graphical method for evaluating mixture designs with respect to the slope. *Comm. Stat. Theory Meth.*, **25**, 1043–1058.

John, P. W. M. (1984). Experiments with mixtures involving process variables. Technical Report 8, Center for Statistical Sciences, University of Texas, Austin, pp. 1–17.

Kamoun, A., M. Chaabouni, M. Sergent, and R. Phan-Tan-Luu (2002). Mixture design applied to the formulation of hydrotopes for liquid detergents. *Chemometr. Intell. Lab. Syst.*, **63**, 69–79.

Kanjilal, P., S. K. Majumdan, and T. K. Pal (2004). Prediction of submerged arc weld-metal composition from flux ingredients with the help of statistical design of mixture experiment. *Scand. J. Metall.*, **33**, 146–159.

Kenworthy, O. O. (1963). Factorial experiments with mixtures using ratios. *Indust. Qual. Control*, **19**, 24–26.

Kettaneh-Wold, N. (1992). Analysis of mixture data with partial least squares. *Chemometr. Intell. Syst.*, **14**, 57–69.

Keviczky, L. (1970). The analysis of first and second order simplex designs. *Period. Polytech.*, **14**, 231–241.

Khuri, A. I. (2005). Slack-variable models versus Scheffe's mixture models. *J. Appl. Stat.*, **32**, 887–908.

Khuri, A. I., and J. A. Cornell (1996). *Response Surfaces: Designs and Analyses*, 2nd ed. Dekker, New York.

Khuri, A. I., J. M. Harrison, and J. A. Cornell (1999). Using quantile plots of the prediction variance for comparing designs for a constrained mixture region: an application involving a fertilizer experiment. *Appl. Stat.*, **49**, 521–532.

Kissell, L. T. (1959). A lean-formula cake method for varietal evaluation and research. *Cereal Chem.*, **36**, 168–175.

Kissell, L. T. (1967). Optimization of white layer cake formulations by a multiple-factor experimental design. *Cereal Chem.*, **44**, 253–268.

Kissell, L. T., and B. D. Marshall (1962). Multi-factor responses of cake quality to basic ingredient ratios. *Cereal Chem.*, **39**, 16–30.

Kojouharoff, V., and H. Ionchev (1986). Experimental-statistical method for investigation of multicomponent yttrium garnet systems. *J. Am. Ceramic Soc.*, **69**, 767–769.

Koons, G. F. (1989). Effect of sinter composition on emissions: a multi-component, highly constrained mixture experiment. *J. Qual. Technol.*, **21**, 261–267.

Koons, G. F., and R. H. Heasley (1981). Response surface contour plots for mixture problems. *J. Qual. Technol.*, **13**, 207–212.

Koons, G. F., and M. H. Wilt (1985). Design and analysis of an ABS pipe compound experiment. In R. D. Snee, L. B. Hare, and R. Trout, eds., *Experiments in Industry: Design, Analysis, and Interpretation of Results*. ASQC Chemical and Process Industries Division, ASQC, Milwaukee, WI, pp. 111–117.

Korn, E. L., and P. Lin (1983). Interactive effects of mixtures of stimuli in life table analysis. *Biometrika*, **70**, 103–110.

Kowalski, S., J. A. Cornell, and G. G. Vining (2000). A new model and class of designs for mixture experiments with process variables. *Comm. Stat. Theory Meth.*, **29**, 312–341.

Kowalski, S. M., J. A. Cornell, and G. G. Vining (2002). Split-plot designs and estimation methods for mixture experiments with process variables. *Technometrics*, **44**, 72–79.

Krol, L. H. (1966). Butadiene and isoprene rubber in giant tire treads. *Rubber Chem. Technol.*, **39**, 452–459.

Kurotori, I. S. (1966). Experiments with mixtures of components having lower bounds. *Indust. Qual. Control*, **22**, 592–596.

Laake, P. (1975). On the optimal allocation of observations in experiments with mixtures. *Scand. J. Stat.*, **2**, 153–157.

Lambrakis, D. P. (1968a). Estimated regression function of the $\{q, m\}$ simplex-lattice design. *Bull. Hellenic Math. Soc.*, **9**, 13–19.

Lambrakis, D. P. (1968b). Experiments with mixtures: a generalization of the simplex-lattice design. *J. Roy. Stat. Soc.*, **B30**, 123–136.

Lambrakis, D. P. (1969a). Experiments with p-component mixtures. *J. Roy. Stat. Soc.*, **B30**, 137–144.

Lambrakis, D. P. (1969b). Experiments with mixtures: an alternative to the simplex-lattice design. *J. Roy. Stat. Soc.*, **B31**, 234–245.

Lambrakis, D. P. (1969c). Experiments with mixtures: estimated regression function of the multiple-lattice design. *J. Roy. Stat. Soc.*, **B31**, 276–284.

Lee, H. H., and J. C. Warner (1935). The system biphenly-bibenzyl-naphthalene: nearly ideal binary and ternary systems. *J. Am. Chem. Soc.*, **37**, 318–321.

Leonpacker, R. M. (1978). The ethyl technique of octane prediction. Ethyl Corporation Petroleum Chemicals Division Report RTM-400. Ethyl Corporation, Baton Rouge, LA.

Lewis, S. M., A. M. Dean, N. R. Draper, and P. Prescott (1994). Mixture designs for q components in orthogonal blocks. *J. Roy. Stat. Soc.*, **B56**, 457–467.

Li, J. C. (1971). Design of experiments with mixtures and independent variable factors. Ph.D. dissertation. Department of Statistics, Rutgers University, New Brunswick, NJ.

Lim, Y. B. (1987). Symmetric D-optimal designs for log contrast models with mixtures. *J. Korean Stat. Soc.*, **16**, 71–79.

Lim, Y. B. (1990). D-optimal design for cubic polynomial regression on the q-simplex. *J. Stat. Plan. Infer.*, **25**, 141–152.

Liu, S., and H. Neudecker (1995). A V-optimal design for Scheffé's polynomial model. *Stat. Prob. Lett.*, **23**, 253–258.

Liu, S., and H. Neudecker (1997). Experiments with mixtures: optimal allocations for Becker's models. *Metrika*, **45**, 53–66.

MacDonald, I. A., and D. A. Bly (1966). Determination of optimal levels of several emulsifiers in cake mix shortenings. *Cereal Chem.*, **43**, 571–584.

Mage, I., and T. Naes (2005). Split-plot design for mixture experiments with process variables: a comparison of design strategies. *Chemometr. Intell. Lab. Sys.*, **78**, 81–95.

Mallows, C. L. (1973). Some comments on C_p. *Technometrics*, **15**, 661–675.

Mandal, N. K., and M. Pal (2008). Optimum mixture design using deficiency criterion. *Comm. Stat. Theory Meth.*, **37**, 1565–1575.

Marquardt, D. W. (1970). Generalized inverses, ridge regression, biased linear estimation and nonlinear estimation. *Technometrics*, **12**, 591–612.

Marquardt, D. W., and R. D. Snee (1974). Test statistics for mixture models. *Technometrics*, **16**, 533–537.

McLean, R. A., and V. L. Anderson (1966). Extreme vertices design of mixture experiments. *Technometrics*, **8**, 447–454.

Menezes, R. R., H. G. Malzac Neto, L.N. L. Santana, H. L. Lira, H. S. Ferreira, and G. A. Neves (2008). Optimization of wastes content in ceramic tiles using statistical design of mixture experiments. *J. Eur. Ceramic Soc.*, **28**, 3027–3039.

Mikaeili, F. (1988). Allocation of measurements in experiments with mixtures. *Keio Sci. Technol. Rep.*, **41**, 25–37.

Mikaeili, F. (1989). D-optimum design for cubic without 3-way effect on the simplex. *J. Stat. Plan. Infer.*, **21**, 107–115.

Mikaeili, F. (1993). D-optimum design for full cubic on q-simplex. *J. Stat. Plan. Infer.*, **35**, 121–130.

Minitab (1999). *User's Guide 2: Data Analysis and Quality Tools, Release 13*. Minitab, Inc., State College, PA.

Mitchell, T. J. (1974). An algorithm for construction of "D-optimal" experimental designs. *Technometrics*, **16**, 203–210.

Mendieta, E. J., H. N. Linssen, and R. Doornbos (1975). Optimal designs for linear mixture models. *Stat. Neerlandica*, **29**, 145–150.

Montgomery, D. C., and S. M. Voth (1994). Multicollinearity and leverage in mixture experiments. *J. Qual. Technol.*, **26**, 96–108.

Montgomery, D. C., E. A. Peck, and G. G. Vining (2001). *Introduction to Linear Regression Analysis*. 3rd ed. Wiley, New York.

Morgan, S. L., and C. H. Jacques (1978). Response surface evaluation and optimization in gas chromatography. *J. Chromatogr. Sci.*, **16**, 500–505.

Morris, W. E. (1975). The interaction approach to gasoline blending. Petrolum Chemicals Division Report AM-75–30. E. I. DuPont de Nemours and Co., Inc., Newark, DE.

Morris, W. E., and R. D. Snee (1979). Blending relationships among gasoline component mixtures. *Auto. Technolo.*, **3**, 56–61.

Moustafa, M. A., A. M. Molokhia, and M. W. Gouda (1981). Phenobarbital solubility in propylene glycol-glycerol-water systems. *J. Pharmaceu. Sci.*, **70**, 1172–1174.

Murty, J. S., and M. N. Das (1968). Design and analysis of experiments with mixtures. *Ann. Math. Stat.*, **39**, 1517–1539.

Murthy, M. S. R., and J. S. Murty (1982). A note on mixture designs derived from factorials. *J. Indian Soci. Agricul. Stats.*, **34**, 87–93.

Murthy, M. S. R., and J. S. Murty (1983). Restricted region simplex design for mixture experiments. *Comm. Stat. Theory Meth.*, **A12**, 2605–2615.

Murthy, M. S. R., and J. S. Murty (1989). Restricted region designs for multifactor mixture experiments. *Comm. Stat. Theory Meth.*, **18**, 1279–1295.

Murthy, M. S. R., and J. S. Murty (1993). Block designs for mixture experiments. *J. Indian Soc. Agricul. Stat.*, **44**, 55–71.

Murthy, M. S. R., and P. L. Manga (1996). Restricted region simplex designs for mixture experiments in the presence of process variables. *Sankhyā*, **58**, 231–239.

Muteki, K., and J. F. MacGregor (2007). Sequential design of mixture experiments for the development of new products. *J. Chemomet.*, **21**, 496–505.

Myers, R. H. (1964). Methods for estimating the composition of a three component liquid mixture. *Technometrics*, **6**, 343–356.

Myers, R. H. (1986). *Classical and Modern Regression with Applications*. Duxbury Press, Boston, MA.

Myers, R. H., D. C. Montgomery, and G. G. Vining (2002). *Generalized Linear Models*. Wiley, New York.

Naes, T., E. M. Faergestad, and J. A. Cornell (1998). A comparison of methods for analyzing data from a three-component mixture experiment in the presence of variation created by two process variables. *Chemometr. Intell. Lab. Syst.*, **41**, 221–235.

Narcy, J. P., and J. Renaud (1972). Use of simplex experimental design in detergent formulation. *J. Am. Oil Chem. Soc.*, **49**, 598–608.

Nardia, J. V., W. Acchar, and D. Hotzad (2004). Enhancing the properties of ceramic products through mixture design and response surface analysis. *J. Eur. Ceramic Soc.*, **24**, 375–379.

Nelson, C., E. V. Thomas, and P. R. Wengert (1986). Compositional quality control study of CON-2 glass. SAND-0430, Sandia National Laboratories, Albuquerque, NM.

Nguyen, N. K., and G. F. Piepel (2005). Computer generated experimental designs for irregular-shaped regions. *Qual. Technol. Quant. Manag.*, **2**, 147–160.

Nigam, A. K. (1970). Block designs for mixture experiments. *Ann. Math. Stat.*, **41**, 1861–1869.

Nigam, A. K. (1973). Multifactor mixture experiments. *J. Roy. Stat. Soc.*, **B35**, 51–66.

Nigam, A. K. (1974). Some designs and models for mixture experiments for the sequential exploration of response surfaces. *J. Indian Soc. Agricul. Stat.*, **26**, 120–124.

Nigam, A. K., S. C. Gupta, and S. Gupta (1983). A new algorithm for extreme vertices designs for linear mixture models. *Technometrics*, **25**, 367–371.

Ochsner, A. B., R. J. Belloto, and T. D. Sokoloski (1985). Prediction of xanthine solubilities using statistical techniques. *J. Pharmaceu. Sci.*, **74**, 132–135.

Oscik-Mendyk, B., and J. K. Rozylo (1987). Analysis of chromatographic parameters in the systems with ternary mobile phases. *J. Liq. Chromatogr.*, **10**, 1399–1415.

Ozol-Godfrey, A., C. M. Anderson-Cook, and D. C. Montgomery (2005). Fraction of design space plots for examining model robustness. *J. Qual. Technol.*, **37**, 223–235.

Paku, G. A., A. R. Manson, and L. A. Nelson (1971). Minimum bias estimation in the mixture problem. North Carolina State University Institute of Statistics Mimeo. Series No. 757. Raleigh, NC.

Park, S. H. (1978). Selecting contrasts among parameters in Scheffé's mixture models: screening components and model reduction. *Technometrics*, **20**, 273–279.

Pal, M., and N. K. Mandal (2008). Minimax designs for optimum mixtures. *Stat. Prob. Lett.*, **78**, 608–615.

Park, S. H., and J. H. Kim (1982). Axis-slope-rotatable designs for experiments with mixtures. *J. Korean Stat. Soc.*, **11**, 36–44.

Park, S. H., and J. I. Kim (1988). Slope-rotatable designs for estimating the slope of response surfaces in experiments with mixtures. *J. Korean Stat. Soc.*, **17**, 121–133.

Pearson, E. S., and H. O. Hartley (1972). *Biometrika Tables for Statisticians,* Vol. **2**. Cambridge University Press, Cambridge.

Piepel, G. F. (1982). Measuring component effects in constrained mixture experiments. *Technometrics*, **24**, 29–39.

Piepel, G. F. (1983a). Defining consistent constraint regions in mixture experiments. *Technometrics*, **25**, 97–101.

Piepel, G. F. (1983b). Calculating centroids in constrained mixture experiments. *Technometrics*, **25**, 279–283.

Piepel, G. F. (1988a). Programs for generating extreme vertices and centroids of linearly constrained experimental regions. *J. Qual. Technol.*, **20**, 125–139.

Piepel, G. F. (1988b). A note on models for mixture-amount experiments when the total amount takes a zero value. *Technometrics*, **30**, 49–50.

Piepel, G. F. (1990). Screening designs for constrained mixture experiments derived from classical screening designs. *J. Qual. Technol.*, **22**, 23–33.

Piepel, G. F. (1990). *MIXSOFT*^{TM} *Software for the Design and Analysis of Mixture Experiments, User's Guide, Version 2.0.* MIXSOFT — Mixture Experiment Software, Richland, WA.

Piepel, G. F. (1991). Screening designs for constrained mixture experiments derived from classical screening designs—an addendum. *J. Qual. Technol.*, **23**, 96–101.

Piepel, G. F. (1997). Survey of software with mixture experiment capabilities. *J. Qual. Technol.*, **29**, 76–85.

Piepel, G. F. (1999). Modeling methods for mixture-of-mixtures experiments applied to a table formulation problem. *Pharmaceut. Dev. Technol.*, **4**, 593–606.

Piepel, G. F. (2006). 50 Years of mixture experiment research: 1955–2004. In A. I. Khuri, ed., *Response Surface Methodology and Related Topics.* World Scientific Press, Singapore, pp. 281–327.

Piepel, G. F. (2006). A note comparing component-slope, Scheffe and Cox parameterizations of the linear mixture experiment model. *J. Appl. Stat.*, **33**, 397–403.

Piepel, G. F. (2007). A component slope linear model for mixture experiments. *Qual. Technol. Quant. Manag.*, **4**, 331–343.

Piepel, G. F., and J. A. Cornell (1985). Models for mixture experiments when the response depends on the total amount. *Technometrics*, **27**, 219–227.

Piepel, G. F., and J. A. Cornell (1987). Designs for mixture-amount experiments. *J. Qual. Technol.*, **19**, 11–28.

Piepel, G. F., G. B. Mellinger, and M. A. H. Reimus (1989). A statistical approach for characterizing chemical durability within a waste glass compositional region. *Nucl. Chem. Waste Manag.*, **26**, 3–11.

Piepel, G. F., and J. A. Cornell (1994). Mixture experiment approaches: examples, discussion, and recommendations. *J. Qual. Technol.*, **26**, 177–196.

Piepel, G. F., and T. Redgate (1996). Mixture experiment techniques for reducing the number of components applied to modeling waste glass sodium release. *J. Am. Ceramic Soc.*, **80**, 3038–3044.

Piepel, G. F., and T. Redgate (1998). A mixture experiment analysis of the Hald cement data. *Am. Stat.*, **52**, 23–30.

Piepel, G. F., and J. W. Shade (1992). In-situ vitrification and the effects of soil additives—a mixture experiment case study. *J. Am. Ceramic Soc.*, **75**, 112–116.

Piepel, G. F., R. D. Hicks, J. M. Szychowski, and J. L. Loeppky (2002). Methods for assessing curvature and interaction in mixture experiments. *Technometrics*, **44**, 161–172.

Piepel, G. F., J. M. Szychowski, and J. L. Loeppky (2002). Augmenting Scheffe linear mixture models with squared and/or crossproduct terms. *J. Qual. Technol.*, **34**, 297–14.

Piepel, G. F., S. K. Cooley, and B. Jones (2005). Construction of a 21-component layered mixture experiment design using a new mixture coordinate-exchange algorithm. *Qual. Eng.*, **17**, 579–594.

Piepel, G. F., and S. K. Cooley (2009). Automated method for reducing Scheffe's linear mixture experiment models. *Qual. Technol. Quant. Manag.*, **6**, 255–270.

Piepel, G. F., and S. M. Landmesser (2009). Mixture experiment alternatives to the slack variable approach. *Quality Engineering*, Vol. **21**, pp. 262–267.

Piepel, M. G., and G. F. Piepel (2000). How soil composition affects density and water capacity—a science fair project using mixture experiment methods. *STATS*, **28**, 14–20.

Prescott, P. (2004). Modeling in mixture experiments including interactions with process variables. *Qual. Technol. Quant. Manag.*, **1**, 87–103.

Prescott, P. (2000). Projection designs for mixture experiments in orthogonal blocks. *Comm. Stat. Theory Meth.*, **29**, 2229–2253.

Prescott, P., N. R. Draper, A. M. Dean, and S. M. Lewis (1993). Mixture designs for five components in orthogonal blocks. *J. Appl. Stat.*, **20**, 105–117.

Prescott, P., N. R. Draper, S. M. Lewis, and A. M. Dean (1997). Further properties of mixture designs for five components in orthogonal blocks. *J. Appl. Stat.*, **24**, 147–156.

Prescott, P., and N. R. Draper (1998). Mixture designs for constrained components in orthogonal blocks. *J. Appl. Stat.*, **25**, 613–638.

Prescott, P., A. M. Dean, N. R. Draper, and S. M. Lewis (2002). Mixture experiments: ill-conditioning and quadratic model specification. *Technometrics*, **44**, 260–268.

Prescott, P., and N. R. Draper (2004). Mixture component-amount designs via projections, including orthogonally blocked designs. *J. Qual. Technol.*, **36**, 413–431.

Prescott, P., and N. R. Draper (2009). Modeling in restricted mixture experiment spaces for three mixture components. *Qual. Technol. Quant. Manag.*, **6**, 207–217.

Price, W. L. (1977). A controlled random search procedure for global optimization. *Comput. J.*, **20**, 367–370.

Quenouille, M. H. (1953). *The Design and Analysis of Experiments.* Charles Griffin and Company, London.

Quenouille, M. H. (1959). Experiments with mixtures. *J. Roy. Stat. Soc.*, **21**, 201–202.

Rajagopal, R., E. Del Castillo, and J. J. Peterson (2005). Model and distribution-robust process optimization with noise factors. *J. Qual. Technol.*, **37**, 210–222.

Rao, P. (1971). Some notes on misspecification in regression. *Am. Stat.*, **25**, 37–39.

Reis, C., J. C. de Andrade, R. E. Bruns, and R. C. C. P. Moran (1998). Application of the split-plot, experimental design for the optimization of a catalytic procedure for the determination of Cr(VI). *Anal. Chimica Acta*, **369**, 269–279.

Rusin, M. (1975). The structure of nonlinear blending models. *Chem. Eng. Sci.*, **30**, 937–944.

Rusin, M. H., H. S. Chung, and J. F. Marshall (1981). A "transformation" method for calculating the research and motor octane numbers of gasoline blends. *I&EC Fund.*, **20**, 195–204.

Sahrmann, H. F., G. F. Piepel, and J. A. Cornell (1987). In search of the optimum Harvey Wallbanger recipe via mixture experiment techniques. *Am. Stat.*, **41**, 190–194.

Sahni, N. S., G. F. Piepel, and T. Naes (2009). Product and process improvement using mixture-process variable methods and robust optimization techniques. *J. Qual. Technol.*, **41**, 181–197.

SAS Institute (1982). *SAS/GRAPH User's Guide.* SAS Institute, Inc., Cary, NC.

SAS Institute (1982). *SAS User's Guide: Statistics*, SAS Institute, Inc., Cary, NC.

Saxena, S. K., and A. K. Nigam (1973). Symmetric-simplex block designs for mixtures. *J. Roy. Stat. Soc.*, **B35**, 466–472.

Saxena, S. K., and A. K. Nigam (1977). Restricted exploration of mixtures by symmetric-simplex designs. *Technometrics*, **19**, 47–52.

Scheffé, H. (1958). Experiments with mixtures. *J. Roy. Stat. Soc.*, **B20**, 344–360.

Scheffé, H. (1959). Reply to Mr. Quenouille's comments about my paper on mixtures. *J. Roy. Stat. Soc.*, **B23**, 171–172.

Scheffé, H. (1963). The simplex-centroid design for experiments with mixtures. *J. Roy. Stat. Soc.*, **B25**, 235–263.

Shelton, J. T., A. I. Khuri, and J. A. Cornell (1983). Selecting check points for testing lack of fit in response surface models. *Technometrics*, **25**, 357–365.

Shillington, E. R. (1979). Testing lack of fit in regression without replication. *Can. J. Stat.*, **7**, 137–146.

Singh, S. P., and M. Pratap (1985). Analysis of symmetric-simplex designs in mixture experiments. *Calcutta Stat. Assoc. Bull.*, **34**, 65–73.

Singh, S. P., M. Pratap, and M. N. Das (1982). Analysis of mixture experiments in presence of block effects. *Sankhyā*, **B44**, 270–277.

Smith, W. F. (2005). *Experimental Design for Formulation*. ASA-SIAM Series on Statistics and Applied Probability. SIAM, Philadelphia, ASA, Alexandria, VA.

Smith, W. F., and T. A. Beverly (1997). Generating linear and quadratic Cox mixture models. *J. Qual. Technol.*, **29**, 211–224.

Smith, W. F., and J. A. Cornell (1993). Biplot displays for looking at multiple response data in mixture experiments. *Technometrics*, **35**, 337–350.

Stat-Ease (1998). *DESIGN-EXPERT Software for Response Surface Methodology and Mixture Experiments. Version 5.0.* (1999) *Version 6.0.* Stat-Ease, Inc., Minneapolis, MN.

Snee, R. D. (1971). Design and analysis of mixture experiments. *J. Qual. Technol.*, **3**, 159–169.

Snee, R. D. (1973). Techniques for the analysis of mixture data. *Technometrics*, **15**, 517–528.

Snee, R. D. (1975a). Experimental designs for quadratic models in constrained mixture spaces. *Technometrics*, **17**, 149–159.

Snee, R. D. (1975b). Discussion of the use of gradients to aid in the interpretation of mixture response surfaces. *Technometrics*, **17**, 425–430.

Snee, R. D. (1979a). Experimental designs for mixture systems with multi-component constraints. *Comm. Stat. Theory Meth.*, **A8**, 303–326.

Snee, R. D. (1979b). Experimenting with mixtures. *CHEMTECH*, **9**, 702–710.

Snee, R. D. (1981). Developing blending models for gasoline and other mixtures. *Technometrics*, **23**, 119–130.

Snee, R. D. (1985). Computer-aided design of experiments: some practical experiences. *J. Qual. Technol.*, **17**, 222–236.

Snee, R. D., and D. W. Marquardt (1974). Extreme vertices designs for linear mixture models. *Technometrics*, **16**, 399–408.

Snee, R. D., and D. W. Marquardt (1976). Screening concepts and designs for experiments with mixtures. *Technometrics*, **18**, 19–29.

Snee, R. D., and A. A. Rayner (1982). Assessing the accuracy of mixture model regression calculations. *J. Qual. Technol.*, **14**, 67–79.

Sommerville, P. M. Y. (1958). *An Introduction to the Geometry of N Dimensions*. Methuen, London, 1929; Dover, New York, 1958.

Soo, H. M., E. H. Sander, and D. W. Kess (1978). Definition of a prediction model for determination of the effect of processing and compositional parameters on the textural characteristics of fabricated shrimp. *J. Food Sci.*, **43**, 1165–1171.

Sorenson, H., N. Cedergreen, I. M. Skovgaard, and J. C. Streibig (2007). An isobole-based statistical model and test for synergism/antagonism in binary mixture toxicity experiments. *Environ. Ecol. Stat.*, **14**, 383–397.

Steiner, S. H., and M. Hamada (1997). Making mixtures robust to noise and mixing measurement errors. *J. Qual. Technol.*, **29**, 441–450.

St. John, R. C. (1984). Experiments with mixtures, ill-conditioning, and ridge regression. *J. Qual. Technol.*, **16**, 81–96.

Stat-Ease, Inc. (2000). *Design-Expert 6 User's Guide*. Minneapolis, MN.

Tait, J. C., D. L. Mandolesi, and H. E. C. Rummens (1984). Viscosity of melts in the sodium borosilicate system. *Phys. Chem. Glasses*, **25**, 100–104.

Tang, M., J. Li, L. Y. Chan, and D. K. J. Lin (2004). Application of uniform design in the formation of cement mixtures. *Qual. Eng.*, **16**, 461–474.

Thompson, D. R. (1981). Designing mixture experiments—a review. *Trans. ASAE*, **24**, 1077–1086.

Thompson, W. O., and R. H. Myers (1968). Response surface designs for experiments with mixtures. *Technometrics*, **10**, 739–756.

Tiede, M. L. (1981). How to use SAS for ANOVA of mixture problems. *Proc. Sixth Annual SAS Users Group International Conf.* SAS Institute, Inc., Cary, NC.

Uranisi, H. (1964). Optimum design for the special cubic regression on the q-simplex. *Math. Rep.*, **1**, 7–12. General Education Department, Kyushu University.

Van Schalkwyk, D. J. (1971). On the design of mixture experiments. PhD dissertation. University of London.

Vining, G. G., J. A. Cornell, and R. H. Myers (1993). A graphical approach for evaluating mixture designs. *Appl. Stat.*, **42**, 127–138.

Vuchkov, I. N., H. A. Yonchev, D. L. Damgaliev, V. K. Tsochev, and T. D. Dikova (1978). *Catalogue of Sequentially Generated Designs.* Department of Automation Higher Institute of Chemical Technology, Sofia, Bulgaria (in Bulgarian and English).

Vuchkov, I. N., and H. A. Yonchev (1979). Design and analysis of experiments for the investigation of properties of mixtures and alloys. *Technika*, Sofia, Bulgaria (in Bulgarian).

Vuchkov, I. N., D. L. Damgaliev, and H. A. Yonchev (1981). Sequentially generated second-order quasi D-optimal designs for experiments with mixture and process variables. *Technometrics*, **23**, 233–238.

Vuchkov, I. N., H. A. Yonchev, and D. L. Damgaliev (1983). Continuous D-optimal designs for experiments with mixture and process variables. *Math. Operations-forschung Stat., Series Statistics*, **14**, 33–51.

Wagner, T. O., and J. W. Gorman (1962). The lattice method for design of experiments with fuels and lubricants. *Application of Statistics and Computers to Fuel and Lubricant Research Problems.* Department of the Army, San Antonio, TX, pp. 123–145.

Wagner, T. O., and J. W. Gorman (1963). Fuels, lubricants, engines, and experimental design. *Trans. Soc. Autom. Eng.*, **196**, 684–701.

Watson, G. S. (1969). Linear regression on proportions. *Biometrics*, **25**, 585–588.

Welch, W. J. (1982). Branch-and-bound search for experimental designs based on D-optimality and other criteria. *Technometrics*, **24**, 41–48.

Welch, W. J. (1983). A mean squared error criterion for the design of experiments. *Biometrika*, **70**, 205–213.

Welch, W. J. (1984). Computer-aided design of experiments for response estimation. *Technometrics*, **26**, 217–224.

Welch, W. J. (1985). ACED: algorithms for the construction of experimental designs. *Am. Stat.*, **39**, 146.

Weyland, J. W., C. H. P. Bruins, and D. A. Doornbos (1984). Use of three-dimensional minimum a-plots for optimization of mobile phase composition for RP-HPLC separation of sulfonamides. *J. Chromatogr. Sci.*, **22**, 31–39.

Wheeler, R. E. (1972). Efficient experimental design. Joint statistical meetings of the American Statistical Association, IMS, and Biometric Society, Montreal, August 1972.

White, D. B., H. K. Slocum, Y. Brun, C. Wrzosek, and W. R. Greco (2003). A new nonlinear mixture response surface paradigm for the study of synergism: a three drug example. *Curr. Drug Metab.*, **4**, 399–409.

White, D. B., H. M. Faessel, H. K. Slocum, L. Khinkis, and W. R. Greco (2004). Nonlinear response surface and mixture experiment methodologies applied to the study of synergism. *Biomet. J.*, **46**, 56–71.

Williams, N. A., and G. L. Amidon (1984). Excess free energy approach to the estimation of solubility in mixed solvent systems. I: Theory. *J. Pharmaceut. Sci.*, **73**, 9–13. II: Ethanol–water mixtures. *J. Pharmaceut. Sci.*, **73**, 14–18. III: Ethanol–propylene glycol–water mixtures *J. Pharmaceut. Sci.*, **73**, 18–23.

Wright, A. J. (1981). The analysis of yield-density relationships in binary mixtures using inverse polynomials. *J. Agri. Sci., Cambridge*, **96**, 561–567.

Wynn, H. P. (1970). The sequential generation of *D*-optimum experimental designs. *Ann. Math. Stat.*, **41**, 1655–1664.

Ying-nan, G. (1981). A computing method of the number of lateral sides and the coordinates of vertices of a mixture convex-polyhedron having lower and upper bounds. *J. Northeast Inst. Technol.*, **28**, 13–22 (in Chinese).

Ying-nan, G. (1983). On some optimalities of simplex-centroid design. Technical Report 5. Department of Management Science and Engineering, Tokyo Institute of Technology, Tokyo, Japan.

Ying-nan, G. (1984). A branch-constructing algorithm for extreme vertices of the mixture convex polyhedron and a modified CONSIM algorithm. *Acta Math. Appl. Sinica*, **7**, 340–346 (in Chinese).

Ying-nan, G. (1985). A branch-boundary algorithm of various dimension bound faces for the mixture convex hyper-polyhedron. *J. Northeast Inst. Technol.*, **2**, 71–76 (in Chinese).

Ying-nan, G. (1987). Division design for mixture convex polyhedra. *Acta Math. Appl. Sinica*, **10**, 68–73 (in Chinese).

Yonchev, H. A. (1982). Optimization of multicomponent systems. *Technika*, Sofia, Bulgaria (in Bulgarian).

Yonchev, H, A. (1985). Optimization of processes and compositions in petroleum processing industry. *Technika*, Sofia, Bulgaria (in Bulgarian).

Zemroch, P. J. (1986). Cluster analysis as an experimental design generator with application to gasoline blending experiments. *Technometrics*, **28**, 39–49.

Zhu, W. Y., C. Hu, and W. Chen (1986). The mixture models with logarithmic terms and their *D*-optimality. *Chinese J. Appl. Prob. Stat.*, **2**, 322–333 (in Chinese).

Zhu, W. Y., C. Hu, and W. Chen (1987a). *D*-Optimality and D_N-optimal designs for mixtures regression models with logarithmic terms. *Acta Math. Appl. Sinica*, **3**, 26–36 (in Chinese).

Zhu, W. Y., C. Hu, and W. Chen (1987b). *D*-Optimal and D_N-optimal designs in five and six components for mixture models with logarithmic terms. *Acta Math. Appl. Sinica*, **3**, 317–329 (in Chinese).

Answers to Selected Questions

CHAPTER 1

1.1. The constraints (1.1) and (1.2) on the component proportions are $x_i \geq 0, i = 1, 2, \ldots, q$ and $x_1 + x_2 + \ldots + x_q = 1$, respectively, implying the proportions are nonnegative and not independent. Furthermore, the shape of the factor space is that of a $(q-1)$-dimensional simplex, or if additional constraints in the form of upper and/or lower bounds are placed on some or all of the component proportions, the shape of the factor space is a subregion inside the $(q-1)$-dimensional simplex. Many scientists are of the opinion that when q is small, say $q \leq 5$ or 6, setting up experimental designs and fitting mixture models is not only easier, but with today's software packages, even highly constrained regions are explorable.

1.3. **(a)** Pure mixtures are single components.

    (b) Binary mixtures are blends of two components.

    (c) Ternary mixtures are three component blends.

      .

      .

    (q) complete mixtures are blends where all components are present.

1.4. Case 1   synergistic because $15 > 27/2 = 13.5$, the latter figure being $(A + B)/2$

    Case 2   additive since $15 = 30/2$

    Case 3   antagonistic since $4 < 12/2 = 6$

    Case 4   antagonistic since $12 < 30/2 = 15$

    Case 5   synergistic since $12 > 21/2 = 10.5$

                Fuels that are neither synergistic nor antagonistic are said to be additive.

A Primer on Experiments with Mixtures, By John A. Cornell
Copyright © 2011 John Wiley & Sons, Inc. Published by John Wiley & Sons, Inc.

1.6. (a) $b_{ij} = 4(\bar{y}_{ij}) - 2(\bar{y}_i + \bar{y}_j)$ where $\bar{y}_{ij}$ is the average of the mid-edge data and $\bar{y}_i$ and $\bar{y}_j$ are the averages at the two vertices at the ends of the edge.

$$b_{AB} = 4(11.95) - 2(10.6 + 9.8) = 47.8 - 40.8 = 7.0, \sqrt{\widehat{\text{Var}}(b_{AB})} = \sqrt{12 \, \text{MSE}} = 3.34$$

(b) $b_{AC} = 4(9.15) - 2(10.6 + 12.6) = 36.6 - 46.4 = -9.8$

(c) $b_{BC} = 4(6.2) - 2(9.8 + 12.6) = 24.8 - 44.8 = -20.0, \sqrt{\widehat{\text{Var}}(b_{BC})} = \sqrt{12 \, \text{MSE}} = 3.34$

(d) $b_{ABC} = 27 y_{ABC} - 12(\bar{y}_{AB} + \bar{y}_{AC} + \bar{y}_{BC}) + 3(\bar{y}_A + \bar{y}_B + \bar{y}_C)$
$$= 27(12.6) - 12(11.95 + 9.15 + 6.2) + 3(10.6 + 9.8 + 12.6)$$
$$= 111.6$$

$$\widehat{\text{Var}}(b_{ABC}) = 958.5 \, \text{MSE} = 958.5(0.928) = 889.49, \sqrt{\widehat{\text{Var}}(b_{ABC})} = 29.82$$

Test statistic: Compute $t_{\text{Cal}} = b_{ij}/\sqrt{\widehat{\text{Var}}(b_{ij})}$ and compare the absolute value of t_{Cal} against Appendix Table A entry $t_{(6 \text{ d.f.}, \, 0.025)} = 2.447$

The test of the null hypothesis $H_0 : \beta_{ij} = 0$ versus the alternative hypothesis $H_A : \beta_{ij} \neq 0$ is as follows:

$$t_{\text{Cal}} = \frac{b_{AB}}{\sqrt{\widehat{\text{Var}}(b_{AB})}} = \frac{7.0}{\sqrt{11.136}} = \frac{7.0}{3.34} = 2.096$$

which does not exceed 2.447 so do not reject $H_0 : \beta_{AB} = 0$ and infer the blending is additive along the A–B edge of the triangle.

$$t_{\text{Cal}} = \frac{b_{AC}}{\sqrt{\widehat{\text{Var}}(b_{AC})}} = \frac{-9.8}{3.34} = -2.93$$

which does exceed 2.447 in absolute value and so we reject the null hypothesis $H_0 : \beta_{AC} = 0$ at the $\alpha = 0.05$ level of significance and declare the blending between A and C to be nonlinear and antagonistic (not beneficial).

$$t_{\text{Cal}} = \frac{b_{BC}}{\sqrt{\widehat{\text{Var}}(b_{BC})}} = \frac{-20.00}{3.34} = -5.99$$

which also exceeds 2.447 in absolute value and we reject $H_0 : \beta_{BC} = 0$ at the 0.05 level of significance and declare the blending between B and C to be nonlinear and antagonistic.

$$t_{\text{Cal}} = \frac{b_{ABC}}{\sqrt{\widehat{\text{Var}}(b_{ABC})}} = \frac{111.6}{29.82} = 3.74$$

which exceeds 2.447 and we reject $H_0 : \beta_{ABC} = 0$ at the 0.05 level of significance and declare the blending among A, B, and C to be nonlinear and synergistic (blending is beneficial when all three components are present in equal proportions).

1.7. The mixture system seems simpler than the nonmixture system because with two components the factor space is a 1-dimensional line while the nonmixture system with two variables or factors requires a 2-dimensional configuration. In addition, the Scheffé-type second-degree model with two components has only three terms while the nonmixture second-degree model with two factors has six terms.

When operating with three components, the mixture factor space is either a regular 2-dimensional simplex or a 2-dimensional region inside the simplex. The second-degree or quadratic Scheffé-type model equation contains only six terms. In the nonmixture case with three factors, the experimental region will be a 3-dimensional configuration and the quadratic or second-degree model will contain ten terms, almost twice as many as the mixture model.

CHAPTER 2

2.1. For the $\{2, 4\}$ lattice, $x_i = 0, \frac{1}{4}, \frac{2}{4}, \frac{3}{4}, 1$ for $i = 1,2$. The $\binom{2+4-1}{4} = 5$ points have as coordinates $(x_1, x_2) = (1, 0), \left(\frac{3}{4}, \frac{1}{4}\right), \left(\frac{2}{4}, \frac{2}{4}\right), \left(\frac{1}{4}, \frac{3}{4}\right), (0, 1)$. For the $\{4, 3\}$ lattice, $x_i = 0, \frac{1}{3}, \frac{2}{3}, 1$ for $i = 1, 2, 3,$ and 4. The $\binom{4+3-1}{3} = 20$ points are

$(1, 0, 0, 0)$	4 of these
$\left(\frac{1}{3}, \frac{2}{3}, 0, 0\right)$	6 of these for $i < j$, where $x_i < x_j$
$\left(\frac{2}{3}, \frac{1}{3}, 0, 0\right)$	6 of these for $i < j$, where $x_i > x_j$
$\left(\frac{1}{3}, \frac{1}{3}, \frac{1}{3}, 0\right)$	4 of these

2.2. (a)

(b) $b_1 = 3.5, b_2 = 4.5$

$$b_{12} = 4(2.0) - 2(3.5 + 4.5) = -8.0$$

$$\widehat{\mathrm{Var}}(b_{12}) = 24s^2/3 = 8(0.14) = 1.12$$

To test $H_0: \beta_{12} = 0$ versus $H_A: \beta_{12} \neq 0$,

$$t_{\text{Cal}} = \frac{b_{12}}{\sqrt{\widehat{\text{Var}}(b_{12})}} = -\frac{8.0}{\sqrt{1.12}} = -7.56$$

From Appendix Table A, $t_{6,\,0.025} = 2.447$ and since $t_{\text{Cal}} = |-7.56| > 2.447$, we reject $H_0: \beta_{12} = 0$ in favor of $H_A: \beta_{12} \neq 0$ and declare the blending of A with B is nonlinear.

2.3. At the centroid, $\eta_{123} = \beta_1\left(\frac{1}{3}\right) + \beta_2\left(\frac{1}{3}\right) + \beta_3\left(\frac{1}{3}\right) + \beta_{12}\left(\frac{1}{9}\right) + \beta_{13}\left(\frac{1}{9}\right) + \beta_{23}\left(\frac{1}{9}\right) + \beta_{123}\left(\frac{1}{27}\right)$ and $\beta_{123} = 27\eta_{123} - 9(\beta_1 + \beta_2 + \beta_3) - 3(\beta_{12} + \beta_{13} + \beta_{23})$. But $\beta_i = \eta_i$ and $\beta_{ij} = 4\eta_{ij} - 2(\eta_i + \eta_j)$, $i < j$. Hence $\beta_{123} = 27\eta_{123} - 12(\eta_{12} + \eta_{13} + \eta_{23}) + 3(\eta_1 + \eta_2 + \eta_3)$. Also the formulas for β_i and β_{ij}, in terms of η_i and η_{ij}, are unaffected by the addition of a centroid point as long as the $\beta_{123}x_1x_2x_3$ term is in the model.

2.4. From Eq. (2.24), $\hat{y}(\mathbf{x}) = 11.7x_1 + 9.4x_2 + 16.4x_3 + 19.0x_1x_2 + 11.4x_1x_3 - 9.6x_2x_3$ so that if $\mathbf{x} = (0.50, 0.20, 0.30)$, then $\hat{y}(\mathbf{x}) = 11.7(0.5) + 9.4(0.2) + 16.4(0.3) + 19.0(0.5)(0.2) + 11.4(0.5)(0.3) - 9.6(0.2)(0.3) = 15.68$. From Eq. (2.25), the estimated variance of $\hat{y}(\mathbf{x})$ is

$$s^2\left\{\sum_{i=1}^{3}\frac{a_i^2}{2} + \sum\sum_{i<j}^{3}\frac{a_{ij}^2}{3}\right\}$$

$$= 0.73\left\{\tfrac{1}{2}(0 + 0.12^2 + 0.12^2) + \tfrac{1}{3}(0.40^2 + 0.60^2 + 0.24^2)\right\}$$

$$= 0.1511$$

since $a_i = x_i(2x_i - 1)$ so that

$a_1 = 0.5(1 - 1) = 0$, and $a_{ij} = 4x_ix_j$; so that $a_{12} = 4(0.5)(0.2) = 0.40$

$a_2 = 0.2(0.4 - 1) = -0.12$, $\qquad\qquad\qquad a_{13} = 4(0.5)(0.3) = 0.60$

$a_3 = 0.3(0.6 - 1) = -0.12$, $\qquad\qquad\qquad a_{23} = 4(0.2)(0.3) = 0.24$

Now $\Delta = \left[t_{(9,\,0.025)} = 2.262\right]\sqrt{0.1511} = 0.879$; hence the 95% confidence interval for the true elongation of yarn at $\mathbf{x} = (0.5, 0.2, 0.3)$ is

$$15.68 - \Delta \leq \eta(\mathbf{x}) \leq 15.68 + \Delta$$

$$14.80 \leq \eta(\mathbf{x}) \leq 16.56$$

2.5. Corresponding to models 1 through 6, we have figures e, a, b, c, f and d. The estimated surfaces are of the second-degree, planar, special cubic, second-degree, planar, and special cubic, respectively.

2.6. (a)
$$\bar{y}_1 = 5.5, \quad \bar{y}_2 = 7.0, \quad \bar{y}_3 = 8.0,$$

$$\bar{y}_{12} = b_{12}/4 + b_1/2 + b_2/2 = 8.75 + \tfrac{1}{2}(5.5 + 7.0) = 15.0,$$

$$\bar{y}_{13} = b_{13}/4 + b_1/2 + b_3/2 = 2.25 + \tfrac{1}{2}(5.5 + 8.0) = 9.0,$$

$$\bar{y}_{23} = 4.5 + \tfrac{1}{2}(7.0 + 8.0) = 12.0.$$

(b) $\widehat{\text{var}}(b_i) = \hat{\sigma}^2/r_i = 3.0/4 = 0.75;$

$$\widehat{\text{var}}(b_{ij}) = 16\hat{\sigma}^2/r_{ij} + 4\hat{\sigma}^2/r_i + 4\hat{\sigma}^2/r_j = (16/3 + 8/4)\,\hat{\sigma}^2$$
$$= (22/3)\,(3.0) = 22.0 \text{ and } \sqrt{22.0} = 4.69$$

$t = b_{ij}/4.69$ and compare with $t_{(15,0.05)} = 1.753$.

To be significantly greater than zero, the value of b_{ij} must be $\geq [4.69 \times t_{(15,\,0.05)}] = 8.22$. All three b_{ij} are greater than zero; certainly $b_{12} = 35.0$ and $b_{23} = 18.0$ are, but what about b_{13}? Yes, $b_{13} = 9.00 > 8.22$.

2.7.

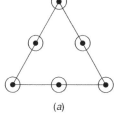

(a)

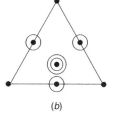

(b)

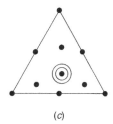

(c)

Source	d.f. (a)	d.f. (b)	d.f. (c)
Regression (fitted model)	5	5	5
Residual	6	6	6
LOF +	0	1	4
pure error	6	5	2
Total	11	11	11

With Design (a):

Good properties = Maximum degrees of freedom for pure-error estimate of σ^2. Nonlinear blending estimates $\hat{\beta}_{ij}$ have minimum variance.

Bad properties = No way to test lack of fit of quadratic model. Also, predictions made inside the triangle are based on information collected only on boundaries.

With Design (b):

Good properties = Almost maximum d.f. for pure-error estimate of σ^2. Higher power than design (c) for the lack-of-fit test of $H_0\!:\!\beta_{123} = 0$. If necessary, can fit the special cubic model.

With Design (c):

Good properties = Space-filling design in terms of extracting information inside the triangle. If necessary, can fit special-cubic model as well as the special-quartic model.

Bad properties = Only 2 d.f. for pure-error estimate of σ^2. Lack-of-fit test for quadratic model is not as powerful as is the test with design(b).

2.8. At $x_i = x_j = \frac{1}{2}$, $b_{ij}x_ix_j = b_{ij}/4$ and $b_ix_i = b_i/2$. Now in order for $b_{ij}/4 = b_i/2$, then $b_{ij} = 2b_i$.

At $x_i = x_j = x_k = \frac{1}{3}$, $b_{ijk}x_ix_jx_k = b_{ijk}/27$ and $b_{ij}x_ix_j = b_{ij}/9$ and $b_ix_i = b_i/3$. In order that $b_{ijk}/27 = b_{ij}/9$, then $b_{ijk} = 3b_{ij}$ or $b_{ijk}/b_{ij} = 3$ and for $b_{ijk}/27 = b_i/3$, then $b_{ijk}/b_i = 9$.

2.9. (a)

$$\hat{y}(\mathbf{x}) = 4.85x_1 + 1.28x_2 + 2.74x_3 - 4.98x_1x_2 - 2.82x_1x_3$$
$$\quad\;(0.11)\quad\;(0.11)\quad\;(0.11)\quad\;\;(0.46)\quad\;\;\;(0.46)$$
$$+1.48x_2x_3$$
$$\;(0.46)$$

(b) Test H_0: Lack of fit of the model in (a) $= 0$ at point 7.

$$\hat{y}(7) = 2.25, \quad \widehat{\text{var}}[\hat{y}(7)] = s^2\mathbf{x}'_p(\mathbf{X}'\mathbf{X})^{-1}\mathbf{x}_p = 0.022(0.216)$$
$$= 0.0048$$

$$t = \frac{2.75 - 2.25}{\sqrt{0.022 + 0.0048}} = \frac{0.50}{0.165} = 3.05 > t_{9,\,0.025} = 2.262$$

Reject H_0: refit model using $y\,(7) = 2.75$. Could you suggest another model?

2.10.

$$C_3 = \frac{4.148}{0.05} - (15 - 6) = 73.96$$

$$C_6 = \frac{0.201}{0.05} - (15 - 12) = 1.00$$

The reduction is due to the addition of the second-degree terms in the quadratic model.

2.11. (a) $\hat{y}\,(\mathbf{x}) = 12.5x_1 + 19.7x_2 + 14.5x_3 + 6.2x_1x_2 - 10.3x_1x_3 + 3.0x_2x_3$

(b) $\quad\quad\quad\quad$ s.e. $(b_1) = $ s.e. $(b_2) = $ s.e. $(b_3) = 0.9$

$\quad\quad\quad\quad$ s.e. $(b_{12}) = $ s.e. $(b_{13}) = $ s.e. $(b_{23}) = 4.5$

Of the quadratic coefficients, only $b_{13} = -10.3$ may be different from zero and only at the 0.10 level of significance since $t = -10.3/4.5 = -2.29$ and $t_{(6,0.05)} = 1.943$.

From the fitted model where $b_2 - b_1 = 7.2$ is the coefficient estimate of the term $\gamma_2 x_2$, the difference $\gamma_2 = \beta_2 - \beta_1$ is probably significant, since $t = 7.2/2.3 = 3.13 > t_{(6,0.025)} = 2.447$ at the 0.05 level. Thus the estimated height of the surface above the $x_2 = 1$ vertex is greater than the estimated height of the surface above the $x_1 = 1$ vertex.

2.12. Shown are data values at the seven blends of a 3-component simplex-controid design. High values are considered more desirable than low values. Based on the data, indicate the type of blending (linear vs nonlinear and if nonlinear, synergistic or antagonistic) that is present between,

(a) A and B: nonlinear and is antagonistic.

(b) A and C: nonlinear and is synergistic.

(c) B and C: nonlinear and is synergistic.

(d) A and B and C: β_{ABC} is not significantly different from 0 at the 0.05 level.

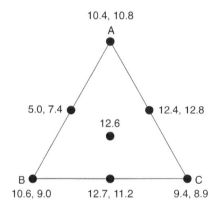

Test H_0: $\beta_{ij} = 0$ versus $\beta_{ij} \neq 0$ at the 0.05 level.

Hint: SST = 63.79, SSR = 58.22, and MSE = 5.57/6 = 0.93.

From the t table (Appendix Table A), $t_{(6,0.025)} = 2.447$.

(a) Recall from Question 1.6

$$b_{AB} = 4(6.2) - 2(10.6 + 9.8)$$

$$= 24.8 - 40.8 = -16.0$$

$$\widehat{Var}(b_{AB}) = 12\,MSE = 11.16$$

$$t_{Cal} = \frac{-16.0}{\sqrt{11.16}} = \frac{-16.0}{3.34} = |-4.79| > 2.447$$

$$\therefore \text{Reject } H_0: \beta_{AB} = 0$$

(b) $b_{AC} = 4(12.6) - 2(10.6 + 9.15) = 10.9$

$t_{Cal} = \dfrac{10.9}{3.34} = 3.26 > 2.447$ Reject $H_0: \beta_{AC} = 0$

(c) $b_{BC} = 4(11.95) - 2(9.8 + 9.15) = 9.9$

$t_{Cal} = \dfrac{9.9}{3.34} = 2.96 > 2.447$ Reject $H_0: \beta_{BC} = 0$

(d) $b_{ABC} = 27(12.6) - 12(6.2 + 12.6 + 11.95) + 3(10.6 + 9.8 + 9.15)$

$= 59.85$

$\widehat{Var}(b_{ABC}) = 958.5 \text{ MSE} = 958.5(0.93) = 891.41$

$t_{Cal} = \dfrac{59.85}{\sqrt{891.41}} = 2.00 \not> 2.447$ Do not reject $H_0: \beta_{ABC} = 0$
                                                at the 0.05 level of significance

CHAPTER 3

3.1.
$$x_1' = \frac{x_1 - 0.10}{0.75}, \; x_2' = \frac{x_2 - 0.15}{0.75}, \; x_3' = \frac{x_3 - 0.0}{0.75}$$

then $x_1 = 0.75x_1' + 0.10, \; x_2 = 0.75x_2' + 0.15, \; x_3 = 0.75x_3'$.

	x_1'	x_2'	x_3'		x_1	x_2	x_3
	1	0	0		0.85	0.15	0
	0	1	0		0.10	0.90	0
	$\frac{1}{2}$	$\frac{1}{2}$	0		0.475	0.525	0
When	0	0	1	then	0.10	0.15	0.75
	$\frac{1}{2}$	0	$\frac{1}{2}$		0.475	0.15	0.375
	0	$\frac{1}{2}$	$\frac{1}{2}$		0.10	0.525	0.375
	$\frac{1}{3}$	$\frac{1}{3}$	$\frac{1}{3}$		0.35	0.40	0.25

3.2.

Blend	x_2	x_3	x_4	y
1	0.20	0	0	6.50
2	0	0.20	0	6.96
3	0	0	0.20	6.00
11	0.10	0.10	0	7.25
12	0	0.10	0.10	6.20
13	0.10	0	0.10	6.47

The fitted second-degree model in x_2, x_3, and x_4 (after each is divided by 0.20) is

$$\hat{y}(\mathbf{x}) = 6.50x_2 + 6.96x_3 + 6.00x_4 + 2.08x_2x_3 + 0.88x_2x_4 - 1.12x_3x_4$$

Since both b_2 and b_3 appear higher than b_4 and since $b_{23} = 2.08$ is positive, try $x_2 = x_3 = 10$ with $x_1 = 0.80$.

3.3. (a)
$$u_1 = \frac{0.92 - x_1}{0.02}, \quad u_2 = \frac{0.06 - x_2}{0.02}, \quad u_3 = \frac{0.04 - x_3}{0.02}$$

(b) $u_1 = 1, u_2 = u_3 = 0 \Rightarrow x_1 = 0.90, x_2 = 0.06, x_3 = 0.04$

$\quad u_1 = u_3 = 0, u_2 = 1 \Rightarrow x_1 = 0.92, x_2 = 0.04, x_3 = 0.04$

$\quad u_1 = u_2 = 0, u_3 = 1 \Rightarrow x_1 = 0.92, x_2 = 0.06, x_3 = 0.02$

$\quad u_1 = u_2 = u_3 = 1/3 \Rightarrow x_1 = 0.9133, x_2 = 0.0533, x_3 = 0.0333$

(c) The experimental region is not an inverted simplex but rather is a convex polyhedron with four vertices and four sides.

(d) Since we need at least six distinct blends, we might use all four extreme vertices, the midpoints of the four edges, and replicate the overall centroid twice.

3.4. (b) $R_2 = 0.57 < R_L = 1 - 0.48 = 0.52$, therefore $U_2 = 0.6$ is unattainable.

(c) One solution would be to leave $L_3 = 0.45$ and reduce U_2 to $U_2^* = 0.55$. Then the constrained region has five extreme vertices and four edges (sides). The vertices are

1. (0.5, 0, 0.5)

2. (0.1, 0, 0.9)

3. (0, 0.1, 0.9)

4. (0, 0.55, 0.45)

5. (0.5, 0.05, 0.45)

Vertices 1 and 5 are close to one another. So, instead of choosing both as design points, we might choose the midpoint of the edge connecting the vertices 1 and 5. This point has the coordinates (0.5, 0.025, 0.475). Then choose vertices 2, 3, and 4 along with the midpoints of the three longest edges, whose coordinates are (0.3, 0, 0.7), (0, 0.325, 0.675), (0.25, 0.3, 0.45). Finally, we might choose to replicate the overall centroid of the region whose coordinates are (0.22, 0.14, 0.64). This is an eight-point design. A nine-point design might be to use the five extreme vertices, the midpoints of the three longest edges, and the overall centroid.

3.5. **(a)** $0.24 \leq x_1 \leq 0.44$, $0.05 \leq x_2 \leq 0.25$, $0.11 \leq x_3 \leq 0.31$, $0.20 \leq x_4 \leq 0.40$
   Yes, the constraints are consistent.

   (b) There are 6 extreme vertices, 12 edges, and 8 two-dimensional faces.

   (c) If we apply the suggestion of Snee (1975) discussed in Section 3.9.1, for
   $q = 4$, we should use the extreme vertices, constraint plane centroids, overall
   centroid, and the midpoints of the longest edges. Since for our region there
   are 6 extreme vertices and 8 faces, one suggestion is to use all 6 extreme
   vertices, the centroids of the 8 faces, and the overall centroid. Another
   suggestion is to use the 6 extreme vertices, the midpoints of four edges,
   the centroids of four faces, and the overall centroid. This second design has
   some appeal because of the shape of the experimental region, which takes
   the form of a double tetrahedron and looks like the figure shown.

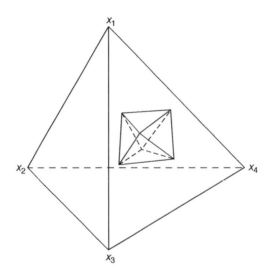

3.6. The coordinates of the extreme vertices (points 1–7) and the midpoints (8–18)
   are

   1. $(0.90, 0.05, 0.02, 0.03)$ **10.** $(0.80, 0.11, 0.06, 0.03)$
   2. $(0.80, 0.15, 0.02, 0.03)$ **11.** $(0.80, 0.09, 0.06, 0.05)$
   3. $(0.80, 0.05, 0.10, 0.05)$ **12.** $(0.89, 0.05, 0.02, 0.04)$
   4. $(0.82, 0.05, 0.10, 0.03)$ **13.** $(0.81, 0.05, 0.10, 0.04)$
   5. $(0.80, 0.07, 0.10, 0.03)$ **14.** $(0.86, 0.05, 0.06, 0.03)$
   6. $(0.88, 0.05, 0.02, 0.05)$ **15.** $(0.84, 0.05, 0.06, 0.05)$
   7. $(0.80, 0.13, 0.02, 0.05)$ **16.** $(0.85, 0.10, 0.02, 0.03)$
   8. $(0.80, 0.14, 0.02, 0.04)$ **17.** $(0.84, 0.09, 0.02, 0.05)$
   9. $(0.80, 0.06, 0.10, 0.04)$ **18.** $(0.81, 0.06, 0.10, 0.03)$

3.7. The $\{2, 3, 2; 1, 2, 2\}$ triple-lattice consists of the 36 combinations formed by

$$
\begin{aligned}
X_{11} + X_{12} &= \tfrac{1}{2}, & \text{where } X_{1j} &= 0 \text{ or } \tfrac{1}{2}, \; j = 1, 2 \\
X_{21} + X_{22} + X_{23} &= \tfrac{1}{4}, & \text{where } X_{2j} &= 0, \tfrac{1}{8}, \tfrac{1}{4}, \; j = 1, 2, 3 \\
X_{31} + X_{32} &= \tfrac{1}{4}, & \text{where } X_{3j} &= 0, \tfrac{1}{8}, \tfrac{1}{4}, \; j = 1, 2
\end{aligned}
$$

such as $X_{11} X_{21} X_{31}, X_{12} X_{21} X_{31}, \ldots, X_{12} X_{22} X_{23} X_{31} X_{32}$.

3.8. Shown are data values at the seven blends of a 3-component simplex-centroid design. High values are considered more desirable than low values. Based on the data, and at the $\alpha = 0.01$ level of significance, indicate the type of blending (linear vs. nonlinear and if nonlinear, synergistic or antagonistic) that is present between:

(a) A and B: linear and is additive.
$$b_{AB} = -0.3, t_{Cal} = -0.09 \quad \text{Do not reject } H_0 \colon \beta_{AB} = 0.$$
(b) A and C: nonlinear and is synergistic.
$$b_{AC} = 14.2, t_{Cal} = 4.25 \quad \text{Reject } H_0 \colon \beta_{AC} = 0.$$
(c) B and C: nonlinear and is synergistic.
$$b_{BC} = 19.7, t_{Cal} = 5.90 \quad \text{Reject } H_0 \colon \beta_{BC} = 0.$$
(d) A and B and C: planar and is additive.
$$b_{ABC} = 5.85, t_{Cal} = 0.20 \quad \text{Do not reject } H_0 \colon \beta_{ABC} \neq 0.$$
Test $H_0 \colon \beta_{ij} = 0$ versus $\beta_{ij} \neq 0$ at the 0.01 level,

Hint: SST $= 63.79$, SSR $= 58.22$, and MSE $= 5.57/6 = 0.93$.

From the t table (Appendix Table A), $t_{(6, 0.005)} = 3.707$.

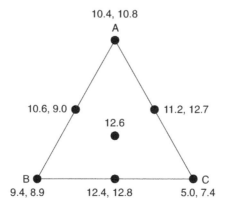

CHAPTER 4

4.1. Fit a special-cubic model to get information on pure blends, on binary blends where x_i and x_j each contribute one-half, and also on the ternary blend where $x_1 = x_2 = x_3 = \frac{1}{3}$. However, the special cubic model is fitted through the data at all seven design points, and thus we have no way of checking the fit at any point other than the design points. Unless some replicated observations are collected, there are no degrees of freedom for estimating the error variance. With a less than special cubic model, an estimate of the error variance is possible.

4.2.
$$\hat{y}(\mathbf{x}) = 2.18x_1 + 0.84x_2 + 1.95x_3, \quad s^2 = 0.025, \quad R_A^2 = 0.8857$$
$$\phantom{\hat{y}(\mathbf{x}) = }(0.09) \quad\; (0.09) \quad\; (0.09) \qquad\qquad\qquad\quad R^2 = 0.9827$$

The null hypothesis $H_0: \beta_1 = \beta_2 = \beta_3 = \beta_0$ is a test that the height of the surface above the triangle is constant (β_0). The fitted model along with the F-test do not agree. In order to acquire a model that describes the shape of the surface better than a plane, one must add some additional terms. Fitting the second-degree model describes the surface above the triangle in greater detail. A test allowing the extra terms (quadratic) to describe the surface shape better than a planar surface is the next step.

$$H_0 : \beta_1 = \beta_2 = \beta_3 = \beta_0, \quad F = \frac{2.57/2}{0.025} = 51.4 > F_{(2, 11, 0.01)} = 7.21$$

Reject H_0 even with an $R^2 = 0.9827$ and fit the second-degree model.

$$\hat{y}(\mathbf{x}) = 2.31x_1 + 0.96x_2 + 2.11x_3 - 0.66x_1x_2 - 1.06x_1x_3 - 0.98x_2x_3,$$
$$\phantom{\hat{y}(\mathbf{x}) = }(0.06) \quad\;\; (0.06) \quad\;\; (0.06) \quad\;\; (0.26) \qquad\; (0.26) \qquad\; (0.26)$$
$$s^2 = 0.006, \quad R_A^2 = 0.9703$$

Increase in R_A^2 value of 0.0846 plus the rejection of $H_0: \beta_{12} = \beta_{13} = \beta_{23} = 0$ by $F = (0.22/3)/0.006 = 12.2 > F_{(3,8,0.01)} = 7.59$ causes us to prefer the second-degree model to the first-degree model.

$$\hat{y}(\mathbf{x}) = 2.32x_1 + 0.97x_2 + 2.12x_3 - 0.76x_1x_2 - 1.16x_1x_3 - 1.08x_2x_3$$
$$\phantom{\hat{y}(\mathbf{x}) = }(0.06) \qquad\qquad\qquad\qquad\quad (0.28)$$
$$+ 2.03x_1x_2x_3, \quad s^2 = 0.006, \quad R_A^2 = 0.9703$$
$$(1.95)$$

Special cubic model is no improvement over second-degree model because $t_{Cal} = b_{123}/\sqrt{\widehat{\text{Var}}(b_{123})} = 2.03/1.95 = 1.04$, and since $t_{Cal} = 1.04 < t_{(7, 0.025)} = 2.365$, we do not reject $H_0: \beta_{123} = 0$ at the 0.05 level of significance. The fitted second-degree model is our choice.

4.3. (a)
$$\hat{y}(\mathbf{x}) = 10.60x_1 + 6.49x_2 + 6.77x_3$$
$$(1.83) \quad (1.83) \quad (1.83)$$

Analysis of Variance Table

Source	d.f.	SS	MS
Regression	2	29.04	14.52
Residual	12	130.64	10.89
Total	14	159.68	

Test statistic:

$$F = \frac{14.52}{10.89} = 1.33 < F_{(2,\,12,\,0.10)} = 2.81$$

Do not reject $H_0: \beta_1 = \beta_2 = \beta_3 =$ constant at this stage. *Note:* $R_A^2 = 0.0450$.

(b) $\hat{y}(\mathbf{x}) = 6.50x_1 + 4.70x_2 + 3.20x_3 + 18.00x_1x_2 + 31.00x_1x_3 + 14.07x_2x_3$
$\qquad\quad (0.74) \quad (0.74) \quad (0.74) \qquad (3.19) \qquad\quad (3.19) \qquad\quad (3.19)$

$$C_3 = \frac{130.64}{1.09} - (15 - 6) \approx 111, \quad C_6 = \frac{9.79}{1.09} - (15 - 12) \approx 6$$

Source	d.f.	SS	MS
Regression	5	149.89	29.98
Residual	9	9.79	1.09
Total	14	159.68	

Increase in regression sum of squares $= 149.89{-}29.04 = 120.85$. Is it a significant increase? Yes, because the increase in MSR of $120.85/3 = 40.28$ relative to MS Residual for the fitted second-degree model MSE $= 1.09$ is

$$F = \frac{120.85/3}{1.09} = 39.96 > F_{(3,\,9,\,0.01)} = 6.99$$

Therefore reject $H_0: \beta_{12} = \beta_{13} = \beta_{23} = 0$ and infer one or more $=$ should be $\neq$. *Note:* $R_A^2 = 0.9044$.

4.4. Shown are data values at the seven blends of a 3-component simplex-controid design. High values are considered more desirable than low values. Based on the data, indicate the type of blending (linear vs nonlinear and if nonlinear, synergistic or antagonistic) that is present between,

(a) A and B: linear and is additive.

$$b_{AB} = 4.6, t_{Cal} = 1.38 \not> 2.447 \qquad \text{Do not reject } H_0: \beta_{AB} = 0$$

(b) A and C: nonlinear and is synergistic.

$$b_{AC} = 9.6, t_{Cal} = 2.87 > 2.447 \qquad \text{Reject } H_0: \beta_{AC} = 0$$

(c) B and C: nonlinear and is synergistic.

$$b_{BC} = 19.7, t_{Cal} = 4.25 > 2.447 \qquad \text{Reject } H_0: \beta_{BC} = 0$$

(d) A and B and C: planar and is not synergistic.

$$b_{ABC} = 5.85, t_{Cal} = 0.52 \not> 2.447 \quad \text{Do not reject } H_0: \beta_{ABC} = 0$$

4.5. Equation (4.37) shows that the total effect upon increasing x_1 by 0.3 along the Piepel direction by starting at $\mathbf{x}_L = (0.10, 0.82, 0.08)$ and ending at $\mathbf{x}_H = (0.4, 0.58, 0.02)$ is

$$TE_1 = \hat{y}(0.40, 0.58, 0.02) - \hat{y}(0.10, 0.82, 0.08).$$

Thus while x_1 increases by 0.3, the quotient "component 2/component 3" or x_2/x_3 increases from $0.82/0.08 = 10.25$ at $\mathbf{x}_L$ up to $0.58/0.02 = 29.0$ at $\mathbf{x}_H$, almost a three-fold or $29.0/10.25 = 2.83$ increase. Might this increase in the ratio x_2/x_3 affect the value of the response as much as the increase of 0.3 in x_1 does?

CHAPTER 5

5.1. Using the values of x_1, x_2, and x_3 in Table 5.5.

(a) *First-degree model:*

$$\hat{y}(\mathbf{x}) = 79.35x_1 + 72.72x_2 + 106.53x_3, \quad C_3 = \frac{28.54}{2.50} - (9-6) = 8$$
$$\quad\;\; (4.01) \quad\;\;\; (7.34) \quad\;\;\; (6.01)$$

(b) *First-degree plus inverse terms:*

$$\hat{y}(\mathbf{x}) = 146.68x_1 + 181.46x_2 + 109.19x_3 - 6.23x_1^{-1} - 12.64x_3^{-1}, \quad C_5 = 4$$
$$\quad\;\; (25.96) \quad\;\; (35.33) \quad\;\; (14.65) \quad\;\; (2.92) \quad\;\;\; (4.59)$$

(c) *Second-degree model:*

$$\hat{y}(\mathbf{x}) = 29.07x_1 - 2.12x_2 - 7.34x_3 + 100.72x_1x_2 + 308.54x_1x_3$$
$$\quad\;\; (29.23) \quad (99.61) \quad (61.78) \quad\;\; (112.81) \quad\;\;\;\; (157.71)$$
$$\qquad + 321.49x_2x_3, \quad C_6 = 8$$
$$\qquad\;\; (291.01)$$

Source	Model (a) d.f.	Model (a) SS	Model (b) d.f.	Model (b) SS	Model (c) d.f.	Model (c) SS
Regression	2	50.69	4	70.87	5	66.85
Residual	6	28.54	4	8.36	3	12.38
		($s^2 = 4.76$)		($s^2 = 2.09$)		($s^2 = 4.13$)
Total	8	79.23				
Note: $R^2 =$		0.6398		0.8945		0.8437
$R_A^2 =$		0.5194		0.7890		0.5830

5.2.

Run:	1	2	3	4	5	6	7
Special cubic:	0.1	−1.0	0.8	0.3	−0.1	0.4	−2.1
Nonpolynomial:	0.6	0.4	0.7	0.1	−0.9	−0.4	−0.7

Run:	8	9	10	11	12	13	14	15
Special cubic:	−1.9	3.2	1.7	−0.5	2.0	0.8	−0.6	−3.0
Nonpolynomial:	−1.9	2.8	0.8	−0.8	0.2	0.0	−0.3	−0.6

$$\# |r_u|_{SC} > |r_u|_{NP} = 10$$
$$\# |r_u|_{SC} < |r_u|_{NP} = 3$$
$$\# |r_u|_{SC} = |r_u|_{NP} = 2$$

Along the x_3 axis at points 2, 12, 13, 14, 7, and 15, the residuals reflect the buckling effect of the special-cubic model.

5.3. The fitted first- and second-degree models and the associated SSE along each ray are

	First Degree	SSE
Ray 1:	$\hat{y}(\mathbf{x}) = 20.56 + 28.76(1 - x_3)$	2.27
	(0.68) (1.11)	
Ray 2:	$\hat{y}(\mathbf{x}) = 21.46 + 61.92(1 - x_3)$	12.19
	(1.56) (2.55)	
Ray 3:	$\hat{y}(\mathbf{x}) = 20.06 + 15.74(1 - x_3)$	0.08
	(0.09) (0.16)	

	Second Degree	SSE
Ray 1:	$\hat{y}(\mathbf{x}) = 20.60 + 28.42(1 - x_3) + 0.34(1 - x_3)^2$	2.26
	(1.01) (4.77) (4.57)	
Ray 2:	$\hat{y}(\mathbf{x}) = 21.23 + 63.75(1 - x_3) - 1.83(1 - x_3)^2$	12.00
	(2.31) (10.93) (10.47)	
Ray 3:	$\hat{y}(\mathbf{x}) = 20.10 + 15.43(1 - x_3) + 0.31(1 - x_3)^2$	0.04
	(0.13) (0.56) (0.53)	

On each ray the drop in SSE from fitting the quadratic model is not significant. Hence the quadratic term is not necessary; that is, do not reject $H_0: \gamma_1 = 0$. However, reject $H_0: \beta_1 = 0$.

5.4. (a) Using the fitted Cox model (5.39), then along the x_1 direction in Figure 5.9,

$$\Delta \hat{y}(\mathbf{x}) = \frac{-2.43(0.166)}{0.346} + \frac{1.36(0.166)^2}{0.346^2}$$
$$= -0.85$$

(b) Using the fitted Scheffé model (5.40), the predicted response values are

$$\hat{y}(\mathbf{x}) = 9.96 \quad \text{and} \quad \hat{y}(\mathbf{s}) = 10.81$$

and the difference is $\hat{y}(\mathbf{x}) - \hat{y}(\mathbf{s}) = 9.96 - 10.81 = -0.85$, the same as in (a) as it should be because the two models (5.39) and (5.40) are equivalent.

CHAPTER 6

6.1.

$$\hat{y}(\mathbf{x})_{\text{at } z=-1} = 4.7x_1 + 5.5x_2 - 2.4x_1x_2$$
$$\hat{y}(\mathbf{x})_{\text{at } z=+1} = 2.5x_1 + 3.7x_2 + 14.0x_1x_2$$

Combined model:

$$\hat{y}(\mathbf{x}, z) = 3.6x_1 + 4.6x_2 + 5.8x_1x_2 - 1.1x_1z - 0.9x_2z + 8.2x_1x_2z$$
$$\quad\;\; (0.2) \quad (0.2) \quad\;\; (1.1) \qquad (0.2) \quad\;\; (0.2) \qquad (1.1)$$

(a) There appears to be a significant binary blending effect $(\gamma_{12}^0 > 0)$, since $t = 5.8/1.1 = 5.21 > t_{(6,0.025)} = 2.447$. Although the binary blending coefficient estimate (5.8) is taken as the average over the two levels of z, the real binary blending occurs at the high level of z only.

(b) Raising the level of z produced a significantly lower average response to each of the pure components and produced an increase in the average response to the binary blend $\left(x_1 = x_2 = \frac{1}{2}\right)$. This is reflected in the significantly-different-from-zero estimates $g_1^1 = -1.1$, $g_2^1 = -0.9$, and $g_{12}^1 = 8.2$ whose respective t-values are -4.84, -3.96, and 7.36.

(c) At $z = +1$, $x_1 = x_2 = \frac{1}{2}$.

6.2. (a) Let x_A and x_B represent beef types A and B, respectively, and let c^1 and c^2 represent the linear and quadratic term for the time—temperature conditions of the process variable. For replication 1,

$$\hat{y}(\mathbf{x}) = 2.17x_A + 1.40x_B + 4.28x_Ax_B + 0.98x_Ax_B(x_A - x_B)$$
$$+ 0.75x_Ac^1 + 0.30x_Bc^1 - 2.36x_Ax_Bc^1$$
$$- 2.36x_Ax_B(x_A - x_B)c^1 + 0.08x_Ac^2 + 0.10x_Bc^2$$
$$- 1.01x_Ax_Bc^2 - 1.31x_Ax_B(x_A - x_B)c^2$$

(b)

Source	d.f.	SS	MS	F
Regression (model below)	11	15.50		
Cooking temp. (c)	2	1.39	0.69	$7.0 > F_{(2,24,0.01)} = 5.61$
Components (C)	3	11.11	3.70	$37.4 > F_{(3,24,0.01)} = 4.72$
Cook $\times$ components	6	3.00	0.50	$5.0 > F_{(6,24,0.01)} = 3.67$
Residual	24	2.38	$0.099 = s_e^2$	
Reps	2	0.17		
Reps $\times$ c	4	0.23		
Reps $\times$ C	6	1.39		
Reps $\times$ C $\times$ c	12	0.59		
Total	35	17.88		

Model:

$$\hat{y}(\mathbf{x}, c) = \underset{(0.11)}{2.02x_A} + \underset{(0.11)}{1.49x_B} + \underset{(0.47)}{3.68x_Ax_B} + \underset{(1.06)}{5.03x_Ax_B(x_A - x_B)}$$
$$+ \underset{(0.13)}{0.63x_Ac^1} + \underset{(0.13)}{0.43x_Bc^1} - \underset{(0.58)}{2.74x_Ax_Bc^1}$$
$$+ \underset{(1.29)}{0.11x_Ax_B(x_A - x_B)c^1} + \underset{(0.07)}{0.04x_Ac^2} + \underset{(0.01)}{0.01x_Bc^2}$$
$$- \underset{(0.33)}{0.64x_Ax_Bc^2} - \underset{(0.75)}{1.24x_Ax_B(x_A - x_B)c^2}$$

6.3.

$$\hat{y}(\mathbf{x}, \mathbf{z}) = \underset{(0.12)}{2.86x_1} + \underset{(0.12)}{1.06x_2} + 0.46x_1z_1 + \underset{(0.12)}{0.14x_2z_1} + 0.58x_1z_2$$
$$+ 0.24x_2z_2$$

Component 1 seems to produce firmer patties than component 2. Component 1 appears to be affected more (firmer patties) by the process variables than does component 2 (all positive effects).

ANOVA table

Source	d.f.	SS	MS	F
Regression	5	9.975	1.995	16.5
Residual	6	0.435	0.0725	
Total	11	10.410		

Answers to Questions QA1, QA2, QB1 and QB2

QA1. When fitting a model of the form (6.32), one assumes the nonlinear blending properties of the components are the same at the different coating thicknesses or levels of Z. This is because there are no terms of the form $b_{ij4}x_ix_jZ$ in the model of (6.32), which would in fact measure the effect of changing the level of Z on the nonlinear blending of components i and j. Hence nonlinear blending, if present, is assumed to be present everywhere in the mixture-amount experimental region, and its magnitude is the same everywhere in the region. When the level of coating thickness is varied over the range of $Z = 10$ to 28 μm, the opacity of the coating behaves in a curvilinear (quadratically) manner. The component linear blending properties on opacity are affected, but only in a linear manner, when varying the coating thickness over the range of 10 to 28 μm. The 15-point design consisting of eight, three, and four blends at $Z = 10$, 19, and 28 μm, respectively, merely supports fitting the model and does not allow one to check the adequacy of the model in terms of checking on the constancy of the nonlinear blending properties everywhere in the region. What would be needed to do so is to have at least six different blends either at $Z = 19$ or 28 μm that support the estimation of the nonlinear blending coefficients in the mixture model, which in turn would enable estimates of the three cross-product terms $g_{ij}^1 x_i x_j Z$ to be obtained. If this happened, one could check on whether or not the changing of coating thickness affected the nonlinear blending of the components.

QA2. Yes, since the model in (6.32) does not contain any cross-product terms between the level of coating thickness and the nonlinear blending of the components, then whatever measure or evidence of nonlinear blending is picked up by the three $b_{ij}x_ix_j$, $i < j$, terms, especially at the $Z = 10$ μm level of coating, will be predicted everywhere else in the experimental region that predictions are made.

QB1. Since the constrained mixture region, shown in Figure 6.9b, is irregular in shape and the configuration of the design points is not symmetrical with respect to the axes of the three L-pseudocomponents, then the choice of which pair of nonlinear blending terms to include in the mixture model only matters if tests of hypotheses or tests of significance are to be performed on the individual coefficient estimates. This is because the separate nonlinear blending terms describe different shape characteristics of the fitted mixture surface. The different shape characteristics are defined in Cornell (1995, table 3). If the five-term reduced quadratic model is to be used only for prediction purposes, on the other hand, then it does not matter which pair of nonlinear blending terms is included in the model.

QB2. A model reduced from Eq. (6.33) can only contain a portion of the information that is provided by (6.33). Thus, if predictions of mixture surfaces at the eight factor-level combinations of Z_1, Z_2, and Z_3 are desired, use the complete model (6.33) or fit eight separate mixture models and use them for prediction purposes, one for each of the $2^3 = 8$ combinations. The model reduced from (6.30) should only be used for prediction of the mixture surface at location within the Z-cuboidal region other than at the factor-level combinations, since a reduced model containing only important or significantly-different-from-zero terms will have lower prediction error than the complete model (6.33), possessing one or more unimportant terms, has.

6.4. Shown are data values at the seven blends of a 3-component simplex-controid design. High values are considered more desirable than low values. Based on the data, indicate the type of blending (linear vs nonlinear and if nonlinear, synergistic or antagonistic) that is present between,

(a) A and B: linear and is additive. $t_{Cal} = \frac{-6.7}{3.34} = |-2.01| \not> 2.447$

(b) A and C: nonlinear and is antagonistic. $t_{Cal} = \frac{-17.4}{3.34} = |-5.2| > 2.447$

(c) B and C: linear and is additive. $t_{Cal} = \frac{-4.3}{3.34} = |-1.29| \not> 2.447$

(d) A and B and C: nonlinear and is synergistic. $t_{Cal} = \frac{122.1}{29.85} = 4.09 > 2.447$

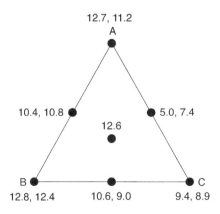

Appendix

Table A Percentage Points of the *t*-Distribution with *v* Degrees of Freedom

v	α										
	0.45	0.40	0.35	0.30	0.25	0.125	0.05	0.025	0.0125	0.005	0.0025
1	0.158	0.325	0.510	0.727	1.000	2.414	6.314	12.71	25.45	63.66	127.3
2	0.142	0.289	0.445	0.617	0.817	1.604	2.920	4.303	6.205	9.925	14.09
3	0.137	0.277	0.424	0.584	0.765	1.423	2.353	3.183	4.177	5.841	7.453
4	0.134	0.271	0.414	0.569	0.741	1.344	2.132	2.776	3.495	4.604	5.598
5	0.132	0.267	0.408	0.559	0.727	1.301	2.015	2.571	3.163	4.032	4.773
6	0.131	0.265	0.404	0.553	0.718	1.273	1.943	2.447	2.969	3.707	4.317
7	0.130	0.263	0.402	0.549	0.711	1.254	1.895	2.365	2.841	3.500	4.029
8	0.130	0.262	0.399	0.546	0.706	1.240	1.860	2.306	2.752	3.355	3.833
9	0.129	0.261	0.398	0.543	0.703	1.230	1.833	2.262	2.685	3.250	3.690
10	0.129	0.260	0.397	0.542	0.700	1.221	1.813	2.228	2.634	3.169	3.581
11	0.129	0.260	0.396	0.540	0.697	1.215	1.796	2.201	2.593	3.106	3.500
12	0.128	0.259	0.395	0.539	0.695	1.209	1.782	2.179	2.560	3.055	3.428
13	0.128	0.259	0.394	0.538	0.694	1.204	1.771	2.160	2.533	3.012	3.373
14	0.128	0.258	0.393	0.537	0.692	1.200	1.761	2.145	2.510	2.977	3.326
15	0.128	0.258	0.392	0.536	0.691	1.197	1.753	2.132	2.490	2.947	3.286
20	0.127	0.257	0.391	0.533	0.687	1.185	1.725	2.086	2.423	2.845	3.153
25	0.127	0.256	0.390	0.531	0.684	1.178	1.708	2.060	2.385	2.787	3.078
30	0.127	0.256	0.389	0.530	0.683	1.173	1.697	2.042	2.360	2.750	3.030
40	0.126	0.255	0.388	0.529	0.681	1.167	1.684	2.021	2.329	2.705	2.971
60	0.126	0.254	0.387	0.527	0.679	1.162	1.671	2.000	2.299	2.660	2.915
120	0.126	0.254	0.386	0.526	0.677	1.156	1.658	1.980	2.270	2.617	2.860
∞	0.126	0.253	0.385	0.524	0.674	1.150	1.645	1.960	2.241	2.576	2.807

Source: Reproduced with permission from W. W. Hines and D. C. Montgomery, *Probability and Statistics in Engineering and Management Science*, Ronald Press, New York, 1972.
v = degrees of freedom.

A Primer on Experiments with Mixtures, By John A. Cornell
Copyright © 2011 John Wiley & Sons, Inc. Published by John Wiley & Sons, Inc.

Table B Percentage Points of the F-Distribution

$$F_{.25, v_1, v_2}$$

v_2 \ v_1	1	2	3	4	5	6	7	8	9	10	12	15	20	24	30	40	60	120	∞
1	5.83	7.50	8.20	8.58	8.82	8.98	9.10	9.19	9.26	9.32	9.41	9.49	9.58	9.63	9.67	9.71	9.76	9.80	9.85
2	2.57	3.00	3.15	3.23	3.28	3.31	3.34	3.35	3.37	3.38	3.39	3.41	3.43	3.43	3.44	3.45	3.46	3.47	3.48
3	2.02	2.28	2.36	2.39	2.41	2.42	2.43	2.44	2.44	2.44	2.45	2.46	2.46	2.46	2.47	2.47	2.47	2.47	2.47
4	1.81	2.00	2.05	2.06	2.07	2.08	2.08	2.08	2.08	2.08	2.08	2.08	2.08	2.08	2.08	2.08	2.08	2.08	2.08
5	1.69	1.85	1.88	1.89	1.89	1.89	1.89	1.89	1.89	1.89	1.89	1.89	1.88	1.88	1.88	1.88	1.87	1.87	1.87
6	1.62	1.76	1.78	1.79	1.79	1.78	1.78	1.78	1.77	1.77	1.77	1.76	1.76	1.75	1.75	1.75	1.74	1.74	1.74
7	1.57	1.70	1.72	1.72	1.71	1.71	1.70	1.70	1.70	1.69	1.68	1.68	1.67	1.67	1.66	1.66	1.65	1.65	1.65
8	1.54	1.66	1.67	1.66	1.66	1.65	1.64	1.64	1.63	1.63	1.62	1.62	1.61	1.60	1.60	1.59	1.59	1.58	1.58
9	1.51	1.62	1.63	1.63	1.62	1.61	1.60	1.60	1.59	1.59	1.58	1.57	1.56	1.56	1.55	1.54	1.54	1.53	1.53
10	1.49	1.60	1.60	1.59	1.59	1.58	1.57	1.56	1.56	1.55	1.54	1.53	1.52	1.52	1.51	1.51	1.50	1.49	1.48
11	1.47	1.58	1.58	1.57	1.56	1.55	1.54	1.53	1.53	1.52	1.51	1.50	1.49	1.49	1.48	1.47	1.47	1.46	1.45
12	1.46	1.56	1.56	1.55	1.54	1.53	1.52	1.51	1.51	1.50	1.49	1.48	1.47	1.46	1.45	1.45	1.44	1.43	1.42
13	1.45	1.55	1.55	1.53	1.52	1.51	1.50	1.49	1.49	1.48	1.47	1.46	1.45	1.44	1.43	1.42	1.42	1.41	1.40
14	1.44	1.53	1.53	1.52	1.51	1.50	1.49	1.48	1.47	1.46	1.45	1.44	1.43	1.42	1.41	1.41	1.40	1.39	1.38
15	1.43	1.52	1.52	1.51	1.49	1.48	1.47	1.46	1.46	1.45	1.44	1.43	1.41	1.41	1.40	1.39	1.38	1.37	1.36
16	1.42	1.51	1.51	1.50	1.48	1.47	1.46	1.45	1.44	1.44	1.43	1.41	1.40	1.39	1.38	1.37	1.36	1.35	1.34

Degrees of Freedom for Numerator (v_1)

Degrees of Freedom for Denominator (v_2)

Degrees of Freedom For Denominator (v_2)

v_2																			
17	1.42	1.51	1.50	1.49	1.47	1.46	1.45	1.44	1.43	1.43	1.41	1.40	1.39	1.38	1.37	1.36	1.35	1.34	1.33
18	1.41	1.50	1.49	1.48	1.46	1.45	1.44	1.43	1.42	1.42	1.40	1.39	1.38	1.37	1.36	1.35	1.34	1.33	1.32
19	1.41	1.49	1.49	1.47	1.46	1.44	1.43	1.42	1.41	1.41	1.40	1.38	1.37	1.36	1.35	1.34	1.33	1.32	1.30
20	1.40	1.49	1.48	1.47	1.45	1.44	1.43	1.42	1.41	1.40	1.39	1.37	1.36	1.35	1.34	1.33	1.32	1.31	1.29
21	1.40	1.48	1.48	1.46	1.44	1.43	1.42	1.41	1.40	1.40	1.38	1.37	1.35	1.34	1.33	1.32	1.31	1.30	1.28
22	1.40	1.48	1.47	1.45	1.44	1.42	1.41	1.40	1.39	1.39	1.37	1.36	1.34	1.33	1.32	1.31	1.30	1.29	1.28
23	1.39	1.47	1.47	1.45	1.43	1.42	1.41	1.40	1.39	1.39	1.37	1.35	1.34	1.33	1.32	1.31	1.30	1.28	1.27
24	1.39	1.47	1.46	1.44	1.43	1.41	1.40	1.39	1.38	1.38	1.36	1.35	1.33	1.32	1.31	1.30	1.29	1.28	1.26
25	1.39	1.47	1.46	1.44	1.42	1.41	1.40	1.39	1.38	1.37	1.36	1.34	1.33	1.32	1.31	1.29	1.28	1.27	1.25
26	1.38	1.46	1.45	1.44	1.42	1.40	1.39	1.38	1.37	1.37	1.35	1.34	1.32	1.31	1.30	1.29	1.28	1.26	1.25
27	1.38	1.46	1.45	1.43	1.42	1.40	1.39	1.38	1.37	1.36	1.35	1.33	1.32	1.31	1.30	1.28	1.27	1.26	1.24
28	1.38	1.46	1.45	1.43	1.41	1.40	1.39	1.38	1.37	1.36	1.34	1.33	1.31	1.30	1.29	1.28	1.27	1.25	1.24
29	1.38	1.45	1.45	1.43	1.41	1.39	1.38	1.37	1.36	1.35	1.34	1.32	1.31	1.30	1.29	1.27	1.26	1.25	1.23
30	1.38	1.45	1.44	1.42	1.41	1.39	1.38	1.37	1.36	1.35	1.34	1.32	1.30	1.29	1.28	1.27	1.26	1.24	1.23
40	1.36	1.44	1.42	1.40	1.39	1.37	1.36	1.35	1.34	1.33	1.31	1.30	1.28	1.26	1.25	1.24	1.22	1.21	1.19
60	1.35	1.42	1.41	1.38	1.37	1.35	1.33	1.32	1.31	1.30	1.29	1.27	1.25	1.24	1.22	1.21	1.19	1.17	1.15
120	1.34	1.40	1.39	1.37	1.35	1.33	1.31	1.30	1.29	1.28	1.26	1.24	1.22	1.21	1.19	1.18	1.16	1.13	1.10
∞	1.32	1.39	1.37	1.35	1.33	1.31	1.29	1.28	1.27	1.25	1.24	1.22	1.19	1.18	1.16	1.14	1.12	1.08	1.00

Source: Adapted with permission from E. S. Pearson and H. O. Hartley, *Biometrika Tables For Statisticians*, Vol. 1, 3rd ed. Cambridge University Press, Cambridge, 1966.

Table B *(Continued)*

$$F_{.10, v_1, v_2}$$

v_2 \ v_1	1	2	3	4	5	6	7	8	9	10	12	15	20	24	30	40	60	120	∞
1	39.86	49.50	53.59	55.83	57.24	58.20	58.91	59.44	59.86	60.19	60.71	61.22	61.74	62.00	62.26	62.53	62.79	63.06	63.33
2	8.53	9.00	9.16	9.24	9.29	9.33	9.35	9.37	9.38	9.39	9.41	9.42	9.44	9.45	9.46	9.47	9.47	9.48	9.49
3	5.54	5.46	5.39	5.34	5.31	5.28	5.27	5.25	5.24	5.23	5.22	5.20	5.18	5.18	5.17	5.16	5.15	5.14	5.13
4	4.54	4.32	4.19	4.11	4.05	4.01	3.98	3.95	3.94	3.92	3.90	3.87	3.84	3.83	3.82	3.80	3.79	3.78	3.76
5	4.06	3.78	3.62	3.52	3.45	3.40	3.37	3.34	3.32	3.30	3.27	3.24	3.21	3.19	3.17	3.16	3.14	3.12	3.10
6	3.78	3.46	3.29	3.18	3.11	3.05	3.01	2.98	2.96	2.94	2.90	2.87	2.84	2.82	2.80	2.78	2.76	2.74	2.72
7	3.59	3.26	3.07	2.96	2.88	2.83	2.78	2.75	2.72	2.70	2.67	2.63	2.59	2.58	2.56	2.54	2.51	2.49	2.47
8	3.46	3.11	2.92	2.81	2.73	2.67	2.62	2.59	2.56	2.54	2.50	2.46	2.42	2.40	2.38	2.36	2.34	2.32	2.29
9	3.36	3.01	2.81	2.69	2.61	2.55	2.51	2.47	2.44	2.42	2.38	2.34	2.30	2.28	2.25	2.23	2.21	2.18	2.16
10	3.29	2.92	2.73	2.61	2.52	2.46	2.41	2.38	2.35	2.32	2.28	2.24	2.20	2.18	2.16	2.13	2.11	2.08	2.06
11	3.23	2.86	2.66	2.54	2.45	2.39	2.34	2.30	2.27	2.25	2.21	2.17	2.12	2.10	2.08	2.05	2.03	2.00	1.97
12	3.18	2.81	2.61	2.48	2.39	2.33	2.28	2.24	2.21	2.19	2.15	2.10	2.06	2.04	2.01	1.99	1.96	1.93	1.90
13	3.14	2.76	2.56	2.43	2.35	2.28	2.23	2.20	2.16	2.14	2.10	2.05	2.01	1.98	1.96	1.93	1.90	1.88	1.85
14	3.10	2.73	2.52	2.39	2.31	2.24	2.19	2.15	2.12	2.10	2.05	2.01	1.96	1.94	1.91	1.89	1.86	1.83	1.80
15	3.07	2.70	2.49	2.36	2.27	2.21	2.16	2.12	2.09	2.06	2.02	1.97	1.92	1.90	1.87	1.85	1.82	1.79	1.76
16	3.05	2.67	2.46	2.33	2.24	2.18	2.13	2.09	2.06	2.03	1.99	1.94	1.89	1.87	1.84	1.81	1.78	1.75	1.72

Degrees of Freedom for Numerator (v_1)

Degrees of Freedom for Denominator (v_2)

ν_2																			
17	3.03	2.64	2.44	2.31	2.22	2.15	2.10	2.06	2.03	2.00	1.96	1.91	1.86	1.84	1.81	1.78	1.75	1.72	1.69
18	3.01	2.62	2.42	2.29	2.20	2.13	2.08	2.04	2.00	1.98	1.93	1.89	1.84	1.81	1.78	1.75	1.72	1.69	1.66
19	2.99	2.61	2.40	2.27	2.18	2.11	2.06	2.02	1.98	1.96	1.91	1.86	1.81	1.79	1.76	1.73	1.70	1.67	1.63
20	2.97	2.59	2.38	2.25	2.16	2.09	2.04	2.00	1.96	1.94	1.89	1.84	1.79	1.77	1.74	1.71	1.68	1.64	1.61
21	2.96	2.57	2.36	2.23	2.14	2.08	2.02	1.98	1.95	1.92	1.87	1.83	1.78	1.75	1.72	1.69	1.66	1.62	1.59
22	2.95	2.56	2.35	2.22	2.13	2.06	2.01	1.97	1.93	1.90	1.86	1.81	1.76	1.73	1.70	1.67	1.64	1.60	1.57
23	2.94	2.55	2.34	2.21	2.11	2.05	1.99	1.95	1.92	1.89	1.84	1.80	1.74	1.72	1.69	1.66	1.62	1.59	1.55
24	2.93	2.54	2.33	2.19	2.10	2.04	1.98	1.94	1.91	1.88	1.83	1.78	1.73	1.70	1.67	1.64	1.61	1.57	1.53
25	2.92	2.53	2.32	2.18	2.09	2.02	1.97	1.93	1.89	1.87	1.82	1.77	1.72	1.69	1.66	1.63	1.59	1.56	1.52
26	2.91	2.52	2.31	2.17	2.08	2.01	1.96	1.92	1.88	1.86	1.81	1.76	1.71	1.68	1.65	1.61	1.58	1.54	1.50
27	2.90	2.51	2.30	2.17	2.07	2.00	1.95	1.91	1.87	1.85	1.80	1.75	1.70	1.67	1.64	1.60	1.57	1.53	1.49
28	2.89	2.50	2.29	2.16	2.06	2.00	1.94	1.90	1.87	1.84	1.79	1.74	1.69	1.66	1.63	1.59	1.56	1.52	1.48
29	2.89	2.50	2.28	2.15	2.06	1.99	1.93	1.89	1.86	1.83	1.78	1.73	1.68	1.65	1.62	1.58	1.55	1.51	1.47
30	2.88	2.49	2.28	2.14	2.03	1.98	1.93	1.88	1.85	1.82	1.77	1.72	1.67	1.64	1.61	1.57	1.54	1.50	1.46
40	2.84	2.44	2.23	2.09	2.00	1.93	1.87	1.83	1.79	1.76	1.71	1.66	1.61	1.57	1.54	1.51	1.47	1.42	1.38
60	2.79	2.39	2.18	2.04	1.95	1.87	1.82	1.77	1.74	1.71	1.66	1.60	1.54	1.51	1.48	1.44	1.40	1.35	1.29
120	2.75	2.35	2.13	1.99	1.90	1.82	1.77	1.72	1.68	1.65	1.60	1.55	1.48	1.45	1.41	1.37	1.32	1.26	1.19
∞	2.71	2.30	2.08	1.94	1.85	1.77	1.72	1.67	1.63	1.60	1.55	1.49	1.42	1.38	1.34	1.30	1.24	1.17	1.00

Degrees of Freedom for Denominator (ν_2)

Table B (*Continued*)

$$F_{.05, v_1, v_2}$$

Degrees of Freedom for Numerator (v_1)

v_2	1	2	3	4	5	6	7	8	9	10	12	15	20	24	30	40	60	120	∞
1	161.4	199.5	215.7	224.6	230.2	234.0	236.8	238.9	240.5	241.9	243.9	245.9	248.0	249.1	250.1	251.1	252.2	253.3	254.3
2	18.51	19.00	19.16	19.25	19.30	19.33	19.35	19.37	19.38	19.40	19.41	19.43	19.45	19.45	19.46	19.47	19.48	19.49	19.50
3	10.13	9.55	9.28	9.12	9.01	8.94	8.89	8.85	8.81	8.79	8.74	8.70	8.66	8.64	8.62	8.59	8.57	8.55	8.53
4	7.71	6.94	6.59	6.39	6.26	6.16	6.09	6.04	6.00	5.96	5.91	5.86	5.80	5.77	5.75	5.72	5.69	5.66	5.63
5	6.61	5.79	5.41	5.19	5.05	4.95	4.88	4.82	4.77	4.74	4.68	4.62	4.56	4.53	4.50	4.46	4.43	4.40	4.36
6	5.99	5.14	4.76	4.53	4.39	4.28	4.21	4.15	4.10	4.06	4.00	3.94	3.87	3.84	3.81	3.77	3.74	3.70	3.67
7	5.59	4.74	4.35	4.12	3.97	3.87	3.79	3.73	3.68	3.64	3.57	3.51	3.44	3.41	3.38	3.34	3.30	3.27	3.23
8	5.32	4.46	4.07	3.84	3.69	3.58	3.50	3.44	3.39	3.35	3.28	3.22	3.15	3.12	3.08	3.04	3.01	2.97	2.93
9	5.12	4.26	3.86	3.63	3.48	3.37	3.29	3.23	3.18	3.14	3.07	3.01	2.94	2.90	2.86	2.83	2.79	2.75	2.71
10	4.96	4.10	3.71	3.48	3.33	3.22	3.14	3.07	3.02	2.98	2.91	2.85	2.77	2.74	2.70	2.66	2.62	2.58	2.54
11	4.84	3.98	3.59	3.36	3.20	3.09	3.01	2.95	2.90	2.85	2.79	2.72	2.65	2.61	2.57	2.53	2.49	2.45	2.40
12	4.75	3.89	3.49	3.26	3.11	3.00	2.91	2.85	2.80	2.75	2.69	2.62	2.54	2.51	2.47	2.43	2.38	2.34	2.30
13	4.67	3.81	3.41	3.18	3.03	2.92	2.83	2.77	2.71	2.67	2.60	2.53	2.46	2.42	2.38	2.34	2.30	2.25	2.21
14	4.60	3.74	3.34	3.11	2.96	2.85	2.76	2.70	2.65	2.60	2.53	2.46	2.39	2.35	2.31	2.27	2.22	2.18	2.13
15	4.54	3.68	3.29	3.06	2.90	2.79	2.71	2.64	2.59	2.54	2.48	2.40	2.33	2.29	2.25	2.20	2.16	2.11	2.07
16	4.49	3.63	3.24	3.01	2.85	2.74	2.66	2.59	2.54	2.49	2.42	2.35	2.28	2.24	2.19	2.15	2.11	2.06	2.01

Degrees of Freedom for Denominator (v_2)

ν_2																			
17	4.45	3.59	3.20	2.96	2.81	2.70	2.61	2.55	2.49	2.45	2.38	2.31	2.23	2.19	2.15	2.10	2.06	2.01	1.96
18	4.41	3.55	3.16	2.93	2.77	2.66	2.58	2.51	2.46	2.41	2.34	2.27	2.19	2.15	2.11	2.06	2.02	1.97	1.92
19	4.38	3.52	3.13	2.90	2.74	2.63	2.54	2.48	2.42	2.38	2.31	2.23	2.16	2.11	2.07	2.03	1.98	1.93	1.88
20	4.35	3.49	3.10	2.87	2.71	2.60	2.51	2.45	2.39	2.35	2.28	2.20	2.12	2.08	2.04	1.99	1.95	1.90	1.84
21	4.32	3.47	3.07	2.84	2.68	2.57	2.49	2.42	2.37	2.32	2.25	2.18	2.10	2.05	2.01	1.96	1.92	1.87	1.81
22	4.30	3.44	3.05	2.82	2.66	2.55	2.46	2.40	2.34	2.30	2.23	2.15	2.07	2.03	1.98	1.94	1.89	1.84	1.78
23	4.28	3.42	3.03	2.80	2.64	2.53	2.44	2.37	2.32	2.27	2.20	2.13	2.05	2.01	1.96	1.91	1.86	1.81	1.76
24	4.26	3.40	3.01	2.78	2.62	2.51	2.42	2.36	2.30	2.25	2.18	2.11	2.03	1.98	1.94	1.89	1.84	1.79	1.73
25	4.24	3.39	2.99	2.76	2.60	2.49	2.40	2.34	2.28	2.24	2.16	2.09	2.01	1.96	1.92	1.87	1.82	1.77	1.71
26	4.23	3.37	2.98	2.74	2.59	2.47	2.39	2.32	2.27	2.22	2.15	2.07	1.99	1.95	1.90	1.85	1.80	1.75	1.69
27	4.21	3.35	2.96	2.73	2.57	2.46	2.37	2.31	2.25	2.20	2.13	2.06	1.97	1.93	1.88	1.84	1.79	1.73	1.67
28	4.20	3.34	2.95	2.71	2.56	2.45	2.36	2.29	2.24	2.19	2.12	2.04	1.96	1.91	1.87	1.82	1.77	1.71	1.65
29	4.18	3.33	2.93	2.70	2.55	2.43	2.35	2.28	2.22	2.18	2.10	2.03	1.94	1.90	1.85	1.81	1.75	1.70	1.64
30	4.17	3.32	2.92	2.69	2.53	2.42	2.33	2.27	2.21	2.16	2.09	2.01	1.93	1.89	1.84	1.79	1.74	1.68	1.62
40	4.08	3.23	2.84	2.61	2.45	2.34	2.25	2.18	2.12	2.08	2.00	1.92	1.84	1.79	1.74	1.69	1.64	1.58	1.51
60	4.00	3.15	2.76	2.53	2.37	2.25	2.17	2.10	2.04	1.99	1.92	1.84	1.75	1.70	1.65	1.59	1.53	1.47	1.39
120	3.92	3.07	2.68	2.45	2.29	2.17	2.09	2.02	1.96	1.91	1.83	1.75	1.66	1.61	1.55	1.50	1.43	1.35	1.25
∞	3.84	3.00	2.60	2.37	2.21	2.10	2.01	1.94	1.88	1.83	1.75	1.67	1.57	1.52	1.46	1.39	1.32	1.22	1.00

Degrees of Freedom for Denominator (ν_2)

Table B *(Continued)*

$$F_{.025, \nu_1, \nu_2}$$

ν_2	\multicolumn{19}{c}{Degrees of Freedom for Numerator (ν_1)}
	1	2	3	4	5	6	7	8	9	10	12	15	20	24	30	40	60	120	∞
1	647.8	799.5	864.2	899.6	921.8	937.1	948.2	956.7	963.3	968.6	976.7	984.9	993.1	997.2	1001	1006	1010	1014	1018
2	38.51	39.00	39.17	39.25	39.30	39.33	39.36	39.37	39.39	39.40	39.41	39.43	39.45	39.46	39.46	39.47	39.48	39.49	39.50
3	17.44	16.04	15.44	15.10	14.88	14.73	14.62	14.54	14.47	14.42	14.34	14.25	14.17	14.12	14.08	14.04	13.99	13.95	13.90
4	12.22	10.65	9.98	9.60	9.36	9.20	9.07	8.98	8.90	8.84	8.75	8.66	8.56	8.51	8.46	8.41	8.36	8.31	8.26
5	10.01	8.43	7.76	7.39	7.15	6.98	6.85	6.76	6.68	6.62	6.52	6.43	6.33	6.28	6.23	6.18	6.12	6.07	6.02
6	8.81	7.26	6.60	6.23	5.99	5.82	5.70	5.60	5.52	5.46	5.37	5.27	5.17	5.12	5.07	5.01	4.96	4.90	4.85
7	8.07	6.54	5.89	5.52	5.29	5.12	4.99	4.90	4.82	4.76	4.67	4.57	4.47	4.42	4.36	4.31	4.25	4.20	4.14
8	7.57	6.06	5.42	5.05	4.82	4.65	4.53	4.43	4.36	4.30	4.20	4.10	4.00	3.95	3.89	3.84	3.78	3.73	3.67
9	7.21	5.71	5.08	4.72	4.48	4.32	4.20	4.10	4.03	3.96	3.87	3.77	3.67	3.61	3.56	3.51	3.45	3.39	3.33
10	6.94	5.46	4.83	4.47	4.24	4.07	3.95	3.85	3.78	3.72	3.62	3.52	3.42	3.37	3.31	3.26	3.20	3.14	3.08
11	6.72	5.26	4.63	4.28	4.04	3.88	3.76	3.66	3.59	3.53	3.43	3.33	3.23	3.17	3.12	3.06	3.00	2.94	2.88
12	6.55	5.10	4.47	4.12	3.89	3.73	3.61	3.51	3.44	3.37	3.28	3.18	3.07	3.02	2.96	2.91	2.85	2.79	2.72
13	6.41	4.97	4.35	4.00	3.77	3.60	3.48	3.39	3.31	3.25	3.15	3.05	2.95	2.89	2.84	2.78	2.72	2.66	2.60
14	6.30	4.86	4.24	3.89	3.66	3.50	3.38	3.29	3.21	3.15	3.05	2.95	2.84	2.79	2.73	2.67	2.61	2.55	2.49
15	6.20	4.77	4.15	3.80	3.58	3.41	3.29	3.20	3.12	3.06	2.96	2.86	2.76	2.70	2.64	2.59	2.52	2.46	2.40
16	6.12	4.69	4.08	3.73	3.50	3.34	3.22	3.12	3.05	2.99	2.89	2.79	2.68	2.63	2.57	2.51	2.45	2.38	2.32

Degrees of Freedom for Denominator (ν_2)

344

v_2																		
17	6.04	4.62	4.01	3.66	3.44	3.28	3.16	3.06	2.98	2.92	2.82	2.72	2.62	2.56	2.50	2.44	2.38	2.32	2.25
18	5.98	4.56	3.95	3.61	3.38	3.22	3.10	3.01	2.93	2.87	2.77	2.67	2.56	2.50	2.44	2.38	2.32	2.26	2.19
19	5.92	4.51	3.90	3.56	3.33	3.17	3.05	2.96	2.88	2.82	2.72	2.62	2.51	2.45	2.39	2.33	2.27	2.20	2.13
20	5.87	4.46	3.86	3.51	3.29	3.13	3.01	2.91	2.84	2.77	2.68	2.57	2.46	2.41	2.35	2.29	2.22	2.16	2.09
21	5.83	4.42	3.82	3.48	3.25	3.09	2.97	2.87	2.80	2.73	2.64	2.53	2.42	2.37	2.31	2.25	2.18	2.11	2.04
22	5.79	4.38	3.78	3.44	3.22	3.05	2.93	2.84	2.76	2.70	2.60	2.50	2.39	2.33	2.27	2.21	2.14	2.08	2.00
23	5.75	4.35	3.75	3.41	3.18	3.02	2.90	2.81	2.73	2.67	2.57	2.47	2.36	2.30	2.24	2.18	2.11	2.04	1.97
24	5.72	4.32	3.72	3.38	3.15	2.99	2.87	2.78	2.70	2.64	2.54	2.44	2.33	2.27	2.21	2.15	2.08	2.01	1.94
25	5.69	4.29	3.69	3.35	3.13	2.97	2.85	2.75	2.68	2.61	2.51	2.41	2.30	2.24	2.18	2.12	2.05	1.98	1.91
26	5.66	4.27	3.67	3.33	3.10	2.94	2.82	2.73	2.65	2.59	2.49	2.39	2.28	2.22	2.16	2.09	2.03	1.95	1.88
27	5.63	4.24	3.65	3.31	3.08	2.92	2.80	2.71	2.63	2.57	2.47	2.36	2.25	2.19	2.13	2.07	2.00	1.93	1.85
28	5.61	4.22	3.63	3.29	3.06	2.90	2.78	2.69	2.61	2.55	2.45	2.34	2.23	2.17	2.11	2.05	1.98	1.91	1.83
29	5.59	4.20	1.61	3.27	3.04	2.88	2.76	2.67	2.59	2.53	2.43	2.32	2.21	2.15	2.09	2.03	1.96	1.89	1.81
30	5.57	4.18	3.59	3.25	3.03	2.87	2.75	2.65	2.57	2.51	2.41	2.31	2.20	2.14	2.07	2.01	1.94	1.87	1.79
40	5.42	4.05	3.46	3.13	2.90	2.74	2.62	2.53	2.45	2.39	2.29	2.18	2.07	2.01	1.94	1.88	1.80	1.72	1.64
60	5.29	3.93	3.34	3.01	2.79	2.63	2.51	2.41	2.33	2.27	2.17	2.06	1.94	1.88	1.82	1.74	1.67	1.58	1.48
120	5.15	3.80	3.23	2.89	2.67	2.52	2.39	2.30	2.22	2.16	2.05	1.94	1.82	1.76	1.69	1.61	1.53	1.43	1.31
∞	5.02	3.69	3.12	2.79	2.57	2.41	2.29	2.19	2.11	2.05	1.94	1.83	1.71	1.64	1.57	1.48	1.39	1.27	1.00

Degrees of Freedom for Denominator (v_2)

Index

A Primer on Experiments with Mixtures, By John A. Cornell
Copyright © 2011 John Wiley & Sons, Inc. Published by John Wiley & Sons, Inc.

WILEY SERIES IN PROBABILITY AND STATISTICS
ESTABLISHED BY WALTER A. SHEWHART AND SAMUEL S. WILKS

The **Wiley Series in Probability and Statistics** is well established and authoritative. It covers many topics of current research interest in both pure and applied statistics and probability theory. Written by leading statisticians and institutions, the titles span both state-of-the-art developments in the field and classical methods.

Reflecting the wide range of current research in statistics, the series encompasses applied, methodological and theoretical statistics, ranging from applications and new techniques made possible by advances in computerized practice to rigorous treatment of theoretical approaches.

This series provides essential and invaluable reading for all statisticians, whether in academia, industry, government, or research.

BECHHOFER, SANTNER, and GOLDSMAN • Design and Analysis of Experiments for Statistical
    Selection, Screening, and Multiple Comparisons
BEIRLANT, GOEGEBEUR, SEGERS, TEUGELS, and DE WAAL • Statistics of Extremes:
    Theory and Applications
BELSLEY • Conditioning Diagnostics: Collinearity and Weak Data in Regression
† BELSLEY, KUH, and WELSCH • Regression Diagnostics: Identifying Influential Data and Sources
    of Collinearity
BENDAT and PIERSOL • Random Data: Analysis and Measurement Procedures, *Fourth Edition*
BERNARDO and SMITH • Bayesian Theory
BERRY, CHALONER, and GEWEKE • Bayesian Analysis in Statistics and Econometrics: Essays
    in Honor of Arnold Zellner
BHAT and MILLER • Elements of Applied Stochastic Processes, *Third Edition*
BHATTACHARYA and WAYMIRE • Stochastic Processes with Applications
BIEMER, GROVES, LYBERG, MATHIOWETZ, and SUDMAN • Measurement Errors in Surveys
BILLINGSLEY • Convergence of Probability Measures, *Second Edition*
BILLINGSLEY • Probability and Measure, *Third Edition*
BIRKES and DODGE • Alternative Methods of Regression
BISGAARD and KULAHCI • Time Series Analysis and Forecasting by Example
BISWAS, DATTA, FINE, and SEGAL • Statistical Advances in the Biomedical Sciences: Clinical
    Trials, Epidemiology, Survival Analysis, and Bioinformatics
BLISCHKE AND MURTHY (editors) • Case Studies in Reliability and Maintenance
BLISCHKE AND MURTHY • Reliability: Modeling, Prediction, and Optimization
BLOOMFIELD • Fourier Analysis of Time Series: An Introduction, *Second Edition*
BOLLEN • Structural Equations with Latent Variables
BOLLEN and CURRAN • Latent Curve Models: A Structural Equation Perspective
BOROVKOV • Ergodicity and Stability of Stochastic Processes
BOSQ and BLANKE • Inference and Prediction in Large Dimensions
BOULEAU • Numerical Methods for Stochastic Processes
BOX • Bayesian Inference in Statistical Analysis
BOX • Improving Almost Anything, *Revised Edition*
BOX • R. A. Fisher, the Life of a Scientist
BOX and DRAPER • Empirical Model-Building and Response Surfaces
* BOX and DRAPER • Evolutionary Operation: A Statistical Method for Process Improvement
BOX and DRAPER • Response Surfaces, Mixtures, and Ridge Analyses, *Second Edition*
BOX, HUNTER, and HUNTER • Statistics for Experimenters: Design, Innovation, and Discovery,
    Second Editon
BOX, JENKINS, and REINSEL • Time Series Analysis: Forcasting and Control, *Fourth Edition*
BOX, LUCEÑO, and PANIAGUA-QUIÑONES • Statistical Control by Monitoring and
    Adjustment, *Second Edition*
BRANDIMARTE • Numerical Methods in Finance: A MATLAB-Based Introduction
† BROWN and HOLLANDER • Statistics: A Biomedical Introduction
BRUNNER, DOMHOF, and LANGER • Nonparametric Analysis of Longitudinal Data in Factorial
    Experiments
BUCKLEW • Large Deviation Techniques in Decision, Simulation, and Estimation
CAIROLI and DALANG • Sequential Stochastic Optimization
CASTILLO, HADI, BALAKRISHNAN, and SARABIA • Extreme Value and Related Models with
    Applications in Engineering and Science
CHAN • Time Series: Applications to Finance with R and S-Plus®, *Second Edition*
CHARALAMBIDES • Combinatorial Methods in Discrete Distributions
CHATTERJEE and HADI • Regression Analysis by Example, *Fourth Edition*

*Now available in a lower priced paperback edition in the Wiley Classics Library.
†Now available in a lower priced paperback edition in the Wiley–Interscience Paperback Series.

*Now available in a lower priced paperback edition in the Wiley Classics Library.

†Now available in a lower priced paperback edition in the Wiley–Interscience Paperback Series.

DUNN and CLARK • Basic Statistics: A Primer for the Biomedical Sciences, *Third Edition*
DUPUIS and ELLIS • A Weak Convergence Approach to the Theory of Large Deviations
EDLER and KITSOS • Recent Advances in Quantitative Methods in Cancer and Human Health
    Risk Assessment
* ELANDT-JOHNSON and JOHNSON • Survival Models and Data Analysis
ENDERS • Applied Econometric Time Series
† ETHIER and KURTZ • Markov Processes: Characterization and Convergence
EVANS, HASTINGS, and PEACOCK • Statistical Distributions, *Third Edition*
EVERITT • Cluster Analysis, *Fifth Edition*
FELLER • An Introduction to Probability Theory and Its Applications, Volume I, *Third Edition,*
    Revised; Volume II, *Second Edition*
FISHER and VAN BELLE • Biostatistics: A Methodology for the Health Sciences
FITZMAURICE, LAIRD, and WARE • Applied Longitudinal Analysis, *Second Edition*
* FLEISS • The Design and Analysis of Clinical Experiments
FLEISS • Statistical Methods for Rates and Proportions, *Third Edition*
† FLEMING and HARRINGTON • Counting Processes and Survival Analysis
FUJIKOSHI, ULYANOV, and SHIMIZU • Multivariate Statistics: High-Dimensional and
    Large-Sample Approximations
FULLER • Introduction to Statistical Time Series, *Second Edition*
† FULLER • Measurement Error Models
GALLANT • Nonlinear Statistical Models
GEISSER • Modes of Parametric Statistical Inference
GELMAN and MENG • Applied Bayesian Modeling and Causal Inference from Incomplete-Data
    Perspectives
GEWEKE • Contemporary Bayesian Econometrics and Statistics
GHOSH, MUKHOPADHYAY, and SEN • Sequential Estimation
GIESBRECHT and GUMPERTZ • Planning, Construction, and Statistical Analysis of Comparative
    Experiments
GIFI • Nonlinear Multivariate Analysis
GIVENS and HOETING • Computational Statistics
GLASSERMAN and YAO • Monotone Structure in Discrete-Event Systems
GNANADESIKAN • Methods for Statistical Data Analysis of Multivariate Observations,
    Second Edition
GOLDSTEIN • Multilevel Statistical Models, *Fourth Edition*
GOLDSTEIN and LEWIS • Assessment: Problems, Development, and Statistical Issues
GOLDSTEIN and WOOFF • Bayes Linear Statistics
GREENWOOD and NIKULIN • A Guide to Chi-Squared Testing
GROSS, SHORTLE, THOMPSON, and HARRIS • Fundamentals of Queueing Theory,
    Fourth Edition
GROSS, SHORTLE, THOMPSON, and HARRIS • Solutions Manual to Accompany Fundamentals
    of Queueing Theory, *Fourth Edition*
* HAHN and SHAPIRO • Statistical Models in Engineering
HAHN and MEEKER • Statistical Intervals: A Guide for Practitioners
HALD • A History of Probability and Statistics and their Applications Before 1750
HALD • A History of Mathematical Statistics from 1750 to 1930
† HAMPEL • Robust Statistics: The Approach Based on Influence Functions
HANNAN and DEISTLER • The Statistical Theory of Linear Systems
HARMAN and KULKARNI • An Elementary Introduction to Statistical Learning Theory
HARTUNG, KNAPP, and SINHA • Statistical Meta-Analysis with Applications
HEIBERGER • Computation for the Analysis of Designed Experiments
HEDAYAT and SINHA • Design and Inference in Finite Population Sampling

*Now available in a lower priced paperback edition in the Wiley Classics Library.
†Now available in a lower priced paperback edition in the Wiley–Interscience Paperback Series.

*Now available in a lower priced paperback edition in the Wiley Classics Library.
†Now available in a lower priced paperback edition in the Wiley–Interscience Paperback Series.

*Now available in a lower priced paperback edition in the Wiley Classics Library.
†Now available in a lower priced paperback edition in the Wiley–Interscience Paperback Series.

*Now available in a lower priced paperback edition in the Wiley Classics Library.
†Now available in a lower priced paperback edition in the Wiley–Interscience Paperback Series.

SCHOUTENS • Levy Processes in Finance: Pricing Financial Derivatives

SCHUSS • Theory and Applications of Stochastic Differential Equations

SCOTT • Multivariate Density Estimation: Theory, Practice, and Visualization

* SEARLE • Linear Models

† SEARLE • Linear Models for Unbalanced Data

† SEARLE • Matrix Algebra Useful for Statistics

† SEARLE, CASELLA, and McCULLOCH • Variance Components

SEARLE and WILLETT • Matrix Algebra for Applied Economics

SEBER • A Matrix Handbook For Statisticians

† SEBER • Multivariate Observations

SEBER and LEE • Linear Regression Analysis, *Second Edition*

† SEBER and WILD • Nonlinear Regression

SENNOTT • Stochastic Dynamic Programming and the Control of Queueing Systems

* SERFLING • Approximation Theorems of Mathematical Statistics

SHAFER and VOVK • Probability and Finance: It's Only a Game!

SHERMAN • Spatial Statistics and Spatio-Temporal Data: Covariance Functions and Directional Properties

SILVAPULLE and SEN • Constrained Statistical Inference: Inequality, Order, and Shape Restrictions

SINGPURWALLA • Reliability and Risk: A Bayesian Perspective

SMALL and MCLEISH • Hilbert Space Methods in Probability and Statistical Inference

SRIVASTAVA • Methods of Multivariate Statistics

STAPLETON • Linear Statistical Models, *Second Edition*

STAPLETON • Models for Probability and Statistical Inference: Theory and Applications

STAUDTE and SHEATHER • Robust Estimation and Testing

STOYAN, KENDALL, and MECKE • Stochastic Geometry and Its Applications, *Second Edition*

STOYAN and STOYAN • Fractals, Random Shapes and Point Fields: Methods of Geometrical Statistics

STREET and BURGESS • The Construction of Optimal Stated Choice Experiments: Theory and Methods

STYAN • The Collected Papers of T. W. Anderson: 1943–1985

SUTTON, ABRAMS, JONES, SHELDON, and SONG • Methods for Meta-Analysis in Medical Research

TAKEZAWA • Introduction to Nonparametric Regression

TAMHANE • Statistical Analysis of Designed Experiments: Theory and Applications

TANAKA • Time Series Analysis: Nonstationary and Noninvertible Distribution Theory

THOMPSON • Empirical Model Building

THOMPSON • Sampling, *Second Edition*

THOMPSON • Simulation: A Modeler's Approach

THOMPSON and SEBER • Adaptive Sampling

THOMPSON, WILLIAMS, and FINDLAY • Models for Investors in Real World Markets

TIAO, BISGAARD, HILL, PEÑA, and STIGLER (editors) • Box on Quality and Discovery: with Design, Control, and Robustness

TIERNEY • LISP-STAT: An Object-Oriented Environment for Statistical Computing and Dynamic Graphics

TSAY • Analysis of Financial Time Series, *Third Edition*

UPTON and FINGLETON • Spatial Data Analysis by Example, Volume II: Categorical and Directional Data

† VAN BELLE • Statistical Rules of Thumb, *Second Edition*

VAN BELLE, FISHER, HEAGERTY, and LUMLEY • Biostatistics: A Methodology for the Health Sciences, *Second Edition*

Printed and bound by CPI Group (UK) Ltd, Croydon, CR0 4YY

16/04/2025

14658368-0005